KENNETH JOHNSTONE is a freelance writer.

In 1898 a group of scientists, working without pay, often under hazardous conditions with only the most primitive equipment, began a systematic study of the fishes in Canada's inland and marine waters. The team operated under the aegis of a board of management which was later to evolve into the Fisheries Research Board, a scientific organization that has placed Canada among the world's leaders in fishery knowledge.

This history of the Fisheries Research Board examines the aims and achievements of its research, its attempts to deal with the often conflicting demands of pure and applied science, and its confrontations with a frequently uncomprehending and dissatisfied government. The people who shaped and sustained the organization figure prominently in the account.

In-depth taped interviews with senior members and employees of the Fisheries Research Board, as well as the Annual Reports and other publications of the Department of Fisheries, have enabled Kenneth Johnstone to produce a rich history of a remarkable scientific organization.

KENNETH JOHNSTONE

The Aquatic Explorers:

A History of
the Fisheries
Research Board
of Canada

Published by the University of Toronto Press,
Toronto and Buffalo,
in cooperation with
the Fisheries Research Board of Canada,
the Scientific Information and Publications Branch,
Fisheries and Marine Service,
Department of Fisheries and the Environment

Co-published through Publishing Centre, Supply and Services Canada

Government Catalogue Number Fs 99-13-1977

Canadian Cataloguing in Publication Data

Johnstone, Kenneth, 1909–
    The aquatic explorers

    Bibliography: p.
    Includes index.
    ISBN 0-8020-2270-7

    1. Canada. Fisheries Research Board – History.
    2. Fisheries – Research – Canada – History.
    I. Title.

SH223.J64   639'.2'072071   C77-001297-3

This book has been published during the
Sesquicentennial year of the University of Toronto

Contents

Illustrations

Tide and wind and crag,
Sea-wind and sea-shell
And broken rudder –
And the story is told
Of human veins and pulses,
Of eternal pathways of fire,
Of dreams that survive the night,
Of doors held ajar in storms.

Reprinted from
*The Collected Poems
of E.J. Pratt* by permission from
the estate of E.J. Pratt and
the Macmillan Co. of Canada

Preface

They were called, simply, investigators. They came from the universities, colleges, high schools, even the elementary schools of Canada, and they sought knowledge of the fishes contained in the streams and lakes of Canada and in the oceans at her door. As scientists, they numbered among them some of the most brilliant figures of their time. Volunteers originally, they served without remuneration beyond bare travel expenses, and they spent their summers in field research and their winters – in their spare time – producing papers on that work; all in the cause of science and the search for knowledge. From their efforts there soon emerged a superb scientific organization which has placed Canada among the world's leaders in fishery knowledge and which has served its country well.

To the spirit that animated those first investigators, who worked often under hazardous conditions, with primitive equipment, and with a meager and grudged budget, this book is respectfully dedicated.

This history of the first 75 years of the Fisheries Research Board of Canada is not a history in the traditional sense. It is not replete with reference numbers and footnotes. It does not attempt deep and penetrating explanations, nor does it offer an interpretation of events, except in a tentative way. Rather, I have endeavored to record the chief events that provided the background to the formation of a quite unusual scientific enterprise, and then to make a necessarily arbitrary selection of recorded events that took place during the 75 years covered here and to provide formal and informal information about the principal people who were involved.

This account is thus more of a reportage than a history. It reports on the successes and failures in the research of Board scientists; the delicate relations of the Board itself with an often uncomprehending and frequently dissatisfied government department; and then, of course, the flesh and blood of any history, the personalities involved, their conflicts and triumphs, failures and disappointments, ambitions and achievements.

The story is written from the viewpoint of an outside observer rather than a participant. It claims no scientific status, apart from the editing to which it has been subjected for accuracy. In fact, it may fairly be said that my chief qualification for tackling the project is an abiding love for seafood and a great sense of gratitude towards a remarkable Canadian scientific institution which has contributed much to ensure the continuing enjoyment of the harvest of the inland and marine waters of Canada for all of us. Without the Fisheries Research Board, there would be fewer Canadian lobsters today and oysters would be much less numerous, reminiscent of the extinction of the Eastern salmon from our inland waters a century ago. There would be less cod and halibut, less haddock and trout, and all would be at luxury prices. We would not know the Queen crab or the ocean perch, and the price of Pacific salmon would be astronomical.

The written source materials used as a basis for this history consist for the most part of the Annual Reports of the Fisheries Research Board of Canada and its predecessors the Biological Board of Canada and the Board of Management of the Marine Biological Station. In addition, I had access to a series of in-depth taped interviews conducted by Dr J.C. Stevenson with a number of senior members and employees of the Board, past and present. Quotations from these tapes have been used by permission of those interviewed, who have in some cases modified their remarks for publication. Exceptions are the quotations from Dr A.G. Huntsman, who died before the manuscript was ready for review, and many of those from Dr J.L. Hart, who was engaged in a review of the manuscript at the time of his death in 1973 but had completed detailed criticisms of only the first dozen chapters.

I was also able to interview senior members and employees of the Board, past and present (many of whom had already been interviewed initially by Dr Stevenson), and benefited greatly from the advice and criticism of Dr Stevenson and other senior employees and Board members. In particular I must single out the thoughtful criticism and great support given the project by Dr J.R. Weir and Dr W.R. Martin. Without their patient and persistent pressure this manuscript would probably never have gotten into print. I also owe a debt of gratitude to Dr Win Billingsley.

Finally I must acknowledge a great debt to Dr W.E. Ricker. He not only undertook a final overview of the manuscript, but pulled together many loose ends and filled in gaps in the story from his own considerable knowledge of the Board's development over the years. It is the simple truth that whatever coherence the history may have in its final form can be attributed in great part to Dr Ricker's fine touch, excellent judgment, and towering scholarship.

Although the pertinent records of the Board and the government have been made available to me, three written sources require particular recognition. These are 'Materials Relating to the History of the Fisheries Research Board of Canada (Formerly the Biological Board of Canada) for the Period 1898–1924,' by Margaret S. Rigby and A.G. Huntsman, 'History of the Fisheries Research Board of Canada' by H.B. Hachey, and *The Fisheries Research Board of Canada, 1898 to 1973* by J.C. Stevenson. These and other sources are listed at the back of the book.

As much as possible, people have been quoted verbatim, not only to give their interpretation of the facts as they saw them and to shed light on the personalities of their contemporaries, but also because of the involuntary self-revelation that is often afforded.

A point has been made, too, of recording, sometimes year by year, the work that went on at the various research stations of the Board. The records gave me a sense of the growth that was taking place, both in the scope of investigations and in the nature of fisheries science itself. What began at the turn of the century as fairly elementary fisheries biology became ever broader and more refined as time went on, and more and more scientific disciplines come within its purview. A quick comparison of one of the reports of the early years with the summary given by the Chairman in his annual report for 1972 (chapter 21) offers a startling contrast and a measure of that growth.

I have tried to present the evidence as objectively as possible, but occasionally, as an observer, I have not been able to suppress a startled comment. Generally, however, I have told the story as I found it, so that the reader might form his own opinions concerning the achievements and failures of the Board, and its significance for Canadians as well as for the rest of the world.

Impartiality is perhaps for angels. But even angels, if they were fish lovers, would find it difficult not to display some partiality for this remarkable institution and the equally remarkable people who created it. To that extent, at least, I find myself on the side of the angels.

Kenneth Johnstone
25 August 1976

THE AQUATIC EXPLORERS:
A HISTORY OF
THE FISHERIES RESEARCH BOARD
OF CANADA

1 Canada's fisheries to Confederation

THE ATLANTIC FISHERIES

It was fitting that Canada's first honorary scientific research board should have developed around the fisheries, for the economic and political history of Canada, the countries of western Europe, and the United States has been intimately connected with the development of the fisheries of the western North Atlantic from the first reports of their existence until the present day.

James T. Shotwell, in his preface to the Harold A. Innis study *The Cod Fisheries*, published in 1940, made this point eloquently: 'A thousand miles of misty sea, yielding a harvest which for a long time rivalled in importance the produce of our opening wildernesses, it evoked the courage and daring of competing nations as did the conquest of the mainland. Its history reflects ... the changing economies of Europe during the rise and growth of capitalism. Even today ... it remains a world of its own in international economy, one in which the interplay of American, Canadian and European interests has created a situation unique in history.' If we include the interests of Japan and the Soviet Union in this interplay, we bring the concluding statement right up to date.

Robert de Loture, in his *History of the Great Fishery of Newfoundland*, speculates on the basis of the Norse sagas that the three sons of Eric the Red, Lief, Thornwald, and Thorstein, from a base in Greenland in the summer of 1001, discovered Labrador (Helluland), Newfoundland (Markland), and the Massachusetts coast (Vinland), and that the Norsemen maintained colonies in Labrador, Nova Scotia, and Massachusetts until the end of the 12th century. The latter claims are disputed, but there is no question about the Norse settlement that has been excavated at Anse aux Meadows at the northern tip of Newfoundland. De Loture also cites a legend that Newfoundland was rediscovered by Basque voyagers as

3

early as 1372, in their search for whales that had disappeared from the Bay of Biscay at the beginning of that century; if so, they failed to take advantage of their initiative for more than 150 years.

But irrespective of who were the first discoverers of North America, it was the report brought back to England by John Cabot from his voyage of discovery in 1497 that galvanized the more daring fishermen of western Europe into action, particularly those from Portugal, France, and England. Through a variety of versions and tongues went Cabot's historic description: 'The sea there is full of fish to such a point that one takes them not only by means of a net but also with baskets to which one attaches a stone to sink them in the water.'

The fishing banks

The prize   The fishing banks referred to, which were to provide a seemingly inexhaustible source of protein for the next 500 years, extend in a rough line from southeast of Newfoundland's Avalon peninsula, where the giant Grand Bank covers some 4 degrees of latitude and 6 degrees of longitude, to the Georges Bank off Nantucket Island: a vast continental shelf almost equal in area to that of France. Lying south of Newfoundland, Grand, Green, and St Pierre banks have an area of 37,000 square miles. Off Nova Scotia, extending from Artimon Bank in the east to Brown's Bank in the Gulf of Maine, are 10 separate banks, and Georges Bank lies to the southwest of Brown's Bank. There are four other banks in the Gulf of St Lawrence, the largest encircling the Magdalen Islands, two others lying between the Magdalens and the New Brunswick coast, and the fourth and smallest, Whittle Bank, lying between Anticosti Island and the Strait of Belle Isle. Another and larger bank lies off the coast of Labrador at Hamilton Inlet.

The location of these banks was to play an important part in shaping the history of Canada both internally and in its relation to other countries.

Although Cabot, in the name of England's Henry VII, formally took possession of all the new lands extending from the regions discovered by the Spaniards to the Arctic Circle, for a great part of the 16th century the English failed to take full advantage of the claim he had established. The English sent a few ships to the new fisheries, but in the main they continued to depend on Iceland for their supplies. The Portuguese and the French were more aggressive in exploiting the new discovery.

Cod, the 'beef of the sea,' was the chief and almost exclusive target of the fishing effort from the beginning. It lent itself ideally to salting, either by a wet or dry-cured process. The French pushed out to the Banks and, having abundant supplies of cheap solar salt required for the wet or green-cured process, harvested the larger cod which abounded there. The Portuguese tended to concentrate around the Avalon peninsula. Both countries pursued the fisheries originally for their own home markets, but later developed an export trade, principally with England when she began to rely less on Icelandic supplies. For Spain, fishing was less important than a great whaling industry that flourished from about 1540 in the 'Granbaya,' the northern part of the Gulf of St Lawrence and Strait of Belle Isle. Large vessels sailed from the Basque ports every summer for shore stations that had been established in several harbors along the Labrador coast. Plants were built for rendering blubber to oil, as well as accommodation for the large labor force that was required. Whales were captured from open boats as they made their annual migration down the strait in late summer and autumn. Several hundred were taken each year, and for half a century this industry made a major contribution to the prosperity of northwestern Spain.

4

Then the curtain went up on a dramatic struggle for the possession of this rich and bountiful self-renewing resource. Attacks and counterattacks, destruction and counterdestruction, pitched battles, piracy, murder, and looting, war on both land and sea marked the course of the conflict, punctuated with brief periods of truce, signalized by solemn treaty. The intermissions were short, designed to provide the warring sides with just enough respite to gather forces for a renewal of the struggle. Such was the scenario for the next 350 years. Only with the dawn of Confederation would the confrontation take a less bloody form, though even then it did not at once diminish.

First it was the French and the Portuguese who most actively exploited the fisheries of the western Atlantic. They were preyed upon in turn by British free-booters and the British navy, which took a toll of fish and funds impartially, even from their own fishing vessels.

Then Spain, in the second half of the 16th century, made its major effort to control the resources of 'la Provincia de Terranova.' This ended only in 1588 with the destruction of the Invincible Armada, for which all the best vessels of the fishing and whaling fleets had been requisitioned as transports. While the Spaniards were active, they provided fair game for the frequently indistinguishable British pirates and British navy. The French, while attacking and driving off British fishing vessels, were, in turn harassed to some effect by the Spanish. In 1554, for instance, 10 French fishing vessels were attacked by a Spanish squadron, and in 1555 the Spaniards captured 48 French vessels and conducted a successful raid on St John's.

By the end of the century the field had effectively narrowed down to the British and the French, whose struggle went on for another 150 years, interrupted from time to time by conventions and treaties.

The Convention of 1635 gave the British theoretical sovereignty in Newfoundland but this was ignored by the French, who from their base at Plaisance attacked the British at St John's and repeatedly defeated British efforts to take Plaisance.

The Treaty of Utrecht in 1713 finally gave Plaisance to the British, who forthwith named it Placentia; the French lost everything but four islands: St Pierre, Miquélon, Cap Bréton, and St Jean (now Prince Edward Island). Most of the French in Newfoundland migrated to Cape Breton and the protection of Louisburg, the great fortress which for half a century would be the buttress of French fishing effort in the New World. The treaty conceded the French the right to fish for and dry cod along the northern coast of Newfoundland from Cape Bonavista in the east to Point Riche in the west, and this became known as the 'French shore.'

Meanwhile, just as the French had moved towards the Gulf of St Lawrence and Cape Breton in search of new fishing grounds, by the end of the 16th century the English had started to explore the grounds farther south in the Gulf of Maine and New England. The monumental importance of discovering a winter fishery close to the continental shore, in terms of American history, was pointed out nearly 300 years later by the historian C.L. Woodbury: 'It was the winter fishery that placed on our coasts a class of permanent consumers, and gave to agriculture the possibility of flourishing. The lumber trade marched beside it. In these pursuits they who tilled the land during the short summer could find profitable employment in the winter on the ocean or in the forest near their home ... It was the discovery of the winter fishery on its shores that led New England to civilization.'

More treaties punctuated the long-drawn-out battle between the British and the French for supremacy in the New World. The British captured Louisburg, but the Treaty of Aix-la-Chapelle in 1748 returned it to the French, much to the disgust of the New Englanders who had participated in its capture. They took it again in 1756, and the Treaty of Paris in 1763 saw French possessions in the New World reduced to the islands of St Pierre and Miquélon, with the restoration of rights to the French shore as the sole concession to the losing side. This concession was strongly resented in Newfoundland, and was maintained only by means of a French naval patrol. New England fishermen promptly moved into the fisheries vacated by France.

The expanding New England fishery (appraised by Adam Smith as 'one of the most important perhaps in the world') and the trade flowing from it made that colony prosperous, and developed a breed of hardy and enterprising men who were soon to challenge the authority of the mother country. It was not surprising that the American Revolution began with the Boston Tea Party.

New contestants   The struggle for supremacy in the western Atlantic still went on, but now the protagonists were changed. With the outbreak of the American Revolution in 1775, Nova Scotia and Halifax quickly replaced New England and Boston as the outposts of Old England in the New World. The colony experienced a rapid expansion in the fisheries and in trade as the war excluded New England from both the fisheries and the highly profitable trade which it had enjoyed with the West Indies. New England fishing vessels were converted to privateers and their fishermen to seamen, who harried the fishing vessels which were replacing them on the Banks.

The Treaty of Versailles in 1783 allowed American fishermen back on the Grand Bank and other banks of Newfoundland and in the Gulf of St Lawrence, and France made a strong effort to revive her fisheries with a system of bounties, but the effort collapsed with the outbreak of war with Britain again in 1793. French installations at St Pierre and Miquelon were destroyed by the British and the people there migrated to the Magdalen Islands. Although the Treaty of Amiens in 1802 restored the islands to France, they were seized again the following year and remained in British possession until 1814.

The renewed hostilities between Britain and France in 1793 enabled New England to expand her fisheries rapidly, and the effect of this expansion was felt immediately by Nova Scotian and British fishermen. In 1792 American fishermen were accused of taking unwarrantable liberties on the Labrador coast and of driving British fishermen from the fishery. It was claimed that New England sent 300 ships and 10,600 men to the Gulf of St Lawrence in 1804 and 900 ships the following year. A large portion of the catch was sold on the West Indies market, where the subsidized American fish easily undersold its competition, and on the expanding home market, where it was protected by a tariff.

Britain endeavored to protect the West Indies market for its North American colonies, Newfoundland and Nova Scotia in particular, and actions to restrict American trade there led to the passage of the Embargo Act by the United States in 1807. This act, however, backfired on the Americans for it cut off their ships from fishing in British colonial waters, particularly in the Gulf of St Lawrence, and it led to widespread smuggling and exchange of goods at sea between British colonial and American ships.

The outbreak of war between Britain and the United States in 1812 further demoralized New England's fishing effort, and the enterprising Nova Scotians were quick to take advantage of the opening in foreign markets. Newfoundland also benefited from American difficulties.

The Treaty of Ghent in 1814 brought peace with the United States, and the end of the Napoleonic Wars saw St Pierre and Miquelon and the privileges of the French shore of Newfoundland restored again to France, the latter over bitter Newfound-

6

land protests. Settlers had moved onto the shore, and difficulties between Britain and France on this score would continue until 8 April 1904, when France ceded its ancient right to the area. Supported by bounty and protected by tariff, the French fishery made a rapid recovery. The re-entry of the French into the North Atlantic fishery struck mainly at the New England operators, who had moved into the markets vacated by France during her long war with Britain.

The Treaty of Ghent had pointedly not covered the fisheries, which were subject to a separate convention in 1818. This convention granted fishing rights to American fishermen on the southern coast of Newfoundland between Cape Ray and the Ramea Islands, on the western and northern coasts between Cape Ray and the Quirpon Islands, in the waters surrounding the Magdalen Islands, and on the coast of Labrador from Mont Joli east to the Strait of Belle Isle and north on the Labrador coast. Importantly, the United States renounced 'any liberty heretofore enjoyed to take, dry and cure fish on or within three marine miles of any of the coasts, bays, creeks, or harbours ... Provided however that the American fishermen shall be admitted to enter such bays or harbours for the purpose of shelter and of repairing damages therein, of purchasing wood, and of obtaining water, and for no other purpose whatever.' They were also forbidden to fish, after settlements became established, on those portions of the coastline on which fishing rights were granted. The inshore fishery had been definitely forbidden to Americans.

The colonies versus New England

This exclusion was received bitterly by the New England fishermen, who claimed that their government had betrayed them. They proceeded to defy the ban, a move which, in turn, brought on a demand from the colonies' fishermen that the British government enforce it. Meanwhile, American fishing vessels flooded back into the Gulf of St Lawrence and the Bay of Chaleur, and along the Labrador and Newfoundland coasts, all of which had been out of bounds for the previous decade.

Newfoundland

During the long period of wars ending in 1815, Newfoundland became increasingly a center of settlement, and the fishing vessels from England practically disappeared. Territory north of Cape Bonavista was occupied. The seal fishery emerged and expanded rapidly. Until New England and France were able to restore their fishing industries, Newfoundland possessed an important market advantage, which was reflected in the high prices she was able to obtain on the world market.

A new factor

The growth of settlement in Newfoundland created demands for the extension of political institutions, and in 1832 provision was made for representative government. In 1854 Newfoundland finally achieved responsible government, and immediately began to press for the exclusion of French, British, American, and Nova Scotian activities from her inshore fisheries.

The French fishing effort depended heavily on the availability of large quantities of bait, usually purchased in Newfoundland, and Newfoundland tried to block this supply by imposing restrictions on the export of bait. When Great Britain reached an agreement with France in 1857, designed to settle difficulties arising from the conflicts between English and French fishermen, Newfoundland protested strongly and drew British recognition of the fact that the consent of the community of Newfoundland was regarded by Her Majesty's government as the essential preliminary to any modification of territorial and maritime rights.

Nova Scotia

The early history of the Nova Scotian fishery dates back to the French regime at the beginning of the 17th century, centered in the Ingonish, Louisburg, and Canso

A vigorous fishery

areas, but it was the last half of the 18th century that ushered in its modern history. Halifax was established in 1749 on a site which the French had briefly settled as Chebuctou back in 1698. Then followed the expulsion of the Acadians in 1755 and the Seven Years' War, which led to the downfall of New France and the Treaty of Paris in 1763. Pre-Loyalist immigrants from New England as well as the British Isles saw a rapid growth of the colony and the fishery.

With the outbreak of the American Revolutionary War, the Nova Scotian fishing industry was given a tremendous impetus, and the province itself experienced a new influx of Loyalist settlers from the United States. This new flood of immigrants spread right up into the Bay of Chaleur and resulted in the formation of New Brunswick as a separate colony in 1784.

In the half-century between 1783 and 1833, the Nova Scotian fishing industry boomed, and Nova Scotia took a leading part in endeavoring to check the American fishing threat. That struggle would finally bring the colony into Confederation.

Expansion was interrupted briefly by the war of 1812, but when the Convention of 1818 limited the activities of American fishing vessels in inshore waters, Nova Scotia enforced the regulations by the seizure of American ships within the three-mile limit, as much to check their smuggling as to prevent their purchase of bait and catches of spawning herring. At the same time, she endeavored to maintain her grip on the West Indies trade which she had secured during and since the period of the American Revolutionary War.

In the middle of the 19th century Nova Scotia reached a peak of prosperity, based upon the expanding fisheries, the timber trade, and the building of wooden ships. 'Nova Scotia ship-building is conquering the world' was the cry in 1846 when the colony turned out 2,538 vessels, as compared with 605 for Canada.

During the period of reciprocity, 1854 to 1866, Nova Scotia penetrated the rapidly expanding American market, but the abrogation of the treaty slammed this door in her face, turning her to the other colonies for joint action in the face of the new emergency. It may well have tipped the scales in favor of Confederation when Charles Tupper secured his bare majority in the Nova Scotian Legislative Assembly.

At the same time, Nova Scotia sharply resisted Newfoundland's moves to restrict her fishery. She protested Newfoundland's attempt to impose an export tax on bait in 1846, and Newfoundland had to make special provisions for the British colonies. Newfoundland's insistence on collecting duties on Nova Scotian goods traded in Labrador again turned Nova Scotian fishing interests toward Confederation.

Prince Edward Island

It was the mackerel fishery pursued by the New Englanders which caused Prince Edward Island to develop a fishery. American vessels came there to obtain supplies and to hire labor. Reciprocity resulted in increased sales of agricultural products to the United States. The settlements developed an interest in the fishery, and in 1851 created a ship-building industry and bounty support for a vessel fishery. By 1859 Prince Edward Island ships were in strength around the Strait of Belle Isle.

The termination of reciprocity abruptly ended this development. The vessels were sold and a boat fishery took over. But customs regulations, introduced and enforced in 1868 by the newly confederated governments of New Brunswick, Nova Scotia, and Quebec, diverted the trade of the American fishing vessels to Prince Edward Island, which found it more profitable to transship American fish and engage in American trade than to fish herself.

Canada endeavored to plug the loophole by persuading the British government to enforce customs regulations against the Americans in Prince Edward Island. However, Prince Edward Island successfully resisted the pressure and continued her lucrative trade with the Americans. Just as the conflict of interests between Newfoundland and Nova Scotia was to keep Newfoundland out of Confederation until the middle of the next century, so Prince Edward Island's divergent interest delayed her entry into Confederation until 1873, when, according to Lord Dufferin, she entered 'under the impression that it was the Dominion that has been annexed to Prince Edward Island.'

New Brunswick

Although New Brunswick did not become a separate colony until 1784, when it ceased to be part of Nova Scotia, a successful fishery had long been pursued off Miscou, in the Gulf of St Lawrence, first by the French and Basques, and then, when New France fell, by fishing houses from the Island of Jersey. The curious fact is that New Brunswick settlers seemed to have no real attraction for the sea, certainly not to the extent shown by Nova Scotians. Those who did engage in the fishery worked mainly for one of the several Jersey firms which efficiently and successfully exploited both the resource of the sea and the manpower of the local population.

The reluctant fisherman

Moses H. Perley was Her Majesty's Immigration Officer at Saint John, New Brunswick. A remarkable man, gifted with acute perception, he was commissioned by the legislature to make a report on the fisheries of New Brunswick during the years 1849, 1850, and 1851. His three reports were subsequently published in one volume (*The Sea and River Fisheries of New Brunswick*) in 1852. The reports provided a penetrating analysis of the state of the New Brunswick fisheries, together with some far-sighted proposals for their conservation and more effective exploitation. In the light of later developments, they make fascinating reading today.

Perley's penetrating reports

Perley minced no words. In the preface he noted that the people of Prince Edward Island failed to take advantage of their favorable position in the rich fishing area of the Gulf of St Lawrence. Then he added: 'but their efforts can scarcely be described as more limited or more feeble than those of the people of New Brunswick, who dwell upon its shores from Baie Verte to the western extremity of the Bay of Chaleurs ... those shores commanding as great an extent and variety of fishing grounds, and as abundant supplies of valuable fish of every description, as can be found in any other part of the unrivalled Gulf of St Lawrence, while they possess equal, and perhaps superior, facilities for prosecuting its fisheries, both extensively and profitably.'

Perley's first report covered the Gulf of St Lawrence; the second covered the area from north of the Miramichi to the Nova Scotian border. His summary showed the great neglect of the sea fisheries and the rapid decay of the river fisheries, now threatened with total extinction. The sea fisheries of the gulf coast were almost wholly in the hands of the Jersey merchants, who conducted their business admirably, but solely with a view to their own profit regardless of New Brunswick interests. They spent their profits in Jersey or elsewhere; they made no investment in the colony and did not aid its advancement.

Though the mackerel fishery was flourishing near the New Brunswick shores, its people ignored it and left it to the Americans, who pursued it very profitably. Resident fishermen were deliberately discouraged from it by those interested in the cod fishery.

American fishing vessels fished with impunity within three miles of the coast and within bays where they had no right to fish. By their manner of fishing they damaged the cod fishers, by depriving them of bait and by throwing overboard the offal of the fish they cleaned.

The herring fishery was most valuable and abundant, but lack of skill in fishing and ignorance of the best method of curing made the fishery scarcely profitable. Perley recommended that the Scottish experience be studied and that a fishery board be established similar to that which regulated the Scottish herring industry. He proposed that fishing stations be established, first at Caraquet, with inspectors and teachers of the herring fishery and with expert curers.

The laws for the regulation of the inland fisheries appeared to be well devised, but there had been a total failure to enforce them everywhere. 'Hence, the decay of these once valuable and prolific fisheries now hastening rapidly to their termination.' He called for immediate steps, such as the inauguration of a closed season for salmon fishing by any means, the elimination of spearing, restrictions on nets, and cooperation with the Lower Canada government to police the Restigouche.

He also called for a law to prevent the use of fish as manure, because the practice was both destructive to the fisheries and, in the long run, injurious to the land. He recommended that no pickled fish whatever should be exported unless the casks bore the brand of an official inspector. Only in this way could a market for the product of the province be built up in foreign centers. Finally he proposed that the revenues derived from the fisheries should be employed in extending and improving facilities to enable the industry to expand effectively.

Perley's third report was devoted to the fisheries of the Bay of Fundy, and again this remarkable observer did a thorough survey and came up with a series of comments which left little room for complacency. They began: 'It is quite clear from the foregoing Report, that the imperfect and careless manner of curing the fish caught in the Bay of Fundy, whether from neglect or want of skill, is such as to prevent those fish obtaining the best prices, and prohibits their being sent to distant foreign markets, for which they would be otherwise well adapted; thereby preventing an extension of the foreign trade of the Province, and diminishing its general prosperity.'

He said that the laws existing for inspection of fish were ignored except when 'it is convenient to affix what purports to be an official brand, for the purpose of giving character to articles which are short of weight, and oftentimes worthless.' He deplored the enormous destruction of herring and spawn at Grand Manan, and predicted that unless it was checked the herring fishery would eventually fail altogether. He mentioned, as in his previous report, 'the closing of the various rivers flowing into the Bay, and their tributaries, by mill-dams; the injuries arising from saw-dust, and mill-rubbish, being cast into rivers and harbours; and the wholesale destruction of salmon on their spawning beds far up the rivers.' Again he commented on the intrusion of American fishing vessels within the three-mile limit and the need for patrols to enforce the 1818 Convention.

Most of Perley's recommendations, however, were aimed at raising the fishermen's level of competence through the provision of effective inspection, better laws, better curing and packing, better facilities, and superior schools where young fishermen could be taught book-keeping, navigation, astronomy, and other subjects that would be useful in their calling.

It is clear from Perley's reports that New Brunswick lagged far behind Nova Scotia both in the degree that she was using the fisheries and in the level of competence exercised, apart from that of the Jersey houses.

Reciprocity stimulated the fishery of New Brunswick, but when it ended, she was in trouble again. However, it was the threat from the south more than any desire for unity of action on the fisheries that provided the major impulse for New Brunswick to join Confederation.

Quebec

Quebec's history followed a different course. The Treaty of Paris in 1763 had seen the government of Quebec succeed New France. Then, in 1774, the Quebec Act set up the Province of Quebec, to be followed in 1791 by the Constitutional Act, which created Upper Canada west of the Ottawa River and Lower Canada east of the Ottawa, including the Magdalen Islands, Anticosti, and Quebec-Labrador. Finally, the Union Act of 1841 changed the respective names to Canada West and Canada East, and these terms remained until Confederation.

Fisheries slow to start

Jersey firms had been quick to fill the gap in the French fisheries that followed the fall of New France, moving into the Gaspé as well as New Brunswick.

The resident fishery seems to have been officially ignored by the government of Canada until the early 1850s when, in response to pleas for protection against the American fishermen, who were operating unmolested along the Canadian shores, the legislature formed a committee to inquire into the state of the fisheries in the Gulf of St Lawrence and along the Labrador coast. At the same time they engaged the services of a young, recently graduated medical doctor from McGill; put him in charge of the coast guard schooner *Alliance*, together with a captain, a first and second mate, and a crew of 10, all smartly dressed in seamen's uniforms; and sent him off to protect the fisheries of the gulf.

The admirable Pierre Fortin

The young doctor's name was Pierre Fortin, and his career over the next 16 years offers material for a great Canadian epic. Each year he would set out from Quebec City in May or June, as the weather and the progress of winter repairs on the ship would permit, and his patrol would conclude in late autumn. His annual reports (1852–67) duplicated for the Quebec fisheries those of Moses Perley for New Brunswick. Together, the two men might be called spiritual forefathers of the Fisheries Research Board, for they were at heart scientific investigators of the fisheries, at a time when fisheries science did not exist.

But Fortin was all of that and more: as a physician he tended the sick whom he encountered on his patrol; as Justice of the Peace he administered the law, enforced the new Fisheries Act when it was passed, collected payments for the sale of crown lands, and occasionally acted as a customs officer. He constituted a one-man government in an area which had seen little or no government before. He also reported meticulously on the state of the fisheries, and even compiled a detailed study of the various fishes of the gulf, recording one species never reported before. In all, he was a thoroughly versatile and astonishing person, and his annual reports are completely engrossing even today.

Fortin's first report, for 1852, recorded that the schooner left Quebec on 15 June, carrying two guns and 'otherwise armed.' He noted that the fishermen in the Bay of Plaisance at the Magdalen Islands were pleased to learn that the Canadian government had at last decided to offer them 'the protection they had long solicited, on account of their remote situation and the absence of any military or police force whatever at these different places.'

He reported that at the Magdalen Islands the herring fishery was already over, and that more than 100 American schooners had been engaged in it. There had been 'many disorders, foreign fishermen had committed many depredations, had made themselves masters of everything, and frequently drove away our fishermen from

11

the stations they had made choice of in Amherst Harbour, which is very small; so that many schooners are compelled to remain outside, where they are exposed to accidents.' He said that the fisheries carried on at the Magdalen Islands included herring, cod, and seal, and that the fish nearly always abound, though one might occasionally fail. He also reported that the inhabitants, almost all fishermen, were poor despite the ideal location of the islands.

From the Magdalens he sailed to the Labrador coast and visited two seal fisheries and one salmon fishery. In Bradore Bay he counted 49 schooners, seven from the Magdalens, ten Americans, the rest Nova Scotian. There were 29 American schooners at Salmon Bay. The American mackerel fishing had been conducted within the three-mile limit.

Near Mingan he met the British sloop-of-war *Devastation* and the brigantine *Arrow*. The captain of the *Devastation* advised him to hoist a pennant such as vessels of war carry to 'inspire more respect.' At Seven Islands the Hudson's Bay agent told him that more than 50 American schooners had anchored in the bay to fish for mackerel. The Americans had insulted the agent, carried off his firewood, set fire to his fences, and he had saved the store only by a giant effort. Returning later to the Bay of Seven Islands Fortin learned that an American schooner had been there but had left when the captain heard that the *Alliance* was on patrol in the vicinity. It was the general opinion that the presence of the *Alliance* had resulted in driving away the American mackerel fishermen.

Among Pierre Fortin's general observations on the conclusion of his first trip was a recommendation for organizing the seal fishery on a license basis to give security to the fishermen. He noted that the seal and salmon formed just a small part of the riches, and that the cod, mackerel, and herring fisheries of the Gulf of St Lawrence attracted more than 1,000 schooners from the United States and the British provinces, with only 10 from the Magdalen Islands and none from Quebec. Yet 'we build vessels cheaper than either the United States or Nova Scotia. We needed to induce the fishermen to venture into schooner fishing.'

He made specific suggestions about his own patrol. It should begin on the coast by the first of June to protect the eggs of the auks and smaller divers which were being looted each year by some 15 schooners, each with about five and six men. They were armed and had been driving off the local people for 20 years. He suggested that the *Alliance* should arrive on the first of June to protect the eggs, then proceed to the Magdalen Islands to patrol the herring fishery, move on from that point to issue seal and salmon licenses, and finally take up the mackerel patrol.

The first report of the committee to inquire into the state of the fisheries in the Gulf of St Lawrence and the Labrador coast noted that little or no help was given to the fishermen of the British colonies in terms of bounties or other aid. Yet France and the United States furnished very substantial assistance to their fisheries. It stated: 'The liberal encouragements granted by the governments of France and the United States of America to their Fisheries have the effect of sending, year after year, during the fishing season fleets of their bankers and fishing craft into the gulph, where they carry on extensive, and no doubt, profitable Fisheries, constituting nurseries for their respective navies, and enabling them to compete with, and underselling us, in the article of fish, the products of our own waters, in foreign and even in our own, Upper Canada, markets, as the returns hereto appended show.'

The report recommended tax exemptions, bounties, and premiums for the fisheries, but two members of the committee dissented on the grounds that 'an expression of opinion by the Provincial Legislature of Canada, adverse to the surrender of the Fisheries, [is] at this moment premature.' At the same time the dissenter noted that the artificial aid conferred by the American government in the form of bounties to the cod and mackerel fisheries amounted to $1,627,505 from 1844 to 1848, plus the 20 percent duty on fish imported from British North America. They caught and exported fish to us to the amount of $11,156,170 in 1851 and of $13,231,000 in 1852. The minority report recommended a direct bounty to Canadian fishermen.

The report observed that the fishery along Canadian shores was being wiped out by heavy American fishing and that Gaspé fishermen were compelled to go to the Banks in small, open boats, spending two days getting there, two days fishing, and two days returning.

Firm action to assist the Canadian fisheries was undoubtedly delayed by hope of the impending Reciprocity Treaty, which would open American markets to Canadian-caught fish. This was realized in 1854, and a rapid expansion of the gulf fisheries followed.

Meanwhile Pierre Fortin continued his summer patrols. His reports for 1853 and 1854 were not published, but by 1855 he had persuaded the Canadian government to build him another and speedier schooner, *La Canadienne*. It was with obvious satisfaction that he reported *La Canadienne* to be much faster than any American schooner he encountered.

The fastest ship in the east

In his report for 1855 he noted that the fisheries had been successful and fewer American schooners were fishing for mackerel than in previous years. The herring fishery at the Magdalen Islands started on 10 May and lasted for two or three weeks. The total catch was estimated at 50,000 barrels.

He defined his function during the mackerel season there: 'besides the protection necessary to guard property from the depredations of foreigners, consisting for the most part of the scum of the seaport towns, it is necessary to prevent vessels from anchoring in the middle of the nets, from throwing out the refuse of the fish upon the fishing grounds, and from obstructing by means of their nets the entrance to the Bay [of Plaisance], a thing very often done when there is not sufficient force present to compel the observance of regulations having for their object the protection of these fisheries, to the great detriment of our fisheries whose nets are laid out nearer to the shore.'

He went on: 'I have been informed that the inhabitants of the Magdalen Islands have complained very strongly of the conduct of foreigners during last spring. Robberies have been committed in many places. The fishermen had forced entrance into houses in order to pass the night in dancing and the commission of every species of insult, without any fear of being made to answer for their actions in a Court of Justice, inasmuch as they are aware that the authorities of the Magdalen Islands cannot find throughout their entire population any force whatever to carry their order into effect.' He reported that the quantity of cod fishing on the Canadian coast was computed at 90,000 quintals, 'of which our fishermen take not more than 20,000 quintals.'

The following year, 1856, Fortin was apparently able to persuade the government to have *La Canadienne* ready for him early, for he arrived at the Magdalen Islands by 19 May. The herring catch had been a failure, with only 16,000 barrels netted compared to an average of 50,000 barrels. The herring had spawned before the fishermen were ready. There were 60 schooners from the Gut of Canso on hand, and only a few American vessels. He enforced regulations. There was not a single complaint. Public order was maintained. He was pleased to report that the presence of *La Canadienne* had prevented the cutting of crown timbers in the Restigouche by foreigners.

13

He commented at length about the fishing establishments of Robin & Co. and Le Boutillier Brothers at Paspébiac, the two largest in Canada, both Jersey-based firms. They impressed him deeply with their efficiency.

Fortin looks ahead

Fortin had some bitter remarks to make about the new Congreve rocket gun, which was used in whale fishing. It killed some 30 to 40 whales annually, but only six to eight of these were recovered; the others sank. He said:

It is true that the evil carries its own cure; that is to say, in a few years the abortiveness of this method of conducting the fishery will be generally felt, and it will be discontinued; but in the meantime there will be made, almost without any profit, so great a havoc among the whales that they will ultimately disappear altogether from the Gulf of St. Lawrence, as within about fifty years the walrus or sea-horse has disappeared. This animal, at the time of the discovery of Canada and even eighty or a hundred years ago, swarmed in immense herds on the coasts of the Magdalen Islands and of the Bay of Chaleurs.

It is, in my opinion, of some importance, if not a violation of treaties existing between Great Britain and the United States, that a law should be passed prohibiting, under heavy penalties, the use of Congreve rocket guns, or of any other such instrument in fishing which may cause the destruction of a great number of animals without a profitable result.

In succeeding years Fortin proposed that the size of mesh in fishing nets be regulated, that Canadian fishermen equip their ships with iron plates to follow the seal fishery in the ice, and that an act be passed to protect the rapidly vanishing salmon. He proposed that artificial oyster beds be developed, and worked out a very modest budget of $600 to accomplish this. He also developed a plan for a school of navigation aboard *La Canadienne*, a three-year course under his direction.

He was given permission to proceed with his oyster project, and made several plantings around the Gaspé peninsula over a period of years, but finally had to admit that the problem was not quite as simple as he had visualized.

In 1859 he made a detailed report on the various fisheries in the gulf: herring, mackerel, cod, seal, whale, salmon, trout, haddock, halibut, eel, capelin, and lobster. He pointed out that the mackerel fishery was almost completely neglected and was left to the Americans to exploit. The cod fishery was the most important, but it was virtually ignored by Canadians; it was carried on by Newfoundland, France, Nova Scotia, and the United States. Not one large ship owner from Quebec or Montreal was interested, yet 250 to 300 schooners from Nova Scotia and 200 to 300 from the United States were engaged in this fishery.

A lengthy study of the fishes found in the Gulf of St Lawrence was appended to Fortin's report of 1864.

Fortin applies the law

It was with some obvious pleasure that Fortin was able to record in 1865 that he had seized the Nova Scotia schooner *Ocean Bride* engaged in collecting murre and auk eggs on the Murre Islands. The captain and crew got two months in jail and the schooner was confiscated. He found 914 dozen eggs on board and 200 dozen eggs collected on Goose Island. At Boat Island he apprehended another three men with 780 dozen eggs, and two more at Studdard's Harbour with 600 dozen eggs; all were part of the *Ocean Bride*'s crew and all got two months in jail. He netted 11 prisoners. The action gave Fortin particular satisfaction, for the law under which the egg poachers were apprehended was one which had been passed at his suggestion in 1858.

In 1866 Fortin set sail on 7 May with new instructions covering American schooner fishing in Canadian waters, for the Reciprocity Treaty had been abro-

14

gated. He deplored the unlicensed sale of liquor to fishermen, and offered to enforce the regulations. He reported that the wild fowl had not been disturbed that season. He turned away four French schooners fishing illegally at Bradore. He issued fishing licenses for mackerel to American schooners and visited a total of 360 vessels during the season for this purpose.

The year 1867 was Pierre Fortin's last in command of *La Canadienne*. On 1 September he turned over his command to his assistant, Théophile Tétu, who had served in that capacity for six years and whom he had warmly recommended.

Off the Magdalen Islands Fortin encountered a large fleet of American mackerel fishing schooners and made a point of outsailing every one of them, just to let them know that they could be caught if necessary. He noted that the Americans were shifting from cod to mackerel fishing, and that there were 400 schooners from Gloucester alone during the season. He remarked that they were the finest ships in the world for their purpose, speedy and beautiful to watch, ideally adapted to follow the shoals of mackerel.

In his last report, Pierre Fortin was still puzzled over the fate of the oysters. They had lived successfully at Gaspé basin for three years, and then, four years later, had completely disappeared. Four schooners had returned to the Murre Islands to steal eggs for the Halifax market. He hoped that Confederation would end this illegal trade.

His successor, Théophile Tétu, died of a heart attack the following year.

With Confederation, Pierre Fortin ran for federal parliament, as a Conservative representing Gaspé. He was active in the parliamentary Fisheries Committee, and in 1882 he provided his own personal collection for Canada's display at the International Fisheries Exhibition in London. But nothing he did afterwards surpassed his 16-year patrol of the fisheries, for which he earned the title 'Le Roi du Golfe.' During that time he had brought law and order to an area that had been terrorized by the depredations of foreign fishermen. With his handful of well-drilled seamen he didn't hesitate to board vessels whose masters had violated the law and hold them to account, or appear at stormy town meetings when a show of authority was required to restore order. He also made sure his men attended Sunday mass 'to set an example' for the local population. He tracked down poachers and violators of the new Fishery Act, handed out penalties, and administered an impartial justice. And, above all, he showed a deep concern for the preservation and development of a fishery which, he felt sure, could mean a better life to the hard-working people who pursued it.

Upper Canada

No similar detailed record exists for the development of the fisheries in Upper Canada, although following the passage of the Fishery Act of 1858 Superintendent of Fisheries John McCuag reported on unsuccessful efforts to lease fishing stations in Lake Ontario (they had been worked previously without fee). He recorded the fact that the catch had been falling off because of the unrestrained methods of fishing, the use of fine mesh, and spearing. There was widespread hostility towards the government on the part of the fishermen because of this new effort to regulate the fishery.

He noted that the salmon, which had been thick in the streams feeding into Lake Ontario from 1812 to 1815, were now entirely vanished from some. In 1856 Lake Huron yielded 3,244,520 fish, and at Port Credit 470,000 were taken, two-thirds of them 'salmon,' which term must have included 'salmon trout,' now known as lake trout. Lake Huron and Georgian Bay had been similarly productive that year.

The fishermen resist regulations

15

In the same year, Richard Nettle was appointed superintendent of fisheries for Lower Canada and began a systematic effort to enforce the laws against spearing salmon, netting, and out-of-season fishing. The law, he noted, was disregarded in particular around the Saguenay. Local overseers were appointed in each area to enforce the Fishery Act.

Canada's freshwater fish had played an important role in the early days of the fur trade, and had formed a vital part of the food supply for explorers as well as for the native population. A fishing station at Sault Ste Marie for whitefish had existed long before Confederation. However, it was the Reciprocity Treaty that spurred the commerical fishery in the freshwater lakes and streams of Canada, particularly in the Great Lakes. There Canadian fishermen competed directly with American fishermen for the catch and, after the abrogation of the treaty, faced the same difficulties encountered by the Atlantic fishermen.

THE PACIFIC FISHERIES

Although the early history of the Atlantic fisheries covers a span of 500 years, that of the Pacific fisheries covers barely a century. And although the history of the Atlantic fisheries includes wars, piracy, and violent confrontations, that of the Pacific generally takes its nature from the name of the ocean which provides Canada with 7,000 miles of coastline. It is in consequence brief.

History has recorded that some time between 1587 and 1596 one Apostolos, better known as Juan de Fuca, explored the California coast for the Spanish viceroy of Mexico and reached latitude 45°N. Proceeding in a northerly direction, after entering a wide expanse of sea between 47° and 48°N, he reached what he called the 'North Sea,' now Puget Sound or possibly the Strait of Georgia. He believed that this was the entrance to the much-sought-after Northwest Passage.

Although the Spanish laid claim to the Pacific coast of America from Cape Horn to the Arctic, it was the Vitus Bering expeditions of 1725–30 and 1733–43 that brought the Russians first on the scene of northwestern America to trade for furs. A generation later the British, Spanish, Portuguese, French, Swedes, and Americans had joined in the trade.

The first European credited with landing on what is now the west coast of Canada was Captain Cook, who reached Nootka Sound in 1778 and claimed the region for Britain. Russia, too, thought to occupy the coast from Alaska southward to Mexico. In 1795 Governor Baranov wrote: 'Nothing is in Nootka; neither English nor Spanish. It is an abandoned place, when the time comes to extend our trade and occupy it on our side ... This is easy at the present time when Nootka is unoccupied by the English and they are engaged in war with the French.' Baranov eventually established a settlement and farm at Fort Ross north of San Francisco, which was occupied for about 30 years, but apparently no other permanent station was set up south of Dixon Entrance.

In 1792 Captain Vancouver with two ships, the *Chatham* and the *Discovery*, explored the mainland coast from the Columbia River to western Alaska. He recorded the melancholy facts of one death and four illnesses among his crew from eating toxic mussels taken from a cove in Mathieson Channel on the central coast of British Columbia in 1793. This disaster was exceeded by that of a Baranov expedition in 1799, when 100 men died at Peril Strait near Sitka, Alaska, from eating mussels.

In 1793 Alexander Mackenzie of the Northwest Fur Company entered 'New Caledonia' by way of the Peace River and followed the Bella Coola River to salt water at the head of a long fjord. Fort George, now Prince George, was founded by Simon Fraser in 1807. In 1808 he went down what is now the Fraser River to its mouth, and was doubly disappointed: the river was not tributary to the Columbia, as he had expected, and the rapids in its canyon were much too violent for it to become a water route to the coast.

From the first the traders at the interior posts subsisted to a large extent on salmon, as did the Indian tribes around them. Moreover, as white contacts spread along the coast it became apparent that fish, shellfish, and sea mammals constituted the principal economic base supporting large indigenous populations with rich and varied cultures. The Hudson's Bay Company established a post at Fort Langley on the lower Fraser River in 1827. Fort Victoria was opened in 1843, and became the company's central depot on the coast when the international boundary was moved north from the Columbia River in 1849. Salting of salmon obtained from the Indians began at Fort Langley in 1829, and soon there arose an export trade in salt-cured salmon with the Hawaiian Islands and the continent of Asia, which continued for many years.

Just as cod had formed the foundation of the fisheries in the Altantic, salmon were to build up the fisheries in the Pacific. However, those fisheries developed slowly. The influx of population to British Columbia following the gold rush of the late 1850s created the first important local market for salmon, and with the decline of production in the gold fields the new population began to look to other resources, such as the fisheries. Expansion of the industry was, however, held up by the need to develop production techniques and to find new markets for the tremendous resource available. Commercial canning did not begin until about 1870. That story belongs properly to the following period.

2 Prelude to the Board: the political background

Two problems for
Confederation

It requires no great feat of historical scholarship to assert that the political, economic, and social problems of the British colonies had created the need for Confederation as a national instrument. The problem that loomed largest at the time was the threat of invasion by disbanded veterans of the United States Civil War, and the Fenian raid of 1866 had shown that this was not entirely hypothetical. But another problem, and a tricky one, was that of the Atlantic fisheries. It was like a double-barreled shotgun with a hair trigger. In the hands of a tyro, either barrel could blow the head off Confederation. One barrel was the need to organize the fisheries on a rational basis. There was no tradition of effective law enforcement or observance of the law to persuade Canadian fishermen that their best interests lay in that direction. On the contrary, they had the daily witness of American fishermen, literally getting away with murder.

The other barrel was the relations with those New England fishermen who had fished Canadian waters traditionally, and who had no intention of changing their habits, no matter what treaty or abrogation of treaty. Canadian fishermen had to compete directly with them, both for the catch and for the nearest and best market, which was the expanding American market.

Mitchell bites
the bullet

The man who had the primary responsibility for tackling the problem was Canada's first minister of marine and fisheries, Peter Mitchell. A forthright, blunt-speaking native of Newcastle, New Brunswick, he had become premier of that province in 1866 and was a Father of Confederation. In his annual report for 1868, he summed up the situation:

At the date of confederation of the Provinces, the official business relating to the fisheries had

18

been for several years organized and managed as a branch of the department of Crown Lands, for the united provinces of Upper and Lower Canada. In the sister provinces of Nova Scotia and New Brunswick, no similar organization existed. There were, however, in these latter provinces, certain statutory and municipal regulations existing; but owing to the want of effective machinery to enforce them, and a proper system under which the restrictions thus provided could be applied, they produced scarcely any practical benefits, consequently the fisheries were subject to serious abuses that in many respects had already reduced them almost to exhaustion ...

The very extensive fishing interests of the maritime population of these provinces, and the great commercial importance of their coast and river fisheries, rendered it highly desirable that some uniform and efficient system should be devised under which the general 'Sea Coast and Inland Fisheries', as placed under the control of the federal government, might be regulated, protected and developed.

That same year, he brought in a Fisheries Act which consolidated and amended existing fishing laws, including those of both Nova Scotia and New Brunswick. The most important clauses in the act covered the appointment of federal fisheries officers, with magistrate's powers, to enforce it; federal fishery licenses and leases; closed seasons for salmon, whitefish, lake trout, and other important species; fish-passes and clear passage for named species; prohibition of the capture of their young; free passage of fish on Sundays and prohibition of Sunday fishing; prohibition of pollution in fishing waters; and the establishing of fish sanctuaries and fish reserves. Oyster and shellfish fisheries were included in the act.

Equally important was a companion act 'respecting fishing by foreign vessels,' which provided for the licensing of vessels to fish within three miles of the shore, the boarding by crown officers of any ship in harbor or hovering within three miles of shore, the fining of the masters of any foreign ships found fishing, or preparing to fish, or to have been fishing without license within three miles of shore, and the confiscation of the vessel, its stores, and cargo. License fees were increased from 50 cents to $2.00 a ton. The larger American vessels accepted the challenge to run the risk of capture, and smuggling increased.

Soon, however, the license fee system was dropped, and an Order in Council of 8 January 1870 forbade all foreign fishermen from fishing in Canadian waters. A fleet of six cruisers was assembled; about 400 American vessels were boarded and 15 were seized.

This, of course, was the preliminary sparring leading up to the negotiation of a new treaty that would reopen the American market to Canadian-caught fish in return for allowing access to Canadian inshore fishing waters by American fishermen. Britain was the essential go-between, as Canada then lacked the authority to sign an agreement directly with the United States. Canadian leaders strongly suspected that Britain was more interested in reaching an amicable agreement with the United States than in defending Canadian rights.

Peter Mitchell expressed himself with typical bluntness in a 'Memorandum on the Position of the Fishing Question,' dated 4 July 1870. Britain, he said, intended to give privileges to the American fishermen which they had not previously enjoyed and he protested against the 'concessions being made and the policy which Her Majesty's Government have pursued with reference to the Fisheries.'

The Americans, he went on, were insisting that they had the right to enter any Canadian bay and to fish within three miles of the inner shore rather than from a headland-to-headland line, something that had been established by international law throughout the civilized world. The Americans had specifically relinquished

19

liberties previously enjoyed by the first article of the Convention of 1818; they had been granted certain privileges in the Reciprocity Treaty of 1854, for which we were to obtain recompense. It was the American government that ended that treaty, in March 1866. The practical effect of that action was that they had terminated their concessions though they insisted upon maintaining those we had given them. Britain had told the provinces not to enforce their rights; a spirit of concession had characterized British actions since 1865, some of which were taken without even consulting the Canadians. Mitchell concluded: 'The active protection of our Fisheries was the first step in our National Policy.' He called for the appointment of a joint commission to settle the matters in dispute and, meanwhile, demanded the exclusion of all foreigners from fishing in Canadian bays, the exclusion of foreign fishing within the three-mile limit from headland to headland, and the exclusion of foreign fishing vessels from the Gut of Canso.

Mitchell's strong position reflected the Maritimes' point of view in the Canadian government, and was endorsed by Nova Scotia's spokesman in the Canadian government, Sir Charles Tupper. A joint high commission was appointed by Great Britain and the United States, and Canada's prime minister, Sir John A. Macdonald, became a member of it in February 1871. Although he supported Tupper's view that 'the Dominion cannot agree to the sale of fishing rights for money consideration,' a softer line was taken at home by his finance minister, Sir Francis Hincks, who wrote to Macdonald: 'We have no object in refusing the fisheries and the St. Lawrence – on the contrary, the fisheries are a mere expense – but we can't yield the fisheries without at least free importation of our fish and free or low duty on coal, lumber and salt, particularly the first.'

Imperial versus colonial interests

Macdonald pressed the Maritime position to no effect. The British government authorized the commission 'to negotiate on the basis of free fish and arbitration for an additional sum.' Macdonald was tempted to resign from the commission, and Sir George Etienne Cartier, militia minister and Macdonald's closest collaborator, wrote: 'The Queen's government, having formally pledged themselves that our fisheries should not be disposed of without our consent, to force us now into a disposal of them, for a sum to be fixed by arbitration and free fish would be a breach of faith, and an indignity never before offered to a great British Possession.'

This indignity was nothing compared to what was in store. The Washington Treaty was signed to become effective on 1 July 1873, permitting American fishermen access to the inshore waters of the Canadian fisheries in return for the (largely meaningless) similar Canadian access to American shores and the (important) free entry of Canadian fish to the American market, together with a sum in compensation to the Canadians, to be determined by arbitration.

Three separate but interconnected events produced the final indignity. First, the defeat of Macdonald in that year's election brought in a new Liberal administration under Alexander Mackenzie, who endeavored to widen the free entry of Canadian products into American markets in place of the monetary settlement, but the draft reciprocity treaty was rejected by the American Senate. Second, the Fisheries Commission, sitting in Halifax, finally settled on a sum of $4.5 million as a payment for the excess value of Canadian inshore fisheries over those of the United States; $4 million of this amount was set aside to provide an annual bounty to Canadian fishermen of $150,000, a sum later raised to $160,000. Third, this award drew such a howl of protest from New England fishing interests that a demand to abrogate the Washington Treaty resulted in just that. Three years after the Canadian bounty system went into effect, in July 1885, the fishery clauses of the treaty were terminated.

Canada waited for six months, meanwhile permitting American fishermen free entry into the shore fishing. Then, failing any corresponding concessions from the United States, Canada began enforcing the Treaty of 1818, the Fisheries Act, and the Canadian Customs Act, which led to the boarding of 700 American vessels in 1886 and 1,362 in 1887.

American reaction was swift. A non-intercourse act was passed by the American government, authorizing the president to deny entry of Canadian vessels to United States ports and to prohibit the entry of fish or any other product or goods coming from the Dominion. The proclamation of a Canadian act to enforce customs regulations, issued on 24 December 1886, produced a retaliatory bill from the United States government, approved by the president on 3 March 1887. Outgunned, the Canadians capitulated. America cracks the whip

Sir Charles Tupper was appointed high commissioner and sent to Washington to work out a new treaty with the British minister there and the United States secretary of state. A 'modus vivendi' agreement was reached on 15 February 1888, pending the ratification of the treaty. But the treaty was blocked in the Senate by Massachusetts, and the 'modus vivendi' agreement continued to operate until 1918.

The 'modus vivendi' gave American fishermen just about all they wanted, and the Canadian fishermen got nothing in return. The Americans could enter the bays and harbors of the Atlantic coasts of Canada and Newfoundland, upon payment of an annual license fee of $1.50 per ton, to purchase bait, ice, seines, lines, and other supplies and outfits and to transship catches and take on crews. If they did not remain more than 24 hours, they could enter without clearing customs, providing they did not communicate with the shore. Eventually, changes in the fishing industry itself would render these concessions irrelevant.

Behind the barrier of an effective tariff wall, the American fishing industry concentrated on the expanding domestic market for fresh fish, aided by the development of refrigeration and speedier transportation. The American fishing effort stayed closer to home, withdrawing almost entirely from the Labrador coast and concentrating on Bank fishing. In 1888, the Grand and Western Bank fleet of the United States numbered 339 vessels, and the Georges and New England shore fleets, 284.

The American tariff on Canadian fish struck heavily at the Canadian fishing industry and resulted in a migration south of Canadian fishermen. By 1886 nearly 20 percent of the labor force in the New England fisheries was Canadian. At the same time, there was a decline in the dried fish and pickled fish industries, as fresh fish replaced them in the American market, and the West Indies market diminished when the price of sugar there fell in the face of new competition from beet sugar.

New techniques had to be developed in catching, processing, and marketing if the Canadian industry was to remain viable. A domestic market for fresh fish was opening up in Canada, as it had in the United States, with the improvement in refrigeration processes and faster transportation to convey the product to the inland cities.

As the century moved to its end, new products of the sea began to find markets, particularly oysters and lobsters. Soon canned lobster would surpass salt cod in value, though 50 years before, as Moses Perley had noted, lobsters were being used as fertilizer on potato fields.

THE PACIFIC FISHERIES

British Columbia did not enter Confederation until 1871, by which time the first Rapid expansion

21

salmon cannery was in operation on the Fraser River, near New Westminster. Opened in June 1870, the cannery was run by a group which included Alexander Loggie and David Hennessy who had learned lobster canning in New Brunswick. A second cannery was opened at New Westminster that same year, and in the succeeding years further canneries developed all along the coast.

The industry expanded rapidly. Completion of the transcontinental railway gave impetus by providing access to the markets of eastern Canada and the United States. In 1877 the salmon pack amounted to 67,387 cases, and by 1901 it had increased to 1,247,212 cases.

Cold-storage plants for handling salmon were established on the Fraser River in 1887. Because of advantages in freight rates, American lines carried most of the traffic until 1892, when specially designed refrigerator cars were provided for moving fish over the CPR system.

The new technique for handling refrigerated fish spurred the development of the halibut fishery, which had previously been confined to local outlets. By the turn of the century a year-round fishery had developed and was expanding to the offshore grounds north of Cape Flattery. Bait requirements in turn stimulated an important herring fishery as well.

From the completion of the railway until the end of the century, the fishing industry on the Pacific coast consisted mainly of small regional units based chiefly on the Fraser River. The depression of the 1890s resulted in many bankruptcies, followed by amalgamation and integration of operations.

In contrast to developments in the Atlantic, no serious friction occurred between Canada and the United States in the Pacific fisheries in the period following Confederation, and a spirit of cooperation rather than conflict was to mark relations in the succeeding years.

HART'S SUMMARY

Summarizing the situation with regard to the fisheries on both coasts, John Hart has said:

It is clear that before Confederation, several important stocks of Atlantic fishes had been heavily exploited and the accumulated stocks removed. As a result, by 1910 people engaged in the fisheries on the Atlantic coast had no clear recollection of former conditions of excessive abundance or of sharp declines in the reward for fishing effort. Supplies of older fishes had already declined. Acceptance of the situation, of operating in a stabilized fishery, was accordingly current in the early years of the 20th century.

On the Pacific coast, where heavy exploitation began much later, and where the mainstays of the inshore fishery were the anadromous salmons and the beach-spawning strains of herring, attitudes were quite different, and fears of depletion stalked the minds of reasonable people.

Until the turn of the century, research on fishes in Canada, as elsewhere in the world, was restricted to describing and identifying the various species of fishes and determining their distribution. Some work was done on life histories. The studies were carried out principally by scientists from Britain, France and the United States and, increasingly in the latter part of the century, by Canadian scientists. Their contributions have been assessed by [J.R.] Dymond, [W.B.] Scott, [W.A.] Clemens and [G. Van] Wilby, and [J.L.] Hart. Research on ecology, details of life history, physiology, and population dynamics [was] still to come. [Correspondence with the author, 1972]

As elsewhere, application of biological observations to fisheries in Canada was at first unsophisticated, and little consideration was given to population structure or ecological needs. At the 1920 meeting of the American Fisheries Society Dr E.E. Prince, chairman of the Biological Board, recounted that salmon fry were raised in a small hatchery in Quebec City in 1859 by Richard Nettle, superintendent of fisheries for Lower Canada, and that in 1866 Samuel Wilmot attempted to improve salmon runs in Lake Ontario by capturing and stripping fish and carrying eggs through to the parr stage.

Biological observations primitive

Wilmot became the first superintendent of fish breeding in Canada in 1876, and established several hatcheries in Ontario, Quebec, and the Maritimes. The resulting successes and failures of this work made the necessity for objective research increasingly evident. Thus, Dr A.G. Huntsman pointed out in 1942 that in spite of a keen desire by those in authority for more knowledge, fisheries research did not develop in the uncongenial atmosphere of administration prior to 1900, apart from sporadic gestures. One such was the employment of the naturalist and geologist H.Y. Hind to prepare the government's case before the Halifax Commission of 1877.

Other forces were at work. Rapid increase in the population of the continent produced the illusion of diminishing national resources, there being an actual decrease in the amount per person. This illusion was fortified by the undoubted disappearance of certain animals, or their increasing scarcity: in fisheries, by the early extermination of walrus in the Gulf of St Lawrence and the spectacular declines in the Great Lakes sturgeon and New England and Ontario salmon. Certain other valuable species were observed to be subject to fluctuations of unknown origin, and each drop in abundance was viewed as the prelude to disaster.

Such factors brought about the establishment of a federal fishery service in the United States and the appointment of Spencer Baird of the Smithsonian Institution as commissioner of fish and fisheries in 1871. The still largely illusory decrease of fish stocks was seen to have a remedy, which proved equally illusory. Artificial hatching and planting of fry, which originated in Europe, developed simultaneously and rapidly in both Canada and the United States, resulting in the setting up of extensive governmental fish culture services and the formation of the American Fisheries Society in 1870.

Laws and regulations, however necessary, were not sufficient to protect the fisheries without knowledge concerning the fish, their spawning, food, habits, etc. Scientific societies with interests in natural history had been established all over Canada. Founded in 1862, the Natural History Society of New Brunswick included two men among its members who would later become members of the Board of Management for the Marine Biological Station and the Biological Board of Canada, and indefatigable workers in their respective fields: Dr L.W. Bailey (diatoms) and Dr Philip Cox (ichthyology). The redoubtable Moses Perley was the first vice-president, and Professor W.F. Ganong took an active part in the society.

The need for scientific knowledge

In that same year the Nova Scotian Institute of Science was formed. Some of its earliest publications dealt with the food fishes of that province, fishery laws, and other fishery matters. Among early investigators in Nova Scotia were Dr H.R. Storer, T.F. Knight, Dr J.B. Gilpin, Dr A. Gesner, and Reverend John Ambrose.

It was through the efforts of the Natural History Society of Montreal, and its president, Principal William Dawson of McGill University, that the cooperation of

23

the minister of marine and fisheries was obtained to permit a marine dredging expedition. This was carried out in the Gulf of St Lawrence by the society's curator, J.F. Whiteaves, on board the schooners *La Canadienne* and *Stella Maris* during the summers of 1871 and 1872. The following year Whiteaves again carried out investigations on the oyster fisheries in the gulf, and in 1887 a commission was appointed, with Edward Hackett as chairman, to inquire into conditions related to this fishery as well as the lobster fishery. The commission brought in a report on 5 November 1887, which included a number of recommendations to improve the fishery. Meanwhile, Professor W.F. Ganong published his monograph on *The Economic Mollusca of Acadia*, in which he urged government regulation of the oyster fisheries and appointment of a commissioner 'who could not only scientifically investigate the needs of our own region, but also direct the work of private culturists.' Sir Charles Tupper, who was then minister of marine and fisheries, was greatly impressed with the publication, and commented favorably on it.

Prince appointment

It was clear that the government was more and more inclined to turn to science for aid in dealing with the problems of the fishery. So, in 1893, a specialist in fish embryology, Dr E.E. Prince, was appointed commissioner of fisheries. He came to the post from a position as professor of zoology and comparative anatomy at St Mungo's College of Glasgow, and was a disciple of Professor W.C. McIntosh, one of the leading fishery scientists of the world.

Prince immediately advocated 'a marine scientific station for Canada' in order to develop a seaside laboratory similar to that organized by McIntosh in Scotland. The idea was already a popular one in the scientific community. The first laboratory of this nature, the Naples Zoological Station, had been established in 1872, and the example had been quickly followed in England, Scotland, and the United States.

McMurrich's proposed research station

A young graduate of the University of Toronto, James Playfair McMurrich, while continuing his studies at the Johns Hopkins University in Baltimore, had had first-hand experience in such a station, for he had spent several summers at its Zoological Laboratory on the Atlantic coast of the United States. There he met American and English biologists actively engaged in ichthyological and marine biological research. In 1884 McMurrich, then a professor of biology at the Ontario Agricultural College in Guelph, published in *The Week* an article entitled 'Science in Canada.' After dealing with science in general and its application to agriculture, he said:

Another source of revenue to the country is in great need of encouragement and protection by scientific investigation. The Department of Fisheries has become of great importance to Canada and something has already been accomplished by the establishment of fish hatcheries, etc., but this affects only our inland waters, the sea fisheries receiving little or no benefit. The life histories of our various food fishes, their manner and times of migration, their spawning localities, their food, their personal enemies, the destroyers of their food, all these should be properly investigated. True the Americans have done much for us in this line, but there is much yet to be done; in fact, the entire fisheries of the western coast are yet to be studied. Stations established on Vancouver Island, on the Gulf of St. Lawrence and on the Nova Scotia coast with facilities for investigations in the form of small steamers provided with dredging apparatus and hatching banks would in a very short time repay the expenditure made upon them by the important aid they would afford by enabling us to adopt measures for the increase of our fisheries by informing us of their real extent, of which we are yet comparatively ignorant, and by preventing their wanton destruction.

Apart, however, from the practical value the establishment of such departments would have, the scientific importance of their work should not be overlooked. Generalizations of which at present we have not the slightest inkling, might be arrived at; all departments of science would receive encouragement; a new stimulus to scientific work would be aroused in our country and the present ban under which science lies would be removed.

But in this search for practical discoveries let not pure science be neglected. Though apparently valueless at the time, it will yield abundant fruit in the future, not only by becoming in its turn capable of direct application, but also by establishing a starting point where new investigation may branch out in the yet undiscovered realms.

McMurrich's idea was to lie fallow for a few years, but he continued to demonstrate his belief in biological stations by his research work in the Marine Biological Station at Woods Hole, Massachusetts, where he also served as a librarian and evening lecturer in 1890 and as a trustee from 1892 to 1901.

The next pressure for a biological station came, surprisingly enough, from Newfoundland, which, politically, was still feuding with Canada over the fisheries. There, through the influence of Reverend Moses Harvey and others, a fisheries commission, with Dr Harvey as secretary, was appointed by the Newfoundland government. Adolph Neilson, a former inspector of fisheries in Norway, was appointed superintendent of fisheries and produced extensive reports on the Newfoundland fisheries. But Dr Harvey's interests in the fisheries was far-ranging, and on 1 June 1892 he read a paper before Section IV of the Royal Society of Canada. It concluded: A call from Newfoundland's Harvey

In closing this paper the writer wishes to point out the desirability of establishing a Biological Station for the study of Ichthyology and Marine Biology in all their branches. This is a work for the Dominion of Canada, whose fishing interests are so extensive, but, if established at some eligible locality on the shores of the Lower Provinces, such an institution would equally benefit the great fisheries of Newfoundland, and that colony might be expected to share in the expenses of its erection and working ... The scientific and practical should be so combined to render it a Fishery School. It would include a laboratory in which the structure and habits of all kinds of marine life would be studied, especially the life, conditions, food, mode of propagation, movements, etc., of such fishes as possess an economic value ...

The interests of pure biology, as a science, would be served by such an institution ... If we want to increase the quantities of our food fishes, our lobsters and oysters, all our operations must rest on a scientific foundation, and all our regulations of our fisheries must have their basis in a scientific study of fish-life. Failing such accurate knowledge, our legislation regarding the fisheries will be largely groping in the dark; and all efforts for their preservation and improvement will come short of the objects aimed at. A thorough knowledge of the mode of life, development, etc., of those fishes which constitute such a large portion of the national wealth of British North America, is essential to their preservation and the extension of these great industries.

After considering Dr Harvey's paper, the Royal Society decided to appoint a committee to consider the whole subject of marine and inland fisheries from a scientific point of view, to collect information on the subject and to take such steps as the committee might think proper to diffuse information, and to bring the subject before the government in order that means might be taken for effectively promoting research in this important department. However, there is no record of any report by this committee.

Meanwhile the government had appointed Prince as commissioner and general inspector of fisheries for Canada, and he began almost immediately to call the attention of his department to the need for a biological station in Canada. Requested Prince goes into action

25

by the minister of marine and fisheries to prepare a report on the subject, Prince wasted no time. The Special Appended Report no. 2 to the 26th Annual Report of the department for the year 1893 stated:

There is a growing feeling prevailing that our country, which in so many respects has taken a leading place among the nations in regard to fishery matters ... should take a position of equality with other countries in the furtherance of marine and fresh water biological research. Proposals have from time to time been made in this direction, and professors in our universities, as well as practical fishing authorities, have given strong expression to views in favour of a biological station for Canada, on the lines of such institutions in other countries. A period has now been reached, it may be justly claimed, when such a suggested scheme should assume practical shape ... The foundation of a marine station upon the coast would render possible the prosecution of ... necessary researches. The individual efforts of naturalists can never lead to the rapid accumulation of facts necessary to a science of the Canadian Fisheries ... A complete biological survey of the coastal waters of the Dominion ... would fall within the operations of a marine station, and would be gradually pushed forward season by season until the physical conditions, the biological characteristics, the fauna and the flora of every area, wherein the fishing industry is prosecuted, are made known and are available for the guidance and information of those actively engaged in fishing pursuits ... Methods of preserving and transporting fish, improved means of drying, salting, canning and refrigeration ... would be thoroughly tested and new improvements, or novel and unsuspected methods made known ... They all end in supremely practical results, and bear directly upon the welfare and prosperity of the great fishing industries ... Legislation has often been hazardous on account of this lack of ascertained fact and the existence of contradictory opinions. Primarily, a marine station would be a centre for investigation and research for the promotion and diffusion of knowledge. Without interfering with this first and most important work, such a station might also be a school for teaching and for scientific study ...

It is not too much to anticipate that the benefits resulting from the establishment of a marine station at some central point as indicated, would make obvious the necessity of others ... Certainly a more northern and a more southern marine station in the future would promote the great work of thorough investigation. The value and extent of the lake fisheries, in a similar way, would call for an inland station, in order that the conditions of life in these vast inland seas might be better understood. Certainly the practical benefits of a more trustworthy knowledge of our marine and fresh water fisheries can alone lead to their prosperity and growth in the future. Holland has established a floating station which can be moved season by season from one point of the coast to another, and with one permanent marine station as a central institution, such subsidiary stations, migratory or otherwise, might be found useful as secondary adjuncts in a work so extensive.

Curious coincidence

The report appeared in 1894, and it is highly likely that between the date of its publication and 6 May 1895 Professor Prince and Professor A.P. Knight of Queen's University had some serious conversations. In any event, on the latter date Professor Knight sent off the following letter to the honorary secretary of the Royal Society of Canada:

I venture to call the attention of the Royal Society of Canada to the desirability of having either a lake or a seaside laboratory in Canada, to which our naturalists could resort for some months every summer and undertake research work. I have myself felt the need of such an institution, and I know of other biologists in Ontario who have felt it also. Last summer, for example, there were seven Canadians working at the Marine Laboratory at Woods Hole,

Mass., and I have no doubt that more would have been there if they had known of the advantages offered for study and investigation.

A beginning could best be made in Canada in connection with the government fish hatcheries, and I am sure that Professor Prince, the fish commissioner, would willingly co-operate with the Royal Society in formulating a plan for such work, and submitting it to the Minister of Marine and Fisheries ... Canada ought to make a beginning soon. It seems too bad that her biologists should be compelled to expatriate themselves in order to gratify so harmless an ambition as that of adding a little to the sum of human knowledge.

A committee was appointed to study the proposal. It was made up of Messrs D.P. Penhallow, T.J.W. Burgess, and J.F. Whiteaves, who at the 1896 meeting of the society presented this brief report: 'The Royal Society of Canada, fully appreciating the great advantages of an economic and scientific character to be derived from the maintenance of a marine biological station, desire to recognize the importance of the report on this subject presented by the Commissioner of Fisheries in 1894, and to urge upon the Government the desirability of taking steps at an early date to bring these recommendations into force.'

A more effective committee

The report was approved by the society, along with a resolution that it be adopted and transmitted to the minister of marine and fisheries for such action as might be possible.

The next move was worthy of a Machiavelli. Someone (was it Prince?) noted that the 67th meeting of the British Association for the Advancement of Science would be held in Toronto in 1897. The matter was brought before the association, which in 1896 promptly appointed a committee to consider the question of establishing a station to investigate the marine fauna of the Atlantic coast of Canada, with Professor L.C. Miall as chairman. At the 1897 Toronto meeting the association replaced this committee with a Canadian committee charged with the task of establishing a biological station in the Gulf of St Lawrence. Dr E.E. Prince was named chairman and Professor D.P. Penhallow secretary. Other members included Professor J. Macoun, Dr T. Wesley Mills, Professor E. MacBride, Dr A.B. Macallum, and Mr W.T. Thiselton-Dyer.

An astute move

The committee met in Montreal and decided to bring the project before the government during the 1898 session. Accordingly, a memorial was prepared and presented to the minister of marine and fisheries, Sir L.H. Davies, on 20 April 1898. The committee was supported by delegates from major Canadian universities and scientific societies. But, above all, it spoke for the world-famous British Association for the Advancement of Science: and Sir L.H. Davies had been knighted by the queen precisely for his service to the fisheries.

The memorial made the following recommendations:

The proposal

1 That a floating station be established in the Gulf of St Lawrence for a period of five years.
2 That this station be established first on the southern coast of Prince Edward Island and that it be moved each year to a new location according to requirements.
3 That the various universities and scientific bodies of Canada be granted certain privileges with respect to opportunities for qualified investigators as may hereafter be determined.
4 That the scientific work of the station be executed so far as possible by experienced investigators connected with the various universities.

5 That although the station remain a government institution, the administration be vested in a special board consisting of one or more representatives from the Department of Marine and Fisheries and one representative from each of the universities reported in support of this petition.

6 That an appropriation of $15,000 be made for this purpose, of which $5,000 shall be applied to construction and outfitting and $10,000 to maintenance for a period of five years.

Not too surprisingly the minister approved the memorial, and the matter was passed by Parliament on 10 June 1898. An appropriation of $7,000 was made to meet the cost of construction of the floating laboratory and running expenses for one year. It was anticipated that the station would be completed for use in the 1899 season.

The minister stated that he would appoint Commissioner of Fisheries Prince to represent the department on the special board to be appointed. The committee met, and formed a Board of Management of the Marine Biological Station of Canada. In London, at the meeting of the British Association for the Advancement of Science in 1898, the committee reported upon the success of its mission and, mission accomplished, was accordingly dissolved.

3 The first Board and the movable station, 1898–1907

The Board of Management of the Marine Biological Station, which was created in 1898, represented a remarkable cross-section of Canadian scientific talent, most members being men in the full stride of their academic careers. Originally, seven members were appointed, and two more were added in 1899 to make a nine-member board.

The Board consisted of Dr E.E. Prince, representing the Department of Fisheries, as director; Professor D.P. Penhallow of McGill University, as secretary-treasurer; and seven trustees: Professor R. Ramsay Wright of the University of Toronto, Professor L.W. Bailey of the University of New Brunswick, Professor A.P. Knight of Queen's University, Dr A.B. Macallum of the University of Toronto, Canon V.-A. Huard of Laval University, Dr A.H. MacKay of Dalhousie University, and Professor E.W. MacBride of McGill University. (Brief biographies of the nine members appear at the end of the chapter.)

Oldest, at 59, was Bailey, and youngest, at 32, was MacBride. The rest were in their 40s. Most had been involved in the early efforts to create a marine biological station. Prince, immediately after his appointment as commissioner of fisheries in 1893, had recommended the station. MacKay and Ramsay Wright had been part of the abortive committee that had been set up in 1892 by the Royal Society following Dr Harvey's proposal. Knight had written his letter to the Royal Society in 1895; Penhallow, as secretary of the committee in 1892 and 1895, had been involved in both efforts, and was part of the 1896 committee which strongly recommended government action; and Prince, Penhallow, MacBride, and Macallum had formed part of the committee of the British Association for the Advancement of Science which presented the memorial to the government in 1898. When that committee

29

E.E. Prince, member of the Board
1898–1936, chairman 1898–1921, editor
1901–18

A.P. Knight, member of the Board
1898–1926, chairman 1921–5

The Board's tasks

appeared before the minister of marine and fisheries, Ramsay Wright and Bailey were among the supporting delegates from the universities.

As it prepared to launch its investigations into the fisheries of Canada, the Board was faced from the start with two major tasks: it had to prove its value to the Canadian government as an instrument of research in aid of the Canadian fisheries, and it had to prove to the scientific community that it could operate a valuable laboratory for biological and fisheries research. Launched on an appropriation of $15,000, of which $5,000 was to cover the construction and outfitting of a floating laboratory and $10,000 the cost of its maintenance for five years, at a rate of $2,000 per year, the venture was hardly overcapitalized. The Board members served without remuneration.

Prince made himself chief propagandist with the government for the work of the Board, through his frequent departmental reports on its activities, and he performed a similar role with the Royal Society by providing a report at each annual meeting from 1900 to 1913. But it was the scientific papers which proceeded to flow from the summers at the movable station that persuaded the scientific community that it was a valid and important instrument in the development of the science of ichthyology. Similarly, many of the subjects of the papers were matters of practical importance dealing with problems that faced the Canadian fishing industry and thereby justified the enterprise in the eyes of Parliament and successive administrations.

THE MOVABLE STATION

According to the first issue of *Contributions to Canadian Biology*, which appeared in 1901, the first meeting of the Board of Management took place on 10 February

30

A.B. Macallum, member of the Board
1898–1919

A.H. MacKay, member of the Board
1898–1926

1898, but the date was more likely 1899 since the establishment of the Marine Biological Station was authorized by Parliament only in June 1898. It is recorded that plans and specifications for the station were considered at this meeting and by May 1899 notices were placed in newspapers at Saint John, New Brunswick, calling for tenders 'for the construction of a wooden building to be used as a marine biological station.'

The contract was awarded to Messrs D.W. Clark & Sons of West Saint John for construction during the early part of the summer, and Prince came down from Ottawa to select the best site. A location at St Andrews, New Brunswick, was chosen, on the eastern shore of Indian Point near Tongue Shoal and facing Malloch's weir, and work on the station began quickly.

The selection of St Andrews as the first site of the station did not go unquestioned by the auditor-general when the expenditure for the building came to his vigilant eye. He promptly wrote the deputy minister of marine and fisheries, pointing out that authorization for the appropriation was for a building in the Gulf of St Lawrence, and the only St Andrews that he could find on the map was 'not on the Gulf of the St. Lawrence, but on the southwest coast of New Brunswick.' He demanded an explanation, and was apparently satisfied when Deputy Minister F. Goudreau wrote back and politely pointed out that it was a floating station, capable of being towed anywhere in the Gulf of St Lawrence.

Choice of
St Andrews site
challenged

Prince, in his report to the Royal Society of Canada in 1901, gave this explanation for the choice of initial site: 'The waters of the Gulf of St. Lawrence invited from the first the efforts of the scientific staff who proposed to take advantage of the marine biological station, but it was deemed desirable to undertake preliminary work in the prolific waters of the Bay of Fundy, where problems of urgency and importance awaited attention.'

31

L.W. Bailey, member of the Board
1898–1923

Canon V.-A. Huard, member of the Board
1898–1928

The choice was a happy one for the scientists. Dr Joseph Stafford, who served as curator of the station for several years, wrote in *Contributions to Canadian Biology* for 1902–5:

The first location of the Station at St. Andrews presented many special advantages. Its southerly sheltered situation implied, close at hand, a rich and varied fauna while, further out, deep-water forms were also obtained ... The tide rises and falls about 28 feet, making enormous differences in the appearance of the shore and exerting a vast influence, not only upon the habits of many marine animals, but even extending to the inhabitants of the coast ... Where the laboratory stood, on the east side of 'the point' ... the lowest tides receded nearly 400 yards. With the rising tide strong currents are swept inwards, between the islands, carrying hosts of marine animals. When the tide falls again, numbers of these are left stranded on the beach, or confined in small pools easily accessible to the collector ... Turn what way he will, an observer is likely to come upon the common star-fish in many colour varieties, the sea-urchin and the sea-cucumber, among echinoderms. The Mollusca are abundantly represented ... *Nereis, Arenicola* (etc.) are common representatives of the worms; while crabs, hermit crabs, barnacles and sand-hoppers are the commonest types of Crustacea. A good many hydroids, Polyzoa, and sponges may also be easily procured along the shore.

The station in operation

The station was designed in the shape of an ark or oblong building, to be placed on a scow for movement from one location to another along the coastline. It could be either moored or hauled up on dry land above the high-water mark. During the first two years it was not placed upon the scow at all, but was erected on the shore at St Andrews. Prince described it effectively in his report:

32

R.R. Wright, member of the Board 1901–12
B.A. Bensley, director of the Georgian Bay Station 1902–13

The laboratory was completed in June, 1899, and is a neat one-story structure of wood, well-lighted from the roof and sides, and somewhat resembling a Pullman car, with a row of eight large windows along each side, and a door with sash provided with plate glass at either end. Its total length is 50 feet, the principal room, or main laboratory occupying the central part of the structure and forming a well-lighted and cheerful work-room, measuring 30 feet in length and 15 feet in breadth. Two tank and store-rooms are at the anterior end, each room six feet by six feet, while at the opposite end are four rooms, one reserved for the director, and another adjacent to the director's, devoted to the use of the attendant, and provided with a sink and spacious shelving, and certain kitchen appliances, while on the opposite side of the passage, are two rooms, one used as a tank room and the other as a chemical room.

The scow was obtained later. It was larger than the laboratory, measuring 60 by 19$\frac{1}{2}$ feet, with about nine feet vertical depth and providing a narrow platform around the sides of the laboratory and spacious platforms at each end. Both salt and fresh water could be pumped to the workers' stations, which were equipped with individual porcelain wash basins, and a drain beneath the laboratory carried away waste water.

First to arrive that summer of 1899 were Professor Prince and two brothers from the University of Toronto, R.R. and B.A. Bensley. They were followed shortly afterwards by another University of Toronto scientist, Joseph Stafford. It was the first contingent of an army of University of Toronto scientists who, over the years, worked closely with the Board and contributed substantially to its achievements. (Brief biographies of the three scientists are given at the end of the chapter.)

During the first season at St Andrews, research was conducted with the help of a rowboat and a hired sailboat. R.R. Bensley studied the food of fishes and B.A.

The first St Andrews season, 1899

33

The movable Marine Biological Station, Canada's first fisheries research facility, operative from 1899 to 1906

Bensley concentrated on the relation of the sardine industry to the herring industry, the decline of which was attributed to the taking of the young for sardines. Stafford began his study of the clams and the clam industry. His report and that of B.A. Bensley appeared in the first issue of *Contributions to Canadian Biology* in 1901. Stafford also began a study of the fauna of the region, which would continue along the coast as the station was moved to new locations. Knight joined the team later and began an investigation of the eggs and larvae of pelagic fishes and the morphology of the lobster. Macallum also arrived to study the chemistry of the Medusae, and both Bailey and Dr F.S. Jackson of McGill came to the station, though the extent of their studies was not reported.

According to Prince the subjects covered during the first season were 'largely faunistic' but they also included the food of fishes, the sardine industry, catches in sardine weirs, the clam fishery, the spawn of fishes, early stages in the life history of the lobster, and 'research in physiological chemistry, dealing with the analysis of the constituent matters in *Aurelia* and Medusae generally.'

Prince had a happy knack for making a little sound like a lot. Knight was brutally realistic. He reported: 'Last summer we did little more than make a beginning. The carpenters were not out of the building until August, and even at the end of the season the equipment was not complete. No regular servants were employed. The workers were sailors, fishermen, laborers and scientists by turn. We swept floors, washed glassware, delved for worms, waded in mud, hauled up boats on rocky beaches, argued, grumbled, swore, but all were determined to return next season.' (See Rigby and Huntsman 1958:31)

Second season more active, 1900

Operations really got into gear during the second season. The Board of Management met at St Andrews in June 1900, with Prince, Knight, Bailey, and MacKay in attendance, and discussed moving the station to Nova Scotia. It was finally decided to spend a second season at St Andrews to complete work already started. Prince, the director, had arrived in May to reopen the station and he made a study of the young stages of salmon. Dr Jackson was back and acted as curator for a month, Stafford taking over for the final two months of the season. The latter continued his faunal studies, with particular attention to parasitic worms, and completed his clam study.

Five other Board members were there during the season. Macallum continued his

34

End view of the movable station, on scow beached at Canso, 1901–2. Group of Board members and research associates includes (l. to r.) G.A. Cornish, ?, A.P. Knight, J. Stafford (curator 1901–7), J. Fowler, C.M. Fraser, E.E. Prince, R.R. Wright, ?, A.B. Macallum

study of the chemistry of the Medusae; Knight studied the blood of fishes and made experiments on the effects of polluted water on fish life; and MacBride undertook a study of the starfish, assisted by MacKay and Bailey.

Four other workers also came: Professor James Fowler of Queen's to investigate the flora of the region and the marine algae of Passamaquoddy Bay; F.H. Scott of Toronto to study the food of sea urchins and other echinoderms; T.T. Bower of Queen's to assist Knight and make a collection for the university museum; and Susan Ganong of Ladies' College, Halifax. A paper on the paired fins of the mackerel shark was written by MacKay and Prince and appeared in the first edition of *Contributions*.

The gift of a complete set of 'Challenger' reports, 50 large volumes, was presented by the British government to the station's library. A note on the flyleaf explaining torn pages and other damage stated that the set had formed part of the cargo of a ship wrecked off the Lincolnshire coast, which had subsequently been salvaged.

The station obtained its own 22-foot gasoline launch at a cost of $650, but it lasted only a few weeks as the engine had been designed for fresh water and soon became corroded and useless in salt water. The steamer *Annie* was chartered for the balance of the season.

In 1901 it was decided to move the station to Canso, Nova Scotia, which was an important fishing center. After loading it aboard the scow, the station was put in tow of the government cruiser *Curlew*, whose captain J.H. Pratt subsequently reported:

The move to Canso, 1901

On May 31, we returned to St. Andrews ... and on June 2 moored alongside of Biological Station ... Next morning in a strong gale, testing towing appliances by towing station to

35

Campobello, and finding everything working satisfactorily, steamed next day across the Bay of Fundy to Brier Island. The second day we succeeded in safely getting our tow around Cape Sable and anchored in Shelburne. The next morning at daylight, we made another start, and the breeze sprung up from the southwest, freshing during the day and raising up a nasty sea. At 10 a.m. off Liverpool, the towing gear on board the scow broke, but we succeeded with little trouble in picking her up again. At 2 p.m. off LaHave the gear on the scow broke again, and although a heavy sea was running, we managed to pass a hawser to her and steamed into Lunenburg to repair damages.

A heavy sea and fog compelled us to put into Halifax on the 8th, but on the 11th the weather cleared up, and we proceeded towards Canso, arriving there on the morning of the 12th, and we then placed the station in safe quarters. [See Rigby and Huntsman 1958:33–4]

Thus the movable station survived its maiden voyage.

At a meeting of the Board on 20 February 1901, Ramsay Wright was nominated assistant director of the station. The nomination was approved by Order-in-Council, and authority was given to engage a senior curator for the whole season. Stafford was appointed curator for the 1901 season and, a month before the station finally arrived at Canso, he was there, beginning work which lasted for 17 weeks that year. Prince was there for only 20 days in August, the work for the most part being supervised by Wright.

Before leaving for Canso early in the season, Wright spent some time securing new equipment, which included nets, dredges, beam trawls, deep sea thermometers, sounding apparatus, and other gear required for the new location and different conditions at the new site. He also obtained a set of the Reports and Bulletins of the United States Commission of Fish and Fisheries to add to the station's growing library.

Knight described conditions at the new site: 'The Assistant Director, Professor Ramsay Wright, assumed charge as soon as the laboratory arrived. Much work had to be done to prepare it for the scientists who were expected the middle of June. The building was dragged up on shore, high above tide mark ... A large tank for fresh water was built upon the roof, and connected with the wash basins inside. A hot-air engine was installed to pump fresh water from a neighbouring well; and a hand force-pump to draw salt water from the sea. As we were cramped for accommodation, part of an adjoining building was rented for a reference library and reading room.'

Canso had been chosen as the new location for the station because it was the largest fishing center on the east coast and also because the coast there was completely different from that at St Andrews. The scientists counted on getting species from the fishermen which would be difficult to obtain elsewhere, and they counted on a very different fauna than at St Andrews. Knight noted that they were not disappointed.

Whitman Brothers
support the station

The Whitman Brothers, Canada's largest fish merchants, were the chief fish buyers at Canso and they placed their steam tug at the disposal of the scientists, drawing Knight's warm praise: 'And the Messrs Whitman – who can sufficiently thank them for the unending assistance which they gave to the workers during the whole summer? Time after time a steam tug and crew were placed gratis at the disposal of the workers; and it is only simple justice to say that without the Whitmans, our results would, in some lines of research, have been very meagre indeed.'

That year the workers included Prince, who studied the eggs and life history of clupeoids; Ramsay Wright, who examined the plankton; Macallum, continuing his monumental study of the chemistry of Medusae and making analyses of sea water and studying its relation to jellyfish life; J. Fowler, collecting the flora of Canso; Stafford, studying trematodes of fishes and the fauna of the region; Knight, examining the effect of dynamite explosions on fish life; and two more University of Toronto recruits, C. McLean Fraser (of whom more in later chapters), collecting and preparing specimens for the University of Toronto Museum of Zoology, and George A. Cornish, studying fishes of the region.

During 1900 Penhallow had made arrangements with Dr O.C. Whitman, director of the Marine Biological Station at Woods Hole, for the exchange of scientists between Woods Hole and Canada's Marine Biological Station on a seasonal basis. Dr Linville of New York was the American nominee at Canso in 1901, and a Mr Shearer of McGill, making an investigation of annelids, was the Canadian at Woods Hole. *Exchange of biologists*

Knight's experiments on the effects of dynamite on fish life were carried on in a number of locations: the open sea off Canso harbor, in Saint John harbor in a salmon weir, in Lake Ontario at Kingston harbor, and on the Ottawa River. The results were published in *Contributions to Canadian Biology* for 1902–5.

At the annual meeting of the Royal Society of Canada held in May 1902, Prince reported on the work of the station. A committee was appointed, consisting of MacKay, Macallum, Wright, and Penhallow, along with Sir James Grant and Dr William Saunders, to consider increased facilities for biological work in Canada. At a general meeting of the society a resolution was passed: *Royal Society calls for more government funds*

The Royal Society of Canada desires to express its appreciation of the great scientific and economic value of the work at present conducted at the Marine Biological Station of Canada, and in view of the great activity at present manifested in Europe with respect to international co-operation in the prosecution of studies relative to the fishing industry, it takes this opportunity to express to the Honourable the Minister of Fisheries the hope that the work which is designed to contribute so largely to the welfare of one of the most important industries may receive from the Dominion Government and from the Department of Marine and Fisheries the most liberal assistance.

There is no indication that this appeal had any immediate effect, because plans for a new launch were shelved that summer. A larger boat than the gasoline launch was needed as the rocky coast of Canso made shore collecting difficult, and material had to be obtained by netting, trawling, and dredging. In an open and often rough sea, this could be dangerous in a small boat. Many years later, Fraser wrote:

As to the personnel, perhaps the most distinct impression then and now, is that they were rather a hopeless lot of seamen. Stafford, of course, was ready for anything at any time ... and in consequence Stafford and I had to do practically all the collecting that was done outside, ... The boat equipment consisted of a launch, about 23 feet, with a two-cycle engine that was as cantankerous as these usually are or were. There was practically no control of the speed and consequently although we used it to go out for shore collecting it was of little use for dredging. Occasionally when there was quite a wind, we dredged by letting the boat drift. We used a rope cable and hauled it by hand. There was a large rowboat with two pairs of oars and a lighter rowboat with one pair. [Letter to Dr A.G. Huntsman, 23 Nov. 1937; see Rigby and Huntsman 1958:37]

The second season at Canso lasted for 19 weeks. Ramsay Wright supervised the work and Stafford was again curator. Investigations initiated in 1901 continued. *Last Canso season, 1902*

These included plankton studies by Wright, faunal studies by Stafford, chemistry of Medusae by Macallum, hydroids by Fraser, fishes of the region by G.A. Cornish. MacKay studied the diatoms; F.R. Anderson of Mount Allison University, the halcarids; and C.B. Robinson of Pictou Academy, Hydrozoa. Fraser also continued his collection for the University of Toronto Museum.

The Board followed up the Royal Society's initiative. In July 1902 it resolved:

That at the meeting of the Board in January a delegation should be appointed to bring before the Government the desirability of continuing the work of the station, and of rendering it more efficient by increasing the annual appropriation from $2,000 to $3,000, also of appropriating a sum of $4,000 for the purchase of a boat suitable for the dredging and other scientific operations of the station, attention being called to the fact that approximately $3,500 of the appropriations made for the early years of its existence had been allowed to lapse. Further it was recommended that the said delegation should request that the amount of the annual appropriation for the conduct of the station should be handed to the Secretary-Treasurer of the Board, to be expended by him as directed by the Board.

It was also agreed to pay $5.00 per week for the living expenses of those workers not covered by government expenses. The director and assistant director undertook to examine suitable locations on or near Northumberland Straits for the next season.

Oyster studies initiated at Malpeque, 1903

The site selected was at Malpeque, Richmond Bay, Prince Edward Island, the center of the oyster fishery. In the spring the station was placed on the scow and left Canso 18 May in tow of the government steamer *Brant*. Chedabucto Bay and the Strait of Canso were crossed safely, but in George Bay a storm developed and heavy seas flooded the scow, permitting the wash to pound against the doors and windows and even the station roof. The laboratory suffered heavy damage, and supplies and equipment were swept overboard. The station was towed to Pictou harbor and beached for repairs, under the direction of Stafford, who had been in Malpeque awaiting its arrival. Three weeks later the voyage resumed, and the station reached Malpeque on 22 June, where it received further repairs: walls were straightened; new partitions, doors, and windows put in; new tables and plumbing provided; and the roof mended. A well was drilled and fresh and salt water connections were made. A storehouse was built to accommodate boats and apparatus.

The state of the oyster fisheries had long concerned both government and industry. In 1887 a commission had been appointed to study it, and on its recommendation 250 acres of water area had been set aside for the study of natural and artificial oyster propagation. British oyster experts Frederick and Ernest Kemp had been brought over to examine the oyster beds, and Ernest Kemp was subsequently hired by the government to carry out a planned program.

Oysters were the main concern of the station in 1903. Wright, Stafford, MacBride, and J.J. MacKenzie of Toronto studied various aspects of their biology, with the assistance of L.C. Coleman of Toronto and R.R. Gates of Mount Allison University. Ernest Kemp, using the government steamer *Ostrea*, closely cooperated by making daily trips to the oyster beds to secure specimens. This work continued to 15 September, when the station closed for the season.

Wright also continued his studies of plankton in relation to the food of the oyster, and Stafford surveyed the fauna of the region. B.A. Bensley collected material for his class work at Toronto, and experiments in hatching lobster eggs were made.

38

Meanwhile, A.N. Whitman and Son, who had reported favorably on the station when it was in Canso, came out with a specific recommendation which would bear fruit six years later. Pointing out that the scientific investigation being carried out thus far had yielded results 'far beyond what might have been expected in view of the inadequate facilities available,' it called for a 'new departure ... in the study of marine life.' It called for a larger budget and a permanent establishment at some central point in close touch with the Atlantic fisheries. The establishment should include a marine museum, a steamboat attached to it, and a summer school established where university students could acquire first-hand knowledge of the fisheries. The proposal concluded: 'This, too, would cost money, but we are convinced that it would be money well spent, as well as any that is spent on experimental farms, agricultural colleges and dairy schools, or colleges of mining and metallurgy.'

A prophetic proposal

One may legitimately wonder at this date whether A.N. Whitman and Son had been subjected to any of the persuasive charm of the genial Professor Prince. In any event the program for the second season at Malpeque, in 1904, seems to have been designed to prove to the government just how vital its experiment in fisheries science had become.

Prepared by the Board at its meeting of 26 May, the program contained no less than 15 separate projects, including: plankton work; study of pelagic ova and young of bony fishes; continuation of the previous season's oyster investigations together with further experiments in raising oysters; studies on the smelts; faunistic studies of fishes at some of the fishing stations; experiments with the otter-trawl in cod, haddock, and hake fishing; experiments in lobster behavior; shad culture; a study of the peculiar structure developed on tidal gauge floats; studies on clam beds, both soft- and hard-shelled; inquiry into the depredations of dogfish; investigation of migrations of striped bass; examination of qualities of frozen and fresh bait (herring and squid); and pollution of streams and fishing grounds. No small program to cover with an appropriation of $2,000.

A formidable program, 1904

Considered in relation to the Royal Society resolution and the recommendation of A.N. Whitman and Son, the program might well have been deliberately devised by the wily Prince to impress his department and Parliament with the very good value they were getting from that slim appropriation, and to justify a loosening of the purse strings.

Kemp and the *Ostrea* were again on hand to aid oyster investigations, and Knight tested varieties of bait suitable for the capture of lobsters and continued his studies on sawdust in relation to fish life. Stafford resumed his faunal study, and MacBride was there, along with four Toronto workers: G.A. Cornish, L.C. Coleman, J.R.K. Murray, and Dr J.H. Faull.

In 1970 A.G. Huntsman, a later star in our story, recalled a celebrated oyster incident:

Eureka!

The investigations were carried on at Malpeque Bay mainly in connection with the oyster industry there, and one of the important things was to find out about the deposition of their young, which are called spat. The eggs of the oyster and the sperm float in the water, where fertilization takes place, and the larvae finally settle as so-called spat on something on the bottom. Professor MacBride was anxious to determine when the spat of these oysters would settle, and brush was placed in the water, on which one would expect the oyster spat to settle.

When it came time to examine the brush to see whether or not the spat had settled, Professor MacBride, of course, went to examine the brush, and he came to the laboratory, and like Archimedes he cried: 'Eureka! I have found it,' which, of course, meant rejoicing for those at the laboratory. But the Curator of the station, Professor Stafford of McGill, when he examined the brush, had a very different report to make. He said that the presumed spat were really coccid insects ... which were on the brush when it was put into the water.

A number of publications resulted from the 1904 program. Wright produced papers on 'Plankton' and 'Natural History of the Oyster.' Stafford reported on 'Fauna of the Region' and 'Larva and Spat of the Oyster.' Coleman produced a paper on 'Fauna of the Oyster Bed'; Cornish, 'Oyster Investigation'; Faull, 'Flora of the Oyster Bed'; Knight, 'Experiments on Bait for Lobsters' and 'Effects of Sawdust on Fish Life'; MacBride, 'Oyster Investigation'; and Murray prepared the illustrations for Wright's paper on the natural history of the oyster. Thus a substantial part of the ambitious 15-point program was in fact accomplished.

The move to Gaspé, 1905

It was decided at the annual meeting of the Board on 7 January 1905 to locate the station that season at the town of Gaspé, Quebec, in the heart of one of the oldest fishing industries on the coast. Stafford arrived in Malpeque on 6 June to make arrangements for the move. A new 5 horsepower engine was installed in the launch, and the scow was thoroughly overhauled before setting out. The government steamer *La Canadienne* picked up the tow on 30 June, arriving at Gaspé next day without incident. The station was located near the entrance to Gaspé harbor, and from this point collecting trips were made into Gaspé basin and dredging and plankton expeditions were made over the fishing grounds as far as Percé.

Workers at the station that season did not include Prince, Wright, or MacBride. In fact, Knight was the only member of the Board to visit the movable station during the last years of its operating life. The other Board members obviously had a larger project in view, and Stafford was left to wind up the career of the station, which had effectively served its purpose.

So, in 1905, Stafford continued his faunal studies and also studied parasites found in bivalve mollusks. Other investigators included Professor James Fowler of Queen's, who collected plants of the area; Dr Frederick Etherington of Queen's, who classified invertebrates with J.C. Simpson of McGill, who, in turn, determined the free-living species of Protozoa for Stafford; and A. Bruce Macallum and J. MacIntosh of Toronto, who classified the invertebrates and made a catalogue of the copepods.

Bait experiments, 1906

In the final Gaspé season Stafford continued his investigations and Knight carried on a series of fishery bait experiments at Griffin's Cove, Grand Grève, and Gaspé Head, chiefly to determine the respective merits of using fresh and frozen herring as bait, and to attempt to educate the fishermen in the use of frozen bait when fresh bait was not available. As always, Knight's activities had practical application.

Other workers included Irving R. Bell and J.G.R. Murray of Toronto, who had been sent by Wright to study parasitic forms. Bell submitted a report on Sporozoa and Murray one on the Haematozoa found in the blood of salt-water fish. The program had called for the marking of migrating salmon and sea-trout schools as well as studies of whales captured at Seven Islands, but there is no evidence that either project was carried out.

A disastrous journey, 1907

In 1907 it was decided to move the station to the north shore of the Gulf of St Lawrence at Seven Islands, which was then the location of a whaling station. However, while the laboratory atop the scow was being towed by the government cruiser *Princess*, the scow developed a leak and filled so rapidly that it had to be beached on the south shore at Grande Vallée for repairs. The station was put on shore, repaired, and as it turned out, eventually abandoned.

Stafford proceeded to Seven Islands, and was joined by P.M. Bayne and W.G. Scrimgeour of Toronto. The attic of a fish house near the water was obtained and cleaned up for use as a laboratory, and two tables were installed. R. Smith, a Cambridge graduate, arrived late in the season to complete the staff. Since the only transportation was a small and badly functioning motorboat and a rowboat with a sail, the main activity consisted of shore collecting. Bayne studied the marine Crustacea, Scrimgeour studied the hydroids, and Smith assisted in general collecting and dredging, which was done by towing the dredge from a rowboat. Stafford collected and studied the fauna of the area. At the whaling station Bayne obtained the ear bones of a whale, which ended up in the Zoological Museum at Toronto. His collection of marine Crustacea formed the subject of his MA thesis. Several new species were described and the distribution of others was extended to Seven Islands.

At the end of the season the motorboat and other effects were left in the care of M. François Galliene at Seven Islands, where they remained until November 1909. The motorboat was then brought to St Andrews and put into service at the new permanent station. The laboratory remained at Grande Vallée until the same date and was then placed under the administration of the St Andrews station. The books were transferred to the St Andrews library, but the laboratory and the scow vanished from mortal sight, though Grande Vallée legend now includes the apparition of a streetcar atop a scow, which makes its appearance on stormy nights, riding high on the crest of an enormous wave, wreathed in foam, and racing full-tilt for that rock-ribbed shore.

BIOGRAPHIES

Loring Woart Bailey
Bailey was the oldest member of the Board at its inception. Born at West Point, Virginia, in 1839, he was the son of Professor J.W. Bailey, a leading microscopist and authority on diatoms and Infusoria. The young Bailey showed an early interest in his father's specialties. He came to the University of New Brunswick at 22 and spent most of his academic career there, retiring in 1907 as professor emeritus at the age of 68. Although Bailey established his first reputation in Canada in geology, after his retirement he returned to his first interest, the microscope and diatoms. Working in conjunction with the Marine Biological Station at St Andrews, he pursued investigations of the diatoms in Canada, and his last paper, 'An Annotated Catalogue of the Diatoms of Canada,' was published in 1924, almost simultaneously with his death, which took place at the age of 86, on 10 January 1925.

V.-A. Huard
Canon Huard was born at St Roch near Quebec City in 1853. He studied at the Seminaire de Québec and was ordained in 1876. He taught biology and other subjects at the Seminaire de Chicoutimi, eventually becoming its superior. In 1901 he moved to Quebec, where in addition to performing his religious duties, he was for 25 years conservateur of the Musée de l'Instruction publique. He received the degree of docteur-ès-sciences from Laval University. Himself a talented naturalist, Canon Huard assumed the editorship of *le Naturaliste canadien* after the abbé Léon Provancher died in 1893, and retained the post until his own death in 1929. He was a strong advocate of a marine biological station, and the activities of the scientists at the station formed the subject of several reports in *le Naturaliste canadien*.

41

Canon Huard was a scholar in the broadest sense. He wrote articles on history, travel, zoology, general science, entomology, and biography. His books on zoology, hygiene, and entomology became school texts. He served on the Board of Management of the Marine Biological Station and on the succeeding Biological Board of Canada until 1928.

Archibald Patterson Knight

Knight was both an educator and a scientist. Born near Renfrew, Ontario in 1849, he graduated from Queen's University in 1872. He was a high-school headmaster and teacher while pursuing studies in medicine at the University of Toronto, and received his MD degree in 1886. He was appointed professor of biology and physiology at Queen's in 1892 and retired in 1919.

Knight had been active in the events leading up to the formation of the Board and he was a member of its executive committee from the outset. Upon his retirement from Queen's, he served as chairman of the Board from 1920 to 1925.

As will be revealed in later chapters, Knight's scientific activities with the Board led him through a variety of subjects, beginning with the pollution of waters by sawdust and an experimental investigation of the efficiency of frozen baits for catching cod. He investigated the problems of lobster hatching and lobster canning, demonstrating the futility of the hatching methods then employed, and saving the taxpayers thousands of dollars by advising the abolition of the hatcheries then in operation. The canneries profited by his insistence on the need for cleanliness and better sterilization techniques. In his later years he became interested in the general problems of fish culture, and his interest continued past his retirement from the Board in 1925 and almost until his death in 1936.

Archibald Byron Macallum

Macallum was born at Belmont, Ontario, in 1858. After graduating from the University of Toronto, for a time he taught high school at Cornwall, Ontario, before returning to the university. From 1890 to 1908 he was professor of physiology at Toronto, and from 1908 to 1916 he held the new chair of biochemistry there. When the National Research Council was formed in 1917, he became its first chairman, and was credited for its development along scientific lines. He served as secretary-treasurer of the Biological Board of Canada from 1912 to 1919, when his long association with the Board came to an abrupt end after he had thwarted an effort to pass a damaging amendment to the Biological Board Act (see chapter 8). Macallum's appointment was not renewed and, coincidentally, he resigned from the National Research Council to return to academic life in 1920 as head of the Department of Biochemistry at McGill. After his retirement there in 1928, he moved to the University of Western Ontario, where he continued his studies and writings until a few months before his death in 1934.

Ernest William MacBride

A member of the Board whose term was relatively short but active was Professor MacBride, who was also, at 32, the youngest member at its inception. Born at Belfast in 1866, he had a brilliant career at Cambridge before coming to McGill as professor of zoology in 1897. For several years he had held the appointment to the Cambridge table at the Naples Biological Station, where he had been a fellow student with the future Norwegian fisheries expert, Dr Johan Hjort. This background made MacBride a valuable member of the Board.

He set religious Montreal on its ear when his inaugural lecture at McGill made it clear that his approach to science was evolutionary. He was denounced for 'denying God' and, not the least disturbed by the charge, he proceeded to give a series of public lectures at the YMCA, patently combating the fundamentalist position concerning the Bible.

He played an active role in the first programs of the movable station, but in 1909 he was lured back to the Old Country with an appointment as assistant professor of zoology at the Imperial College of Science and Technology in London, under his old mentor Adam Sedgwick. Later he succeeded to the professional chair which once was Huxley's. He continued his interest in fisheries research until his death in 1940.

Alexander Howard MacKay

MacKay was born at Plainfield, Pictou County, Nova Scotia, in 1848, and he died on his 81st anniversary in 1929. He served as a member of the Board's executive committee for 30 years. He had been a teacher from the first, and after serving as principal at a series of Nova Scotia academies with exceptional administrative skill, he was appointed superintendent of education for Nova Scotia in 1891. He held that post until 1927, two years before his death. On the executive committee of the Board, his sane and sober judgment was greatly valued, and he made himself an authority on the diatoms, collaborating closely with Bailey in some of his later papers.

David Pearce Penhallow

Penhallow was fated to have the shortest term of membership of anyone on the roster of the original Board of Management. Born at Kittery Point, Maine, in 1854, he came to McGill in 1883 as professor of botany, after spending four years at the Imperial College of Agriculture in Sapporo, Japan, as professor of botany and chemistry. He worked closely with Sir William Dawson in paleobotany, and as an outstanding administrator he held numerous key posts in various learned societies and committees as well as editorial posts on a number of scientific publications.

In addition to all this, Penhallow played a key role in the work of the Board of Management, particularly in the establishment of the Marine Biological Station on a permanent site from 1907 on. In October 1909 he suffered a serious breakdown, and had to drop all academic activities for many months. His recovery was slow, and it was felt that a longer holiday abroad might restore him to full health. On his way to England with his wife for an extended holiday, this gifted man died at sea, on 20 October 1910, in his 56th year.

Edward Ernest Prince

Director of the Board of Management and destined to serve on it and the succeeding Biological Board of Canada for a longer term than any other founding member – 38 years in all – was 'the genial Professor Prince,' as he was widely called. Prince was born at Leeds, England, in 1858, and completed his education at the universities of St Andrews, Cambridge, and Edinburgh. In 1884 he was senior assistant in zoology at Edinburgh, and the following year he worked as a naturalist under the renowned Professor W.C. McIntosh at the Marine Laboratory at St Andrews, Scotland.

When the Canadian fisheries service was reorganized in 1892, combining Fisheries and Marine under a single deputy minister, it was decided to obtain a fisheries scientist. At that time McIntosh was regarded as the world's leading authority on fisheries, and his laboratory was turning out specialists who were sought for responsible positions all over the world. As one of these, Prince was selected scientific advisor to the Canadian fisheries service and was named commissioner and general inspector of fisheries.

As previously noted, Prince played an active role in the formation of the Board. His interest continued until his retirement in 1924, when the following minute was recorded at a Board meeting: 'In connection with the Board's suggestion as to filling Dr Prince's place on the Board, the Secretary explained that he had discussed the matter with the Department, and it was felt that owing to Dr Prince's connection with the institution of the Board and his long association therewith that he should be allowed to remain a member of the Board as long as he chooses to retain that connection with it.' He remained an ex officio member until his death in 1936.

Robert Ramsay Wright

Ramsay Wright was another Old Country transplant who made a formidable contribution to Canadian biology before retiring to Oxford at 60 to study the Greek classics. Born at Aloa, Scotland, in 1852, he gave up an opportunity to become a member of the Challenger Deep Sea Expedition to become professor of natural history at the University of Toronto in 1874. That university was soon to feel the impact of his drive and enthusiasm for science, in particular in marine exploration and the application of biology to medicine.

In 1884 he published a series of studies on the catfish *Ameiurus* with three of his students, A.B. Macallum, J.P. McMurrich, and J.J. MacKenzie, all destined for brilliant scientific careers. In 1892 he made a report on the fishes and fisheries of Ontario. Undoubtedly his greatest and most lasting accomplishment for the fisheries of Canada was, however, the drive and inspiration that he furnished to the Department of Zoology at the University of Toronto, for its students and graduates were to prove a main source of supply for the scientists that powered and built the Fisheries Research Board of Canada during the first 40 years of its existence.

Professor Wright died in England in 1933.

Robert Russell Bensley and Benjamin Arthur Bensley

Each of the Bensley brothers was remarkable in his own right. Robert Russell was a demonstrator at $800 a year in the Department of Biology under Ramsay Wright and practiced medicine part time to avoid starvation at the time that he first came to work at the movable station in 1899. He lectured in botany, invertebrate zoology, comparative vertebrate anatomy, human gross anatomy, and histology, and yet found time for original studies. In 1902 he went to the University of Chicago to head up the new anatomy department there. When he died in 1956, it was acknowledged that his research had left its impress upon investigations all over the world in fields far beyond anatomy, and had helped to bring anatomy to an eminent position in biology and medicine.

Benjamin Arthur was the younger brother. He graduated with honors in natural science at the University of Toronto and was promptly appointed Fellow in the Department of Biology. After earning a PH D at Columbia in 1902 for his studies on the evolution of the dentition of marsupials, he returned to Toronto as a lecturer in zoology, and he replaced Ramsay Wright as head of the department on Wright's retirement. He was also appointed director of the Royal Ontario Museum of Zoology in 1914 and retained both positions until his death in 1934. Under Bensley's direction the Department of Biology of the University of Toronto went through a considerable expansion, and it was through his persistent efforts that the Ontario Fisheries Research Laboratory was organized within the department with the help of a grant from the provincial government.

Joseph Stafford

Stafford, the third of Ramsay Wright's bright young men to arrive at the movable station in its first year of operation (1899), would prove to be the most important recruit for that period of the Board's history. Joseph Stafford was born in 1864 at Goodwood, Ontario. He graduated from the University of Toronto in 1890, received his MA and PHD from the University of Leipzig, and returned to the University of Toronto in 1898 as lecturer in biology. In 1901 he went to McGill as lecturer and assistant professor of biology and remained there until his death in 1925.

From 1899 until 1908 Stafford was associated with the movable station. He was its first curator and served in that capacity until the station was replaced by the permanent Atlantic Biological Station at St Andrews in 1908, whereupon he became curator there until 1910. The following year he was curator of the Pacific Biological Station. His activities with the Biological Board of Canada seem to have ended in 1913 when he prepared a classic on the Canadian oyster, *The Canadian Oyster – Its Development, Environment and Culture*, for the Commission of Conservation. Some 20 published reports partly cover his accomplishments over a 14-year period with the Board.

In his Presidential Address to the Royal Society in 1907, Prince warmly praised Stafford as 'the most devoted member of the staff of the station' who had 'never missed a single season, being the first to arrive and the last to leave.'

Dr J.C. Medcof in his FRB bulletin on oyster farming (1961) wrote: 'Even in pre-Confederation days there was some oyster farming in Prince Edward Island but it, too, was of the rule-of-thumb kind. It was not until the turn of the century when overfishing had done its worst that Dr Joseph Stafford, an early associate of the Fisheries Research Board, laid the basis for scientific oyster farming in Canada ... In the period 1904 to 1913 he conducted investigations that have made his name familiar to oyster biologists the world over. He also gave advice on management of our oyster resources. His advice was sound.'

4　Go Home Bay, 1901–13

A freshwater station The establishment of the Georgian Bay Biological Station, situated on an island in the mouth of Go Home River in Georgian Bay, took place largely through the efforts of the Madawaska Club, although proposals to create such a station had been made previously by E.E. Prince, A.P. Knight, Ramsay Wright, and others. The Madawaska Club had been organized at the University of Toronto in the winter of 1896–7 for the purpose of creating a university summer settlement, and an application to the Crown Lands Department resulted in a grant of land in 1898. The charter of the club listed as one of its objectives 'to conduct experimental work in Forestry, Biology, and other branches of Natural Science.'

Efforts were first made to get support from the Ontario government, but when these proved fruitless a petition was prepared by the Madawaska Club in the autumn of 1900 and presented to the federal minister of fisheries, asking for federal support. The propagation of bass was listed as one of the chief studies proposed. The petition was favorably received and a grant of $1,500 per annum was made, beginning in 1901, for the establishment and maintenance of a biological laboratory at Go Home Bay. Conditions of the grant required that a competent biologist be obtained as a director, that a caretaker be employed on a year-round basis, and that reports be submitted to the Board of Management of the Marine Biological Station.

R.R. Bensley became the first director of the new station, and its management was nominally under the supervision of a committee of Madawaska Club directors. The laboratory building and boathouse were built in the summer of 1901, and the laboratory was equipped. The total expenditure, by some masterpiece of bookkeeping, amounted to just $17.85 less than the annual grant. The station building consisted of a biological laboratory, a tank room equipped with pump and supply pipes, and a meteorological laboratory. The boathouse was located at a nearby wharf, and equipment there included a 35-foot sailboat, a rowboat, a flat-bottom

The Georgian Bay Biological Station, 1901–13

punt, together with floats, leads, plankton and other nets, and other apparatus for biological research. A small pond on a nearby island owned by the Madawaska Club was to be used for experimental fish-rearing.

Prince reported: 'The scientific staff consists mainly of members of the Madawaska Club, but application from special workers who may desire to carry on approved investigations will be given consideration by the Board of Management, to whom, in the first instance, all reports of scientific work conducted at the station are to be submitted.' Then, with typical zeal, Prince proceeded to outline a program of activities for the first year that would surely impress the minister sufficiently to continue the annual grant, though it was hardly realistic as a beginning program. He proposed investigations on: the food and habits of whitefish, lake herring, black bass, sturgeon, and pickerel; the plankton of the adjacent waters; young stages of valuable species of fish; the Entomostraca and other crustaceans in Georgian Bay; fish parasites and diseases; alleged injuries by fish-eating Amphibia, particularly the mud-puppy (*Necturus*); breeding habits of fish in connection with fish culture; and meteorological conditions of the region, with special reference to atmospheric electricity.

The report of the deputy minister of fisheries for 1902 revealed that, because the station had not been sufficiently equipped in 1901, little work had been done. However, in the spring of 1902 systematic investigations were begun. R.R. Bensley's term proved a short one. He left that year to take up an appointment as head of the new anatomy department at Chicago and was succeeded by his brother, B.A. Bensley, who would remain at that post, if sometimes reluctantly, until the station was finally closed in 1913.

A respectable beginning was made in 1902. A complete hydrographic survey of the vicinity was completed under the supervision of Professor C.H.C. Wright of the University of Toronto, including a surface map of Go Home Bay and Island by students Tom Loudon and Ralph and Justin DeLury. A reference collection of the fishes of Georgian Bay was made, under Bensley's direction, from specimens of the previous season as well as new material. Collections of birds, mammals, and plants

47

were made by Fred S. Carr assisted by University of Toronto student R.T. Anderson. Meteorological observations were made during the season under the supervision of Professor W.J. Loudon, and a hatching pond for bass was prepared. The total expenditure provided another neat bookkeeping feat: $1,495.95.

First activities of the 1903 season were directed towards observation of smallmouth and largemouth bass. Bensley reported that the season was well advanced at the time of the first observations (25 May), and the shores of the shallow marshes and inlets were lined with the characteristic nest excavations of centrarchid fishes in from 1 to 4 feet of water. A large number of the nests belonged to the rock bass (*Ambloplites*) and some to sunfish (*Lepomis*). Attempts to hatch and rear samples of the eggs were unsuccessful because of lack of sufficient running water. Later in the season an efficient aquarium table and pump were installed to overcome this difficulty.

During the balance of the season the young of different species were collected to observe their feeding habits at different stages of growth, and an extensive collection of the microscopic organisms which constitute the first food of the young fish was made. The reference collection begun the previous season was completed. Bensley pointed out the danger of depletion of some of the important game fishes in Georgian Bay, and suggested measures for conservation. Other workers at the station again included C.H.C. Wright, W.J. Loudon, the DeLury brothers, Carr, and Anderson. The last-named, unfortunately, was drowned while attempting to swim to the station from the mainland shore.

Bensley was not happy with the supervision of the Madawaska Club. On 1 February 1904 he wrote to Chancellor Burwash of the University of Toronto:

At the time of my appointment as Director of the Station in August, 1902, it was understood that the management of the Station was under the supervision of a committee of the Directorate of the Club consisting of the President, Secretary-Treasurer and advisory members. This committee has not found it necessary or advisable to supervise the equipment or work of the Station so that its duties have been virtually confined to those of the Treasurer in the management of the finances. On account of the authority assumed by the Treasurer, however, this arrangement has not been beneficial and in my opinion its continuance is not in the interest of the Station. [See Rigby and Huntsman 1958:62]

He concluded by requesting that control of the station be transferred from the Madawaska Club to the federal Department of Marine and Fisheries, and the transfer was made for the 1904 season.

Some new faces appeared at the station in 1904. A.P. Gundy, a science master at Brantford Collegiate Institute, studied aquatic insects. J.M. Cole of Woodstock College investigated aquatic mollusks. L.C. Coleman, an assistant in zoology, undertook to complete the work begun by Anderson, assisted by two advanced students from the University of Toronto. One of these was Davidson Black, later to become famous as the discoverer of 'Peking man' (*Sinanthropus*); the other was A.G. Huntsman, of whom more will be said in later chapters. Together they added to the collections of plants and birds begun by Anderson, and investigated the plankton and the feeding habits of black bass. The station was open from 2 June to 5 September. Bensley and Huntsman spent the entire season there; the others stayed for periods ranging from two to seven weeks. The year 1904 also saw the completion of a house at the station to accommodate workers. Bensley listed it on his 1904 inventory as follows: 'Building – Main 16 × 30 – 2 floors; contains 1 living room, 9 bedrooms – walls, California siding, floor white pine, 11 windows, 2 doors, kitchen leanto 8 × 12, 2 small windows, veranda front 8 × 30.'

However, the change in management from the Madawaska Club to the Board of Management failed to end the Georgian Bay station's financial problems. There were long delays in the payment of accounts. A climax was reached in September 1905, when Bensley wrote the minister of marine and fisheries, outlining the situation and stating that if no adjustment of the station's finances were made and no arrangements arrived at for the future, he wished to be informed so that he could submit his resignation and make a public statement of the reasons. Apparently no such assurance was given, for the files contain a copy of his letter of resignation, dated 18 April 1906 and addressed to Prince. However, the letter was not presented, and with the new field season Bensley was again in charge.

Continued financial troubles

In an effort to get closer to current effort in fish management, Bensley made the following proposal to the government of Ontario in his 1905 report:

Proposal for a hatchery

In view of the increasing sentiment in favor of preserving the game fishes, and of the interest already taken by the Provincial Government in restocking operations, will you allow me to refer to the need of a hatchery in the southern section of Georgian Bay. The northeastern shore and the islands embrace large sections at present undisposed of by the Dominion and Provincial Governments, which are only valuable for fishing and tourist purposes. According to the general opinions expressed, the supply both of food and game fishes is being rapidly depleted. It seems therefore advisable that the efforts of the Government in preventing the capture by irresponsible persons of game fishes during the breeding season by seines and traps should be supplemented by constructive efforts towards increasing the supply through hatching operations. In this region bass and whitefish hatching could be profitably combined.

There was no immediate response to this suggestion, and in the light of present knowledge this seems just as well. The proposal was not based on any objective evaluation by the station of the benefits of such a hatchery, but rather reflected the belief that then went unchallenged among scientists and managers alike, that adding more fish fry to a body of water could greatly increase fish stocks. In the decades ahead this view was to become eroded, for one species after another, largely by the Board's own researches.

The 1905 season saw only five investigators at the station: Bensley, Huntsman, A. Pearson (who was a science master at Ingersoll Collegiate Institute), and students I.R. Bell and J.G.R. Murray. Nests of the smallmouth bass had been found, and the development and growth of the young were followed. Other local fishes were similarly studied, and the first contacts were made with commercial fishermen near by. The only untoward incident was a windstorm that extensively damaged the station's dwelling house, necessitating extensive repairs.

Go Home Bay in 1905

In 1906 the main activity at Go Home concerned a study of the conditions relating to the capture of fish in gill nets. Working with a regular fisherman, the scientists observed this operation, concentrating on comparison of average sizes and weights of whitefish and lake trout taken in nets of one mesh, the effect of capture and handling on the vitality of the fish, the probability of recovery after liberation of fish of less than the minimum legal size, and food and parasites. A general study was made of the average size of various shore fishes which might be retained or allowed to escape in nets of graded mesh, as a preliminary to the examination of pound nets. The habits and food of the carp, which had become numerous in lower Georgian Bay, were also studied.

The years 1906–11

Bensley's colleagues included J.W. Firth, a graduate of the University of Toronto, who secured samples of whitefish to determine their keeping qualities when taken from both shallow and deep water; W.A. McCubbin and P.I. Bryce, students from the same university, who took measurements of the whitefish and lake trout taken in gill nets and observed the condition of the fish. Bryce also made two trips to the southern end of Georgian Bay to collect carp and examine their food.

Between 1907 and 1911 work continued along the lines already established, with particular emphasis on black bass, lake trout, and whitefish. The life histories of aquatic insects, fish parasites in relation to their hosts, and microscopic life in the water were objects of study. A small laboratory had been built on a scow in 1906, and this was towed to various sites to facilitate investigations over a wider area of the Georgian Bay island region.

In 1907 Bensley was absent from the station for a long period and Dr E.M. Walker, a lecturer in zoology at the University of Toronto, was in charge from June until the middle of August. Huntsman was also there, along with two students, E.V. Cowdry and W.J. Fraser. In 1908 Bensley was present full time, along with Walker and four other Toronto people. A 'first' for the season was the tagging of small-mouth bass to ascertain their movements. In 1909 Bensley's time was taken up mainly with preparing his research of previous years for publication. Four other workers were there that year, continuing their earlier work. Huntsman was back in 1910 to supervise the work of A.R. Cooper on the life history of fish parasites and A.D. Robertson on a collection of aquatic mollusks. E.M. Coatsworth made a collection of flatworms and leeches and worked out a provisional classification of both parasitic and free-living forms.

Bensley proposes to retireIn 1911 Bensley expressed his decision to retire as director, and arrangements were made by the Board to have Huntsman take over for the season. However, the situation at St Andrews made Huntsman's presence there more urgent (see the following chapter), and Bensley agreed to continue to act for another year. Professor W.T. MacClement of Queen's University studied the plants of Georgian Bay and A.B. Klugh, also of Queen's, studied the algae (both MacClement and Klugh would prove to be long-time workers at the Board's stations). H.T. White of Sudbury High School studied the 'moss animals' or Bryozoa of the region, and Robertson and Cooper continued their respective studies.

But it was in 1912 that the station at Go Home Bay reached a high point in activities. It would prove to be in the nature of a swan song. Lack of a suitable motorboat had hampered the work of the station in previous years, and this was rectified when a 25-foot launch with a 15-horsepower engine was purchased and shipped to Go Home Bay in May. Although Bensley remained technically as director, the work of the station was under the supervision of E.M. Walker, who arrived on 19 May and stayed until 11 September. Bensley reported: 'The Station had an unusually successful season, both in respect of the number of workers and the results obtained. The progress of the investigation of parasitic animals attacking fishes, which has been under way for several seasons, is now such that many points in the identity and life history of these forms have been cleared up. Several new species have been identified. A.R. Cooper, who has had this work in hand, has also made infection experiments with small bass, using parasite-bearing minute Crustacea, which serve as food, as a source of infection.' (See Rigby and Huntsman 1958:221)

In addition to looking after things, Walker completed his investigations of the seasonal and ecological distribution of the dragonflies and other aquatic insects, which in their larval and other stages are a source of food for fishes. He was assisted by T.B. Kurata of the University of Toronto, as curator, and by W.A. Clemens, also of Toronto, who made collections and identifications of the local species of mayflies. Clemens was to figure largely in the Board's subsequent history.

A.D. Robertson continued his work on the Mollusca, covering the whole of Georgian Bay, with the assistance of A.B. Klugh late in the season. White continued his study of Bryozoa.

The botanical work of the station was under the direction of MacClement, who studied water molds, some of which attack living fishes or their eggs. The fleshy fungi of the region were collected and identified by Miss E. Penson and T.H. Bissonnette of Queen's and R.P. Wodehouse of Toronto, under MacClement's supervision. Klugh also studied the submerged and emergent vegetation of the Bay.

Bensley completed his paper on the fishes of Georgian Bay during the year and brought together for publication other papers dealing with investigations over several seasons.

The final season for which there is any record of scientific work at Go Home Bay was 1913. The work was under the supervision of Dr J.W. Mavor of the University of Wisconsin. In fact, he was apparently the only person actively engaged in field work at the station that year. Bensley was there, but eye trouble prevented him from doing much. Mavor studied the protozoan parasites of the fishes, on which he published two papers.

Last season at Go Home Bay

The Annual Report of the Ontario Department of Game and Fisheries stated: 'It is reported that the Department of Marine and Fisheries are directing their attention to other Provinces and discontinuing for the present their important work in this Province.' During the summer of 1916 a conference was held at Go Home Bay between members of the Madawaska Club and Dr A.B. Macallum, representing the Board, to arrange for the sale of the station to the club. This took place in 1917, and Huntsman went to Go Home Bay in June to pack up all the scientific equipment and other articles belonging to the Board and ship them to St Andrews for the Atlantic Biological Station.

The apparent reason for the closing down of the Georgian Bay Station was the result of an agreement between the federal and Ontario government in 1912, which left the responsibility for stocking those waters used by anglers with the provincial government, freeing the federal government to carry on fish-breeding work in connection with commercial species only.

Bensley later appraised the work at Go Home Bay:

Although restricted in some respects by its geographical position and its establishment as a fixed rather than a movable station, the Go Home Bay laboratory accomplished results of far-reaching importance. Specifically, the technical papers are incorporated in a volume of reports published as Sessional Papers of the Department of Marine and Fisheries. More important, however, it served as a training place at a critical period for young men who later rendered valuable service to economic biology elsewhere, while both in material and in personnel it gave direction to a unique development in laboratory practice, namely the foundation of marine and freshwater laboratories in the University of Toronto. [See Rigby and Huntsman 1958:223]

51

5 The permanent station at St Andrews, 1908–11

As early as January 1906 and again in May of the same year the Board of Management considered the need for a permanent biological station for the Atlantic coast. A committee consisting of Ramsay Wright, A.B. Macallum, L.W. Bailey, E.W. MacBride, and D.P. Penhallow was appointed to examine a number of possible locations. They met at St Andrews on 23 July and, after examining several sites near there, went to Nova Scotia, visiting Yarmouth, Port Maitland, Chester, Lunenburg, Digby, and St Mary Bay. They came to the conclusion that the choice of a permanent location lay between Chester and St Andrews, but felt that before the final choice was made a year of further study was required to make a careful faunistic and floristic examination and comparison. They proposed that the movable station, which was then at Gaspé, should be taken to Chester the following year, and that meanwhile Bailey would make a further study at St Andrews.

The proposal to move the station to Chester was not adopted, but Stafford went there in May 1907. His report confirmed the committee's inclination towards St Andrews. Despite the fact that more than two hundred fishing schooners made Chester and nearby Lunenburg their home ports, the region itself was not a major fishing center because these vessels fished off the Grand Banks and their catch generally went directly to a foreign market. Not many fish were caught in adjacent waters, and, on top of that, it was difficult to obtain supplies at Chester. Penhallow's report on behalf of the committee stated:

St. Andrews, on the other hand, is convenient of access to Montreal, St. John, and other centres whence supplies could be drawn without unreasonable delay, and in itself it possesses sufficient resources to meet all ordinary demands. Moreover it is one of the real fishing centres of eastern Canada. The large number of clams, herring and other fish annually taken in Passamaquoddy Bay and adjacent regions make it the immediate centre of a large live fish

52

industry as important as that of any other town on the coast. The great expanse of sheltered waters embraced in Passamaquoddy Bay and the St. Croix River; the great variety of depths to be met there; the presence of large numbers of fish weirs from which material may be readily obtained, and the exceptionally rich feeding grounds around Grand Manan and Campobello, combine to make it an ideal place for the establishment of a station engaged in solving problems associated with the fishing industry ... For these reasons the choice fell to St. Andrews.

Since most of these facts were known to the committee before they sent Stafford to Chester, it is fair to speculate about the political pressure that made them go through the motions of looking for another location when St Andrews was so obviously their first choice. There is no doubt that the Nova Scotian fishing industry would have preferred a station on the Nova Scotian coast, handy to their problems. On the other hand, in addition to the points made by Penhallow, in those days St Andrews had direct Pullman service from Montreal, which was a great convenience to a Board dominated by 'Upper Canadians.' Equally important, 3.5 acres of waterfront property had been obtained from the Canadian Pacific Railway for the nominal sum of $250. Nevertheless the question of the location of the station, which was handled so smoothly by the Board of Management when it first arose, was to be raised more than once on subsequent occasions, but although other units were established later in the Halifax area, the main biological station has remained at St Andrews.

The general plans for the station were approved by the Board of Directors (as the Board of Management was then known; see chapter 7), and funds were granted by Parliament in April 1907. Penhallow was instructed to go to St Andrews and begin construction of the buildings to have them ready for 1 May 1908. The laboratory was completed, with the exception of the water supply, that spring, and the residence was ready on the 8th of July. By this time available funds were exhausted, and further construction was suspended for the year. _The station takes shape_

Although the buildings were not fully equipped, the station was sufficiently ready to carry out a program during the summer. The laboratory, a building 79 by 31 feet, provided for nine junior and three senior investigators, which was considered sufficient for the scientific work of the station for some years. It was gas-equipped for both lighting and heating, and contained one long table for general purposes, provided with fresh water for culture work. Working tables were placed along the north side, where the light was best for microscope work. Each table was furnished with gas and a set of drawers. There were storerooms for supplies and a room for overhauling and sorting of material, and the attic was used for storage.

The station was equipped with two motorboats: the _Sea Gull_, 35 feet, with a cabin and a 10 horsepower engine; and the _Sagitta_, 24 feet, with a 4 horsepower engine. There were also two rowboats.

Penhallow was appointed resident director and Stafford was named curator. Stafford arrived in the third week of May and began extensive examinations by dredging and other means at localities in the St Croix River and Passamaquoddy Bay, areas which had not been examined when the movable station was at St Andrews. Professor J.P. McMurrich of the University of Toronto arrived early in May and spent about four weeks studying sea anemones or Actinaria and examining copepods. This was the first activity, with the Board, of another of the illustrious scientific figures who would play a major role in its development.

McMurrich, who was born at Toronto in 1859, was the first of the fine biologists developed by Ramsay Wright. He graduated from Toronto in 1879 and received his _James Playfair McMurrich_

The *Sea Gull*, 1908–15

master's degree in 1881. In that year he was appointed professor of horticulture, later biology and horticulture, at the Ontario Agricultural College in Guelph. There, in 1884, he wrote the article for *Science in Canada* calling for the establishment of biological stations on Canada's east and west coasts which has been referred to in chapter 2. Shortly afterward he went to Johns Hopkins University as instructor in osteology and mammalian anatomy, remaining there until 1886. Then he held successive posts at Haverford College, Clark University, and the universities of Cincinnati and Michigan, during which period he built up a formidable reputation as an investigator and a teacher in anatomy. In 1907 he returned to the University of Toronto, where he headed the Department of Anatomy, retiring as professor emeritus in 1930. For eight years he was dean of graduate studies.

While at Johns Hopkins he had spent the summers at the Chesapeake Zoological Laboratory, where he came in contact with an active group of American and English biologists. He received his PH D from Johns Hopkins in 1885. The principal scientific interests of his career by that time were expressed in papers on vertebrate morphology, embryology of snails, and sea anemones. In embryology, he produced a series of papers on gastropods, ascidians, fishes, medusae, actinians, and isopods. He was widely recognized as an authority on sea anemones. His work on vertebrate morphology established his authority as a human anatomist, by which he was best known. When he returned to Canada he took up studies of the plankton and the life histories of the fishes, working at both the St Andrews and the Nanaimo stations.

McMurrich wrote several textbooks, which spread his reputation as a teacher. He was active in many scientific organizations. In 1922 he became president of the Royal Society of Canada, and its Flavelle Medal was awarded him posthumously in 1939. In his lifetime he was awarded no less than three LL D's: from the University of Michigan in 1912, the University of Cincinnati in 1923, and the University of Toronto in 1931.

The Atlantic Biological Station complex, St Andrews, from the wharf, 1909. Left to right: water tower, laboratory, residence, and workshop

Huntsman wrote of him:

He was never robust, particularly in later life, and belonged to the study rather than to the forum. Modest, unassuming, and ever gentlemanly, he was universally liked and esteemed. His interest in all around him, the small things as well as the great, made him an ideal companion. With broad sympathy, extensive knowledge, and sound judgment, he was in general demand as a counsellor. When the professional biologists of Toronto organized in 1935, they created the post of Honorary President for him, who was the pioneer in this field. He did not merely receive the honor, for he remained active in this group as well as in other ways until his death on February 9, 1939. [In *Proc. Roy. Soc. Can.*, 3rd ser., XXXIII (1939), 143–6]

McMurrich became a member of the Biological Board in 1912, was chairman from 1926 to 1934, and stayed on the Board until he retired in 1937.

The year that McMurrich first appeared as an investigator at St Andrews, 1908, Dr Edmond Boyd of Toronto was also there, collecting hydroids and studying fish eggs. D.L. McDonald of McGill assisted the curator and studied the rate of growth of the common barnacle. Knight continued his study of factors in the death of fish, and S. Kirsch of McGill studied the algae. Mary and Adaline Van Horne, daughters of CPR magnate Cornelius Van Horne, prepared a list of fungi of the area which was subsequently published in *Contributions to Canadian Biology*, Other investigations begun that season included examination of several harbors as possible lobster-spawning areas, fish migration, and the suitability of local waters for the culture of oysters.

MacBride continued his study of the oyster in Prince Edward Island and was able

The first season, 1908

to secure an experimental bed to conduct investigations over a period of years. He also investigated complaints of damage to the oyster beds caused by hard-shell clams or quahaugs.

Among the distinguished visitors who came to the station that summer was Dr David Starr Jordan and other members of the International Fish Commission. Penhallow was able to hold a conference with the representative of the United States Fish Commission, Barton Evermann, on joint action and the interchange of data on mutual problems. United States Commissioner of Fisheries G.M. Bowers, Dr W. Bell Dawson, engineer in charge of tidal survey, and W.J. Wilson of the Geological Survey were also there.

In his report on the station for the year, Penhallow called for year-round operation and a full-time curator: 'the policy of the Station and the continuity of its work from year to year should be consistently followed. This can be secured only by placing its administration in the hands of one responsible member, who shall be accountable for its scientific management under the general direction of the Board … The Board has directed that the St. Andrews Station should be constructed with a view to winter occupation, in order to preserve perfect continuity of work … When that time arrives, it will be necessary to have in the paid employ of the Board, a scientific man who shall reside at the Station continuously in the capacity of Curator.'

Wright and MacBride were appointed a subcommittee to see if a young Scandinavian could be secured for the post. MacBride, who had left McGill to accept a post in London, wrote to Wright advising that he had contacted the noted Dr Johan Hjort for a likely candidate. He said that he did this as 'one more turn for an ungrateful government (which had never paid my expenses of last year).' One candidate was eventually contacted but no satisfactory agreement could be reached, and the search was dropped.

The second season, 1909

In 1909 Penhallow was again resident director. Further work was done on the building and an artesian well was drilled. When Penhallow fell ill Macallum took over for a period and then was succeeded by Bailey.

Among the senior scientists, Macallum studied the blood composition of cod, pollock, and lobsters and, after returning to Toronto, examined blood serum of dogfish from material sent to him; Wright, in a stay of a few weeks, continued his studies of plankton; and Bailey, there 10 days, collected freshwater and marine diatoms. Stafford spent part of the season aboard the *Ostrea*, continuing his studies of oyster spat and larvae and collecting plankton from a number of points off the shores of New Brunswick and Prince Edward Island to compare with the fauna of Malpeque Bay.

Anatomical and physiological studies of the intestines of the lobster were made by F.R. Miller, a physiology demonstrator at Toronto; A.B. Klugh spent two months at the station working on the flora of the St Croix River valley and Passamaquoddy region as well as the islands of Campobello and Grand Manan; and G.G. Copeland of Toronto investigated the waters of Passamaquoddy Bay and other adjacent areas in search of locations suitable for oyster culture. Other workers included D.L. McDonald of McGill, who assisted Copeland; Robert Chambers of Queen's assisting Klugh; and H.E. McDermott, assistant to the director. Eric Dahlgren of Princeton spent some time at the station studying luminous shrimps.

The two new gasoline launches, the *Sea Gull* and the *Sagitta*, proved their value for investigations carried out at distant points. There were under the supervision of Captain Shepard Mitchell, a veteran of those waters. His assistant, Arthur Calder, was to remain with the station for a total of 41 years, beginning as an engineer for the small boats, then becoming captain of larger vessels, and concluding as collector for the station. He finally retired in 1950.

His memories still fresh in 1972 when interviewed by the author, Captain Calder spoke of the scientists he had met at St Andrews. He himself had been a fisherman and then worked in Toronto in a factory producing gasoline engines. This experience led directly to his appointment as engineer in 1908: 'I was the only person around St Andrews who knew anything about gas engines, so that got me the job.' Calder recalls Penhallow

Calder recalled Penhallow as being very 'British Navy' in his manner. He adopted the navy practice of 'sealed orders' when he dispatched Mitchell on a mission. The first time this happened, Mitchell called Calder and said: 'Arthur, what do you make of this? We have to go somewhere but we don't know where until we go to sea. It says on the envelope, ''Open at Sea.'' ' Calder offered: 'Go anywhere out of sight, around Joe's Point, and let's take a look.' They set out and rounded Joe's Point, just out of view of the station. Mitchell opened his sealed orders, and read: 'Proceed to Grand Manan. Go 10 miles from Big Duck Island to southeast. You'll sound there in 100 fathoms. Dredge. Save full material. Return to the Biological Station by the most direct route. Failure to do so, wire.' Mitchell complained to Calder: 'Where the hell do we find a wire out in the bay?' However, they accomplished the mission without incident.

Thirty years later, when A.H. Leim was director of the station, he wanted to find a certain compound tunicate. Calder was able to locate it: Jordan and Evermann had originally reported it, and it was the accuracy of their report that Penhallow was checking with his first 'sealed orders' mission.

Calder described Penhallow as a man 6 feet 6 inches in height, with a tiny Indian wife. Penhallow strongly believed that people should keep their proper place. One day Calder undertook to help some laborers on the wharf with a piece of heavy equipment. He was summoned to Penhallow's office and told by the director: 'Calder, a good dog does as he is told. That is all. You may be excused.'

Calder had very warm recollections of the rather elusive Joseph Stafford, about whom there is very little material in the Board's files. 'He was a curious and rather secretive man, but he didn't hide anything from me, and we had some pretty good times together,' Calder recalled. Once, returning by boat from St Andrews to the station, as they rounded Joe's Point, Calder and Stafford spied a group of girls bathing in the nude in a sheltered sandy cove. They continued on to the wharf, tied up hastily, and then slipped back through the woods to get a closer look. Finally, cautiously crawling along the ground, they came to a steep ledge overlooking the cove. Stafford inched forward to the lip of the ledge to enjoy an unobstructed view. As he poised there precariously, Calder pushed from behind. Stafford went sprawling down the slope before the girls and they began promptly to pelt him with pebbles while, chuckling, Calder retreated hastily to the station. When Stafford finally got back to the station, he summoned Calder to his office: 'I want an apology for pushing me over that bank.' Calder countered: 'Can you prove it? Maybe you just slipped.' Stafford looked at Calder in deep disgust. 'Listen, Calder,' he said, 'you may make a good lawyer, but you make a damn poor liar.' And the incident ended there. Warm memories of Stafford

On another occasion Stafford took Mitchell and Calder on a week's trip around the Bay of Fundy to make plankton collections. At one point they sighted a group of young people having a picnic on a promontory over a steep cliff that ran along the shore. After exchanging a series of friendly waves and shouts, Stafford turned to Mitchell: 'Put me ashore. I'm going up to join the party.' Both Mitchell and Calder protested that the climb up the cliff face was too steep and dangerous, but Stafford could not be discouraged: 'After you put me ashore,' he directed, 'I'll go up the cliff and you go round the point to the other side, where you can pick me up.' They bowed to his authority and set him ashore.

Stafford made his tricky ascent up the cliff until he was about two-thirds of the way, when he was stopped by a sheer face and no footholds. He couldn't go forward and it was almost sure suicide to climb down. Calder and Mitchell shouted up to the group, who were now gathered at the cliff edge in fearful anticipation: 'Get a rope at the fishing station and help him.' Someone ran to the nearby fishing station and secured a rope. It was lowered over the cliff and Stafford tied it around his waist. They pulled him up to safety amid the cheers of the girls. Later, when Mitchell and Calder picked up Stafford on the other side of the point, Calder told Stafford bluntly: 'If you keep pulling stunts like that, you're going to need someone to look after you.' Stafford said wryly: 'I realized that when I was about two-thirds the way up that cliff.'

Stafford's career with the Board was to terminate abruptly after he went to Nanaimo in 1911, to stand in for its ailing director. The circumstances, according to Huntsman's recollections, were that Stafford became involved in an unstated 'difference' with a visiting British woman scientist, to the point that she demanded an apology. Stafford refused and the Board felt compelled to dispense with his services. The only female scientist at the Nanaimo station in 1911 was Miss H.L.M. Pixell of Bedford College, University of London, whose stay from 30 June to 14 September overlapped with Stafford's stint as acting curator. It is interesting to speculate on what kind of 'difference' made the Board take such action and made Stafford prefer dismissal to an apology.

Macallum's dignity

Calder recalled Professor Macallum as a man of great dignity at all times. He also liked a drink, and one of Calder's jobs was to purchase Macallum's whiskey. 'He never offered me a drink either. I couldn't have accepted it, for I was a teetotaler.'

Calder recalled one occasion when Macallum was at the station and, after several drinks, decided that he wanted to go to St Andrews. There was only one small boat at the wharf and there was a stiff sea running. Calder warned Macallum that it would be rough going, but Macallum insisted on going just the same. Inevitably, as they rounded Joe's Point, they took one wave aboard that drenched them. Macallum received it stoically, but when they reached the wharf at St Andrews, he drew himself up and solemnly observed to Calder: 'Evidently she is just a fair-weather boat.'

The genial Professor
Prince

Calder remembered Prince as a man of great charm. Prince came to St Andrews one winter when the station was closed, though Calder checked it regularly every day. Prince stayed at a boarding house near the Algonquin Inn, and one morning when the thermometer was registering about −20°F, he called Calder on the phone and told him: 'I want to go out to the station with you this morning.' Calder warned Prince that it was 20 below and that he had better dress accordingly. Prince told him not to worry. When Calder stopped at the boarding house to pick him up, Prince emerged wearing a light topcoat, thin kid gloves, a light cap on his head, and low rubbers over his shoes. Calder protested: 'Professor, I think it would be a good idea if you went back and put some warm clothes on.' Prince assured him: 'I'll be perfectly all right the way I am.'

On the way down to the dock, Calder made a slight detour past the home of a friend. As they approached the house, Calder noticed that Prince's nose was dead white. He told Prince: 'Professor, your nose is frozen.' Prince protested: 'It's perfectly all right.' Calder said: 'It's dead white. Can you feel it?' Prince touched his nose, then exclaimed in alarm: 'I can't feel anything!' Calder bundled Prince into

58

his friend's house and thawed him out. After warming up with a cup of tea, Prince said reflectively: 'You know, I don't think we should go out to the station today.'

Professor Penhallow's illness, which eventually proved fatal, prevented him from submitting a published report for the year 1909. His illness continued in 1910 and made it necessary to arrange for other supervision during that season. L.W. Bailey supervised for the first part of the season, arriving on 10 June and remaining until 9 August, and H.E. McDermott was his assistant. Stafford, at the station for four months, carried on experiments with lobster traps with slats spaced from $1\frac{1}{4}$ to $3\frac{1}{2}$ inches apart, to test their efficiency in catching lobsters and holding them after they were caught.

The loss of Penhallow, 1910

Other workers that season included A.G. Huntsman, from 4 July to 2 August, making a collection and study of marine forms, particularly tunicates; A.B. Klugh, for a short period in August; and J.C. Simpson of McGill. In August Bailey and Klugh made a short hydrographic survey of the Kennebecasis and Lower St John River region. On 28 August they attended a meeting of the New Brunswick Natural History Society and gave a short account of the survey.

The well that had been dug the previous year proved inadequate and a new well was started; drilling was suspended at the end of August when it had reached a depth of 450 feet.

The season ended tragically with the death of Penhallow in October. His fellow workers felt that he had not spared himself in the difficult and exhausting task of preparing the land and in the construction of the station's road, wharf, and buildings. Some of the buildings escaped the disastrous fire of 1932 and still stand as evidence of his thorough work.

The work during the 1911 season was under the general direction of Bailey, Macallum, and Wright. As previously noted, Stafford went to the Pacific Biological Station, and Huntsman replaced him as curator at St Andrews. Fresh water, drainage, and sewage disposal were again problems, and Knight was sent to the station early in the year to investigate and make a report to the Board for its May meeting. Arrangements were made to connect up with the CPR's water system that supplied the Algonquin Hotel from Chamcook Lake. A pit was prepared for sewage disposal, and the grounds were graded.

Huntsman takes over, 1911

Prince had suggested 11 different subjects for investigation and Huntsman enthusiastically reported on them all. Plankton was collected as opportunity afforded and was examined by Bailey and Wright, in part. Food of fishes was determined from studying the contents of stomachs and intestines. Experiments were made on the storage and shipping of squid. Sea cucumbers were tested as fish bait, but the one species available was too tough to be of use. Bait trials were made on 36 line-trawl sets, using several different baits in regular rotation; the whelk proved to be the most useful substitute for herring. Fish caught in the weirs were identified and listed. The spawning of herring was investigated. The fish present in Chamcook Lake were reported on. Food available for herring on the 'ripplings' outside of Grand Manan and among the islands of Passamaquoddy Bay was checked using drift nets, and the euphausid shrimp *Meganyctiphanes norvegica* was the most common food object taken. Rearing of lobsters was studied, an examination of Oak Bay by dredging showing that, except for a small area at the mouth, the bay was quite unsuitable for planting lobsters. Finally, collections of marine animals were made for the museum exhibit and for the use of students.

Sixty years later Huntsman commented on his first yearly report as curator: 'A curious thing about my report on the 11 problems he gave; I was so naive as to think that he wanted me to investigate them all. But I learned afterwards that he expected me to investigate only one of them perhaps; they were only suggestions. I was so foolish as to attempt to report on all.'

Bailey continued with his diatoms; Wright, with plankton. Macallum studied the distribution of potassium in certain Protozoa. W.H.T. Baillie assisted the curator and made a collection of polychaetes for study at the University of Toronto. Dredging, seining, and shore collections extended to new areas around Grand Manan, Bliss Island, and The Wolves, and Huntsman made further collections of ascidians for future study.

Visitors at the station during the 1911 season included Board member A.H. MacKay and Andrew Halkett of the Fisheries Museum in Ottawa. Engineer Skelly of the Department of Marine and Fisheries came in connection with the water supply, as well as W.A. Found, assistant deputy minister of fisheries, probably in connection with the accounts. The Board and the department were already at odds over finances, and the grievances of the Board would soon spill into the open.

6 The Pacific Biological Station, 1908–11

Beginnings

During the second half of the 19th century there was sporadic activity in describing the aquatic fauna and flora of what is now British Columbia. A member of the International Boundary Commission, J.K. Lord, made extensive observations and collections and in 1866 produced two entertaining volumes entitled *The Naturalist in Vancouver Island and British Columbia*. These included the first descriptions of several local fishes, 'giant' octopuses, and a hair-raising account of the little Hudson Bay side-paddle steamer *Beaver*'s passage through the swift current of Dodds Narrows near Nanaimo. During the 1880s the United States survey ship *Albatross* passed along the coast several times and took quite a number of dredge samples of invertebrates and fishes.

Between 1893 and 1896 an international committee made a survey of the boundary waters between the United States and Canada. One of its members was Richard Rathbun, assistant secretary of the Smithsonian Institution. Concerning the data then obtained, the work of the *Albatross*, and reports of the Department of Marine and Fisheries of Canada, he wrote: 'The fact most strikingly brought out in the assembling of these data is the great paucity of accurate or detailed information regarding the aquatic products of the region, such as is requisite in providing for their preservation while still permitting them to be utilized without needless interference. With exceptional opportunities for their study ... the field is one that would richly pay the inquiries of the naturalist and fishery expert, if properly directed, in the practical benefits they promise.' (See Rigby and Huntsman 1958:97)

From the start of the 20th century the government of British Columbia began to take an active interest in the fishing industry, and appointed John Pease Babcock from California to keep it informed and to make surveys; at different times he had the title of commissioner or assistant commissioner of fisheries. However, the individual who played the key role in the establishment of the Pacific Biological

In the entrance to the George W. Taylor wing of the Pacific Biological Station, Nanaimo, hangs this portrait of its first director. In the showcase are pieces of historic scientific apparatus, including the Rev. Mr Taylor's microscope, built about 1830

Station was the Reverend G.W. Taylor of Wellington, one of the satellite coal-mining towns near Nanaimo.

George William
Taylor

Born in Derby, England, Taylor was connected with the museum there before coming to Canada in 1882. He took orders in the Church of England in 1884, and began parish work in Victoria, then Ottawa, and finally at Wellington. He was known to be a 'tireless rambler, and collected without cessation.' He enjoyed a high reputation as an entomologist and marine biologist concentrating particularly on molluscs and crustaceans, as a result of which he became a Fellow of the Royal Society of Canada. In 1887 he was appointed Honorary Entomologist for British Columbia. He was a charter member of the Natural History Society of British Columbia, which was organized in 1890. He was keenly interested in the rapidly developing fisheries for salmon, halibut, and other species, and was a constant advocate of the establishment of a biological station on the Pacific coast of Canada.

The Minnesota Seaside Station

Actually there was already a marine station in British Columbia, called the Minnesota Seaside Station. It flourished from 1901 to 1907 on the southwest coast of Vancouver Island near Port Renfrew, at a site still known as Botany Beach. It owed its existence to a remarkably capable and 'liberated' young botanist of the University of Minnesota, Dr Josephine Tilden, who selected the location of the station after she had inspected all of the shoreline adjacent to San Juan Bay in a dugout canoe. She was especially interested in the larger seaweeds, hence her desire to reach the outer coast where these are more numerous and more varied than in sheltered waters.

To this remote spot, accessible only by a weekly steamer from Victoria followed by a two-mile hike through the rain forest, came scientists and students from Minnesota and other universities, including Raymond Osburn of Ohio, the pioneer Illinois ecologist Victor Shelford, and even one or two visitors from Japan. Although concerned with all branches of natural history, the work of this station always had a strong botanical flavor. It twice produced its own publication, *Postelsia* – handsome volumes named after the beautiful palm-tree kelp that still grows abundantly on a reef at Botany Beach. In those years only one university (Manitoba) and a very few two-year colleges in western Canada existed whose staffs might have taken advantage of the station, but apparently none of them did. However, C. McLean Fraser spent some time there in 1903, presumably collecting hydroids.

There is no direct evidence that the existence of the Minnesota station played any important role in prompting Canadians to do something similar, yet so excellent an example right on our doorstep could not help but be a favorable factor. Another nearby station was the University of Washington's Marine Station at Friday Harbor on San Juan Island, not far from Victoria, which was begun informally during the 1890s.

Prince plays both ends against the middle

Taylor's keen interest in the fisheries resulted in his appointment to the British Columbia Fisheries Commission, which was set up by the federal government in 1905. In its Interim Report of December that year, the commission, of which Prince was chairman, stated:

We recommend that, as early as possible, a thorough survey of the fishing grounds of British Columbia be carried out. We have in the course of our enquiry felt the lack of full and exact information respecting the resources of the prolific waters covered by our investigations. Mr. William Sloan, M.P., and Mr. Ralph Smith, M.P., have, we understand, already urged that such investigations should be authorized by the Dominion Government, and the urgency of the work has profoundly impressed us as a Commission. We recommend that this important matter be referred to the Marine Biological Board, so that the details of such fishery researches may be outlined and the necessary apparatus and gear decided upon by the Board mentioned, which the Commission is informed, holds its annual meeting early in 1906.

The Board, of which Prince was also chairman, duly and with a straight face considered the recommendation, and sent the following communication back to the commission:

As Secretary of the Board of the Marine Biological Station, I was instructed at the last meeting of the Board held in Ottawa on March 30th, 1907, to formally intimate to you as the Chairman of the British Columbia Fisheries Commission, that the question of establishing a Biological Station for fisheries researches and connected scientific investigations was duly considered.

The Board in having this matter referred to them by the Members of the British Columbia Fisheries Commission appreciate the fact that the Commission clearly realizes the importance and necessity of a thorough inquiry into the fish-life and marine-life of the Pacific coast-waters of Canada, the results of which will aid in devising appropriate regulations for the preservation and further development of the fisheries.

During the forthcoming meetings of the Royal Society in Ottawa, the Board will hold a further session at which a definite scheme will be arranged, for the immediate foundation of the Station at some suitable point on the British Columbia coast. I shall then have the honour of further communication with you.

The issue was kept warmed up in Parliament by Sloan who, on 18 January 1907, spoke on the subject and referred to the report of a committee of the Vancouver Board of Trade dated 7 February 1905 which endorsed the proposal for the establishment of a biological station on the Pacific coast.

Prince and Taylor combine

Then the many-hatted Prince, in his presidential address for Section IV of the Royal Society of Canada, delivered in May 1907, made a plea for 'A Dominion Biological Station for British Columbia.' It had his usual astute touch, stating, in part:

The project for a marine biological station for British Columbia has never been allowed wholly to remain in abeyance ... But no public statement was made to the country until the able and far-seeing representative for Comox-Atlin, Mr. William Sloan, M.P., in a memorable speech on fishery matters declared, on January 18th, 1907, to the Federal House at Ottawa that there ought to be no delay in founding a scientific laboratory for fisheries research on the coast of British Columbia ... No one who has cast a dredge over the bow of a vessel in these prolific waters, crowded with exuberant life, can doubt that there is no land of promise ... offering greater reward to the biologist. In the course of a day's dredging, as recorded in the Royal Society's Transactions, no less than 150 species comprising 7,000 specimens were taken in Departure Bay, near Nanaimo, many of them new species ... The specialists of the United States have long recognized the peerless nature of our Pacific coast waters, and it may be doubted if the *Albatross* has anywhere secured, in so short a time, and with such ease, a mass of living treasures to compare with those obtained in her course in 1890 along the west shores of our Dominion.

By curious coincidence Taylor also presented a paper at the same meeting, entitled 'A Plea for a Biological Station on the Pacific Coast.' After reviewing events leading up to the establishment of the Biological Station on the Atlantic coast and referring to the plea of Prince in 1893 that Canada undertake her own scientific investigations, he continued:

This plea was entirely successful, and as a result, in 1898, there was established a Biological Station on the Atlantic Coast of Canada. But the curious feature is, that in all the arguments –

all the negotiations – all the appeals – made to the Government of the day, in respect to these stations, Canada is spoken of as though she possessed but one coast. It was a Canada bounded on the east by the Atlantic Ocean – on the west by a vacancy. There is not in any of the papers I have quoted any mention of the Pacific Ocean or of the British Columbia coast.

Now at this time – up to 1898 say – such omission was not, perhaps, much to be wondered at. The only way to persuade a Government to spend money on science was to point out economic advantages and the fisheries of Canada at that time were on the Atlantic almost entirely. The seats of learning from which investigators were to be drawn were also in the east. But now all this is changed. British Columbia, as far as the fisheries are concerned, is recognized as the premier province. The value of its fisheries amounted last year to one-third of the value of the whole Canadian fisheries.

Meanwhile, however, the project for the establishment of a station on the Pacific coast had been brought before Parliament when the acting minister of fisheries presented, on 5 April 1907, an amount of $15,000 to provide for the construction and maintenance of marine biological stations and investigations, mentioning in particular the intention to establish a biological station on the Pacific Coast. This increase in the Vote from $3,000 to $15,000 was approved by the Committee of Supply and authorized by Parliament.

Success

A subcommittee of the Board of Management, consisting of Prince and Ramsay Wright, was appointed to select a site. At first a tract of land on Nanaimo harbor was secured. Then a far more desirable location on Departure Bay became available through the kind offices of the lieutenant-governor of British Columbia, the Honorable James Dunsmuir, whose coal company owned land around the bay, and was conveyed in the form of a lease at a nominal rental to the Trustees of the Board of Management of the Marine Biological Stations. During the fall and winter of 1907 the Pacific Biological Station was erected on the site under the supervision of Taylor, who was named the station's first curator.

C. McLean Fraser has written about the choice of this site:

Departure Bay was chosen on account of some special features. There is no safer harbour on the coast for all weathers, and the station site is conveniently situated and easy of access since it is but three miles by water and four and one-half miles by road from Nanaimo, which provides the greater part of the supplies needed and has good boat service to Vancouver and good train service to Victoria. This means much to a station that is kept open the whole year round. It is favorably situated because it is within easy reach of the large archipelago situated along the east coast of Vancouver Island between Nanaimo and Victoria and not so very far from the equally extensive archipelago lying between the north end of the island and the mainland, while across the Strait of Georgia it is not so far to the mouth of the Fraser and the Point Grey fishing grounds. Consequently, quite a large portion of the 25,000 miles of tidewater line of British Columbia is less than a day away. For that reason it is a good center from which to work at problems concerning all of the food fishes, with the possible exception of the halibut, as well as for the majority of the species of invertebrates of commercial value. It is pre-eminently a herring center and, since the herring serves as a food supply for a great number of marine forms, that in itself is quite an important matter. Without going outside the bounds of the bay much can be learned concerning the life-history of the herring, coho, spring and dog salmon, steelhead, blue or green cod, rock cod, capelin, the white perch and several of the flat fish, not to mention several other forms that as yet are not considered desirable as food, while the common butter clam and the little-neck clam, as well as the common mussel, are found right at the door. [See Rigby and Huntsman 1958:101]

Departure Bay's advantages

The Pacific Biological Station at Departure Bay

The building was ready for operations in the spring of 1908. The *Nanaimo Free Press* of 10 June 1908 carried a description:

The building which has been erected and fitted up at the Bay is, in the opinion of the designer and of those who have seen it, as well suited for its purpose as any similar institution in the world. It consists of, first, a large office for the reception of visitors and the transaction of necessary business. Adjoining the office is a room for the storage of glassware and chemicals, and the reception of specimens when first brought in. Next is the main laboratory, a room 32 feet long by 16 feet wide. It is lighted by 8 windows – 4 facing north and 4 south. Work tables for the accommodation of 8 [are] provided with shelving on which books and specimens are arranged. Salt and fresh water are both convenient for each worker so that investigation of the living organism can be carried on.

Next to the laboratory on the east side are the library and photographic room. Leading out of the library is the dining-room through which one passes into the outer hallway and to the private entrance to the Station. Upstairs are 4 bed-rooms, which will be for the accommodation of the visiting scientists. A separate building contains the kitchen and the Caretaker's apartment.

There are also a stall, woodshed, boathouse and engine house conveniently situated. On the whole, the establishment may be said to be complete in every way, although a good deal of money can be spent in improving the library and the means of research. A steam or gasoline launch will also be required in the near future.

The station opens, 1908

Research work began at the new station in 1908 under the supervision of Taylor. Library facilities for the first few years were slim, but Taylor put his own large scientific library at the disposal of the workers. Together with his extensive zoological collections, this was of great value to the workers in the early years.

In January 1908 E.M. Burwash, Professor of Natural Sciences at Columbian

66

The residence building, as it appeared in 1964

College, New Westminster, began a geological examination of the area around the station and continued this into the summer. He prepared a geological map and general report on the stratigraphy of the area that was published in *Contributions to Canadian Biology*.

A.G. Huntsman spent two months at the station, making a general collection of sea animals for the University of Toronto and a special collection of tunicates for systematic study. Howard C. DeBeck of Columbian College made a study of the Echinodermata of Departure Bay, particularly the starfishes and sea urchins. McLean Fraser examined the hydroids of the vicinity, resuming the study that was to prove his major scientific achievement.

The station provided headquarters for a party of three investigators from the Geological and Natural History Survey of Canada. John Macoun and William Spreadborough made an extended examination of the fauna and flora of Vancouver Island, while Charles G. Young collected starfish, crustaceans, and other sea animals for the new Victoria Museum in Ottawa.

Taylor, in addition to performing his duties as curator, continued his observations and collections of marine animals. He discovered a new species of sculpin which was subsequently named *Asemichthys taylori* in his honor, by Professor C.H. Gilbert of Stanford University. Taylor also made a collection of parasitic copepods from the fish, which were sent for identification to Dr C.B. Wilson of Massachusetts. Eight species new to science were included in the collection.

James Fletcher spent three days at the station examining Taylor's collection of marine forms, and Andrew Halkett made the station his headquarters for dredging operations being carried out by the Department of Marine and Fisheries.

In discussion of the estimates in the House of Commons in March 1909 the member for New Westminster remarked: 'Is it to be part of the purpose of a Station of this kind to pursue investigations and to ascertain whether there are other fish preying upon these salmon in their haunts between the time they leave their hatching grounds on the Fraser River and their return later? If we would find out that fish prey upon these salmon and devise means of dealing with them and guard against the efforts of the hatchery being made nugatory, this Station would be worth many times the money proposed to be spent upon it.'

The second season, 1909

During the 1909 season Huntsman continued his investigations on the tunicates, dredging in Burrard Inlet with Taylor and, through the courtesy of Macoun, collecting also at Ucluelet on the west coast of the island. The scope of his investigation was enlarged to include a collection of marine worms and plankton. McLean Fraser continued his work on the hydroids, giving particular attention to study of the living forms. J.P. McMurrich spent three weeks at the station, examining the possibilities of investigations into the development of food and other fishes. He also made an anatomical study of the ratfish or chimera, and collected and studied isopods and sea anemones. John Macoun spent the latter part of August there, having his collection of molluscs from Barkley Sound identified by Taylor. Andrew Halkett was also there for two days collecting zoological specimens for the department's exhibit of British Columbia fisheries at the New Westminster Exhibition of 1909.

The British Association for the Advancement of Science held its annual meeting in Winnipeg on 25 August to 1 September 1909, and passed the following resolution:

In view of the enormous importance of the Fisheries of Canada in connection with her prosperity and her rapidly developing position as the great resource of the food supply of the Empire, and appreciating the danger of exhaustion which menaces certain of the Fisheries, the members of the Zoological Section of the British Association for the Advancement of Science now meeting in Winnipeg, desires to congratulate both the Dominion and Provincial Governments upon the work already accomplished in connection with the study of food fishes, upon the establishment of a Marine Biological Station on both the Atlantic and Pacific coasts, and upon the co-operation with the Government of the United States in an international commission, from whose labours much may be expected. At the same time the members of the section are of the opinion that further and more extensive efforts in all these directions are urgently needed if certain of the fisheries, notably that of the Pacific salmon, are to be maintained, even at their present condition of productiveness. For the framing of satisfactory and effective regulations for the utilization and conservation of the food fishes, a complete knowledge of their life-history is absolutely necessary, and the section desire to impress upon the Governments concerned the immediate need for an extensive prosecution of investigations along this line, for greater facilities for the scientific study of the fisheries, especially those of the Pacific coasts, and for a continued co-operation of the Dominion Government with the Governments of the Provinces and also those of the United States in all efforts looking towards the conservation of the fisheries, one of the most valuable natural resources of Canada.

After the Winnipeg meeting, some members of the association visited Nanaimo. Both E.E. Prince and A.B. Macallum were there, and the group was accompanied by Prime Minister Sir Mackenzie Bowell. The city council of Nanaimo gave a dinner, the Vancouver Island Coal Company offered a tour of the mines, and Prince and Taylor superintended dredging operations for the edification of the visitors.

The year 1910 was almost a blank in the history of the station, apparently owing to the serious illness of Taylor early in the year. The station register shows no record of work for the year, but this may be misleading because there is evidence that Miss E. McClughan of McGill University made a study of the moss animals or Polyzoa of Departure Bay and other points along the coast, and it is unlikely that she would be the only visitor.

The continued illness of Taylor in 1911 made it necessary to appoint an assistant, and Joseph Stafford was made acting curator from 1 May to 15 August. He carried on investigations with the small native oyster *Ostrea lurida*. J.P. McMurrich determined the age and rate of growth of halibut and salmon by reading their scales, and served as acting director. The station was also the headquarters for two English scientists, both specialists in marine annelid worms. F.A. Potts of Cambridge University made collections of polychaetes and parasitic crustaceans on both coasts of Vancouver Island and near the Queen Charlotte Islands. Miss H.L.M. Pixell of Bedford College, University of London, made a study of polychaetes and collections of protozoans. Taylor was not wholly incapacitated; he took a great interest in the work and, as far as possible, continued his own study of marine forms.

7

Board of Management to Biological Board, 1898–1912

The Board of Management functioned between 1898 and 1912 with little change. In 1901 an assistant director was appointed to fill in for E.E. Prince, whose duties as commissioner of fisheries demanded more and more of his time elsewhere during the summer seasons. Ramsay Wright was named to the post.

On 7 May 1907 a resolution was passed by the Board calling for a change of name from Board of Management to Board of Directors, the director to be known henceforth as the chairman of the Board and the assistant director to be known as vice-chairman, the members to be known as directors rather than trustees.

With the death of Professor Penhallow in 1910 A.B. Macallum was appointed secretary-treasurer, and with the departure of MacBride to London that same year Arthur Willey became the McGill representative. Otherwise the Board continued unchanged until it became the Biological Board of Canada in 1912.

Willey was already a well-known zoologist when he joined the Board. He was born in Scarborough in 1867 and graduated from University College, London, with his B SC in 1894. He was awarded an honorary MA from Cambridge in 1898 and became a Fellow of the Royal Society in 1899. He studied at the Naples Zoological Station in 1891–2 and held the position of tutor in zoology at Columbia University from 1892 to 1894. He was director of the Colombo Museum of Ceylon from 1902 to 1910, when he came to McGill as professor of zoology, a post he held until 1932. In 1912 he was elected a Fellow of the Royal Society of Canada.

Willey became active in marine biology immediately after his arrival in Canada. In 1912 and 1913 he studied marine plankton at St Andrews, and in 1914 he worked at Nanaimo on the growth of salmon and halibut. In 1915 he was a member of the Canadian Fisheries Expedition under Dr Johan Hjort, developing the copepod collection. He continued to study copepods from the Canadian Atlantic, Bermuda, Hudson Bay, and arctic waters as well as those from freshwater areas. Huntsman has said that he contributed steadily to knowledge of Canadian animal life and showed initiative which bore continual fruit in various directions. Professor Willey died in 1942.

During the period 1898–1912, the operations of the stations were controlled by the Board, but the annual appropriation for the stations was administered by the accounting branch of the Department of Marine and Fisheries, which was bound by the strict rules for the spending of money which are government tradition. Since the Board of Directors and the other scientists, with the exception of Prince, were not government employees, and provided their services without remuneration beyond bare expenses, they fumed at the long delays in the payment of accounts when forms were not filled out exactly according to regulations, or frequently even when they were. Individuals and firms that provided services and materials to the stations became exasperated by the usual long interval between their submitting a bill and receiving a cheque for it. If the proposed use of materials happened to puzzle the Accounting Department or the Auditor General's Department, an account could be held up for many months. The red tape crisis

Efforts were made as early as 1902 to have the Board's appropriation turned over to the secretary-treasurer to be expended by him as directed by the Board. Again in 1904 Prince wrote of the need to obtain the minister's approval for the accountant to transfer the appropriation to the Board. He noted that 'this would remove from the Accounts Branch that unnecessary red tape system which has done so much to hamper the work of the Station.'

But even the persuasive Prince was unable to effect any change in the system, and the files reveal repeated complaints of delays in providing advances to curators and directors in charge of the stations and in obtaining reimbursements for money spent. As noted earlier, B.A. Bensley came close to resignation over this situation at Go Home Bay, and MacBride, in May of 1909, complained that he had not been paid for his 1908 expenses.

In 1943, in an article in *Science*, A.G. Huntsman wrote that 'a clerk's veto of a request for purchase of scientific literature because the latter was in foreign languages caused an explosion and resulted in the creation in 1912 by Act of Parliament of the Biological Board of Canada, an independent body under the Minister to have charge of all biological stations.'

Huntsman later expanded on this comment:

The Board at first was under the department; it received a grant through the department. It was the Board of Management of the Marine Biological Station of Canada ... All its requisitions – for money for paying accounts – had to go through the department. On one occasion the Board failed to get a requisition accepted for books and scientific publications in French and German. There was an official in the department who later became deputy minister, W.A. Found, and the rumor was that it was he who turned down the Board's request for these French and German scientific publications with some remark that we really didn't need those. This aroused the Board so much that they took the matter up with the minister, and were able to make a case for the Board being set up as an independent institution. That was in 1912, I believe.

In his later contacts with the Board Found displayed a certain 'love-hate' relationship, which may well have begun with this veto of the book purchase. He was in fact a remarkable man.

Found was born in 1873 at Found's Mill, Prince Edward Island. After graduating from Prince of Wales College in Charlottetown, he taught school for some years William Ambrose Found before joining the Fisheries Service in 1898 as secretary to the head of the Fisheries Branch. In 1911 he was appointed to the newly established position of superintendent of fisheries, and in the same year he was adviser to the British plenipotentiaries who were negotiating the Pelagic Sealing Treaty. Later, in 1918, he was a member of the Canadian-American Fisheries Commission established to consider settlement of outstanding fishery questions between Canada and the United States. He was also, at different times, a member of the International Fisheries (Halibut) Commission, the International Pacific Salmon Fisheries Commission, and the North American Council on Fishery Investigations.

In 1920, when the Fisheries Branch reverted from a short-lived embrace with Naval Services to the Department of Marine and Fisheries, Found was appointed director of fisheries and assistant deputy minister of fisheries. In 1928 he became deputy minister of fisheries, and in the same year his public service was recognized by the University of New Brunswick, which conferred on him an honorary D SC degree. He remained deputy minister until he retired from government service in 1938, and he died two year later.

From 1911 to 1938 Dr Found was a key person in the relations between the Board and the government. More than one confrontation took place between the strong-willed men on either side, and through this dialectical process the Board developed.

The schizophrenia of scientific research But much more than red tape or personality conflict was involved in the change from Board of Management to Biological Board in 1912. The two objectives which both boards undertook to achieve, one of independent aquatic research and the other of providing answers to the practical problems of the fisheries, required that they do a nice balancing act, with the pole tilted now one way, now the other. This situation, of course, has parallels in research setups of all sorts, from the time of Archimedes down to the present day. For example, the senior scientific service of our federal government is the Geological Survey: it was founded as early as 1842, when Canada consisted only of the southern parts of Ontario and Quebec. In its long history it has always had to balance its activities between fundamental and immediately applicable work, and it has more than once been criticized for too much emphasis in one direction or the other. In 1884 its work was reviewed very critically by a committee of the House of Commons, which found that its 'field of operations should be confined to subjects more closely allied … to a Geological Survey.' But somewhat later it was instructed to expand its surveys into other fields of science – botany, zoology, and even anthropology; in fact, for a time its name was changed to Geological and Natural History Survey, much to the disgust of some of its rock-hammer geologists.

In the same way, when science was harnessed to the service of fisheries, it proved a mettlesome and high-spirited steed that must have caused deputy ministers to wonder who was smuggling in the oats. In their own university departments the members of the Board were laws unto themselves, respected for their scholarship and achievements, and not at all prepared to have their decisions reviewed by 'bureaucrats' unfamiliar with biological matters.

However, the opening of the first marine biological station had offered an unequaled opportunity for on-the-spot research that whetted the appetite of scientific curiosity in each marine biologist, and there was a rush of volunteers to greet the event. The movable station, literally and figuratively, had given them an opportunity to get their feet wet, and they seized it. They understood very well that the government would expect to see some tangible results from the sum that it was spending, modest though it was. Some of the scientists plunged directly into the problems that beset the fishing industry, others pursued goals that mainly satisfied

72

their own curiosities, and still other combined pure and applied research with great success.

The first season of the movable station found eight scientists in attendance, and the second year a peak of 12 was reached. Thereafter, between 1901 and 1902, the number varied between nine and seven. In 1902, with the opening of the Georgian Bay station, the University of Toronto contingent was split, but the total number of participants for the two stations was 16. It declined slightly over the next three years and then dropped sharply to eight in 1906 and nine in 1907.

The opening of the two new stations at St Andrews and Nanaimo saw the number of participating scientists increase sharply to 24 at the three stations in 1908, and this remained the high point during the Board of Management period. It dropped to 23 in 1909, 12 in 1910, and 17 in 1911.

In terms of publications, the results were impressive for that period. At the movable station, between 1899 and 1907, 20 different investigators published a total of 58 papers and reports. At the Georgian Bay station, during the period 1902–13 (which includes its last two years, during the succeeding Biological Board regime), 24 scientists produced a total of 49 papers. During the four years that the permanent station at St Andrews operated under the Board of Directors, 1908–11, 15 scientists published 28 papers. For the same period, Nanaimo received 14 investigators who published 33 papers. Impressive publication record

Although many of these papers and reports represented 'pure' scientific research or were accounts of the activities at the various stations, a substantial number related directly to fishery problems. These included the sardine industry, the effects of polluted waters on fish life, sawdust and fish life, effects of dynamite explosions on fish life, fishery bait experiments, a survey of the clam fisheries, and oyster studies, all from the period of the movable station.

However, the main scientific thrust during this early period was to establish the basic facts of the flora and the fauna of the regions, the life histories of the fishes and the other organisms of Canadian waters. There were reports on the marine Polyzoa of Canso, the fishes of Canso and of Tignish, the flora of St Andrews and Canso, lengthy studies of hydroids and medusae, a report on the diatoms of Canso harbor, a study of the eggs and early life history of the herring, gaspereau, shad, and other clupeoids, and even a publication on the paired fins of the mackerel shark. The seaweeds of Canso, the food of the sea-urchin, trematodes from Canadian fishes, the water mites or Halacaridae of Canso, the crustaceans at Seven Islands, and the sporozoans at Gaspé were all subjects of study at that period. 'Pure' research predominated

At Go Home Bay there were reports on the fishes of Georgian Bay, Cladocera of southwestern Ontario, a new tapeworm from a primitive fish, the bowfin (*Amia calva*), the systematic position of *Haplobothrium globuliforme*, trematodes from marine and freshwater fishes, the life history of the bass tapeworm *Proteocephalus*, the crayfish or Malacostraca of Ontario, the algae and plants of Georgian Bay, molluscs, leeches, Entomostraca, Bryozoa, new nymphs of dragonflies, mutual adaptation of the sexes in the damselfly *Argia*, and the nymphs of another damselfly (*Lestes*).

The preponderance of purely scientific publications became even more marked with the establishment of the permanent station at St Andrews. Marine, estuarine, and freshwater plankton diatoms and diatomaceous earths were reported on. Other papers covered marine works, ascidians, the production of light by marine animals, the flora of the region, the inorganic composition of the blood of vertebrates and invertebrates, the action of surface tension in determining the distribution of salts in living matter, the paleochemistry of the body fluids and tissues, a collection of crustaceans made at St Andrews, the sea anemones of Passamaquoddy Bay, and a list of the fleshy fungi collected at St Andrews.

Practical papers few
In contrast, just a handful of papers with a practical application emerged from St Andrews. There was a report on the temperatures, densities, and allied subjects of Passamaquoddy Bay waters and environs and their bearing on the oyster industry, another on oyster culture and clam fishing at Prince Edward Island, contributions to the physiology of the American lobster, a paper on the recognition of bivalve larvae in plankton collections, another on the conservation of the oyster, and a major study on the Canadian oyster, its development, environment, and culture.

The same emphasis on basic scientific research rather than the practical problems of the fisheries was marked in publications for the first four years at Nanaimo. These included notes on a collection of fishes from Vancouver Island, the geological environment at Departure Bay, two papers on hydroids and one on marine zoology, papers on a new genus and species of cottoid fish, five papers on ascidians, and one on anemones. Also reported on were Phoronidea and a new rhizocephalan (*Mycetomorpha*), and a series of papers dealt with marine worms: the swarming of *Odontosyllis*, stolon formation in certain species of *Trypanosyllis*, the Chaetopteridae, and the Nereidae.

The papers that appeared to have a possible practical application were two on the Pacific salmon, two on parasitic copepods, and a preliminary list of 129 species of British Columbia crabs and other decapod crustaceans.

Pure and practical science
If people in the Fisheries Service had their reservations about the practical value of the Board in finding solutions to the problems of the fisheries, there was not a great deal in the record up to this point to make them feel otherwise. However, the solid scientific achievements that were being registered because of the opportunities and facilities provided by the Board won it the confidence and support of the marine biologists who, in due course, would make their own formidable contribution in practical terms. From the beginning, Prince and his colleagues on the Board insisted that pure and applied science went hand-in-hand: that there could be no valid applied science without the basic knowledge furnished by a total study of the environment. This view was to be repeatedly challenged over the succeeding years, but it was never abandoned by the Board.

Huntsman talks about Prince
Huntsman has contributed some fascinating recollections of the early Board members, including the Chairman, E.E. Prince: 'He was called the genial Professor Prince, and he was a delightful companion, not only for myself, but also for my small daughters, to show them things about the station. The brook near the station became known as Spelerpes Creek because he found that kind of salamander in the stream and showed it to my daughters.'

Was Prince a good leader? In answer to this question, Huntsman has said:

It is difficult to answer that question because in the case of the Board in the early days, the situation was that Macallum was the [real] head of the Board, whether or not he was chairman. It did develop that he, as secretary, led the Board. Professor Prince, as chairman, did not; later Professor Prince thought that he'd have a better position as secretary, but the change didn't make any difference ...

We had an interesting situation when we developed a museum in the laboratory ... The new part in the basement was quite a sizable museum with various exhibits of the local forms of life, and there were cases in which these were mounted. Above one of them was placed a skeleton of some animal which had been found along the shore of the St Croix River, which is entirely marine. This partial skeleton had what seemed like perhaps a skull and a vertebral column, and on one occasion the question arose as to what animal did that partial skeleton belong. Professor Prince was interested in it and examined it carefully. He came to the conclusion that it must have been some kind of a small porpoise or whale. I hesitated, but finally felt I had to tell him that it was really the skeleton of a cow, without the parts which would have been significant for recognizing it, as he might have [otherwise done].

He responded to this very nicely because, not many days later, I found in one of the cases in the museum a glass jar of the proper museum type, with a glass top, and suspended from the top on the inside was a curious animal and over it was a legend, *Cornu bovis*. Now, that might deceive the uninitiated into thinking that here was a real animal that might be found in the adjacent waters, because it seemed to have a tail with fins, and a back fin, and what seemed to be its mouth was something white which superficially could be considered as teeth. But if anyone were very careful and if anyone knew the meaning of the two Latin words, *Cornu bovis*, there was no difficulty at all; it was the horn of a cow.

He had a decided sense of humor. And he would tell stories on himself as well as on others ...

Professor Prince was a member of the American Fisheries Society, and as a member he once presented a paper on a perfect fish pass. Now, there being dams for power purposes that might interfere with the ascent of such fishes as salmon, it had become very important to have fishways or fish ladders or fish passes to take them over these obstacles. He wrote a paper on the perfect fish pass, which was to have the fall of water operate an elevator which would take them over the dam. This was published by the American Fisheries Society, and Professor Knight, who was practical-minded, considered that it would be very desirable to see if this could be done at St Andrews. They were both there and he undertook to construct and operate such a fish pass in the Magaguadavic River across Passamaquoddy Bay from St Andrews, where there was a falls and a dam over which the salmon could not leap, but at the base of which they were said to come every summer.

Professor Knight did construct the fishway, which was operated by water power in having an elevator to carry the fish up over the dam. It was constructed at the station and finally set up at St George at the foot of the falls, and was made to operate.

Unfortunately, that summer no salmon came to the foot of the falls as reported ... The only salmon that went up there were ones that had been caught and put in the fishway. I'm afraid that the conclusion reached was that there was no good prospect of finding that the salmon knew enough to enter the fishway. They had to be caught first for it to operate.

Of Wright and Knight, Huntsman has said:

Ramsay Wright was an outstanding personality in the University of Toronto ... he was an all-round man, one of the few all-round men whom I have known. When re retired in 1912, he went to Oxford; although from Edinburgh, he had an Oxford accent. He admired Oxford and he admired the classics. I saw him there in 1914, and he was interested in studying modern Greek.

Dr Knight wasn't [the best scientist] from a strictly scientific standpoint. He did not pretend so much to discover things, or basic things, as Professor Macallum did, as to try to get the results of scientific investigation practically applied to the fisheries. He had previously attempted to get science applied in the health field by writing a human physiology text for the use of students in the schools ... His flair for application can be appreciated from what he did in the early years of the Board's work ... at St Andrews. He undertook to investigate two things: one was the effect of sawdust pollution on water and the fish in it; the other was the usefulness or the danger in using dynamite to get fish.

Today it is not always easy to appreciate the background of such projects. At the turn of the century eastern Canada was at the height of its white pine lumber boom, and the huge volumes of sawdust and slabs that were spilled into the rivers were widely blamed for the decline of fishing success, both in fresh waters and in the sea. The dynamite study was prompted by the fact that some Bay of Fundy fishermen had started using dynamite to capture schooling pollock.

Huntsman has also appraised Macallum and Bailey:

Macallum and Bailey

Professor A.B. Macallum, physiologist and biochemist of the University of Toronto (later McGill), became the first chairman of our Council for Scientific and Industrial Research, which was started in 1918, the beginning of our National Research Council. He was an outstanding scientific investigator and was made a most unusual thing for a Canadian, a Fellow of the Royal Society of London ...

L.W. Bailey was a specialist in microscopic plants called diatoms, and also a geologist. He was very nice indeed. My experience with him as director when I was curator was an entirely agreeable one. Perhaps because he left practically everything to me except recording the expenditures.

On one occasion the station had a sort of expedition to Wolves Islands, outside the mouth of Passamaquoddy Bay in the Bay of Fundy, not very far away, perhaps 20 miles. We went in the old *Sea Gull* ... These islands had no harbor at all, so we had to land on the beach from a rowboat, anchoring the *Sea Gull* offshore, and we had to walk up over the beach to the island itself. On the beach were various stones, small and large, and even boulders. Being a geologist, Professor Bailey was interested in these, which were quite a variety, and he gave one rounded stone which he thought particularly interesting from a geological standpoint to Shepard Mitchell, the captain of the *Sea Gull*, and the captain treasured this. The captain had a bedroom in the residence of the station, and he kept this on his bureau and showed it to anyone interested as being a diatom, that is, a microscopic plant, which was rather absurd.

So Dr Huntsman summarized his recollections of some of the leading figures in the Board of Management as it was going through its metamorphosis into the Biological Board of Canada. Meanwhile, the Board had won an important victory with the granting of a measure of financial autonomy. But it would soon find that the government, in granting this new status, expected a lot more than it had obtained in the past.

8 Atlantic work by the Biological Board, 1912–21

The new act: advantages and problems

'An Act to Create the Biological Board of Canada' was presented to Parliament during the 1911–12 session, passed its third reading in March 1912, and became law on 1 April. The act gave the Board its sought-after independence, including the key provision that 'from the moneys appropriated by Parliament for the work of the Board ... the Secretary-Treasurer, under the direction of the Board, shall expend such sums as are necessary for the work of the Board.' An annual statement of expenses was to be submitted to the auditor general for examination.

While under the act the Board had the advantage of being able to arrange to have work done directly rather than by warrant as previously and of being able to purchase supplies and equipment as required. Economy became vital, for the cost of many items such as printing and publishing reports, which formerly had been absorbed by the Department of Marine and Fisheries, came out of the Board's appropriation. Moreover, some large expenditures from the previous year were still unpaid. Nevertheless the vetoed order for scientific literature was triumphantly reinstated.

Under the new system many of the difficulties and delays previously experienced in getting supplies and equipment vanished and the work of the stations was enormously facilitated, but the Board had problems in meeting the requirements of the Auditor General's Department's annual statements. In the auditor general's report for 1911–12 and again for 1912–13 it was pointed out that details of expenditures and supporting vouchers had not been received from the Biological Board.

New appointments

At the organization meeting of the new Board held on 13 May 1912 an Order-in-Council was read appointing A.B. Macallum secretary-treasurer and E.E. Prince was elected chairman. A.P. Knight, Canon Huard, and Ramsay Wright were appointed members of the executive committee, along with the chairman and secretary-treasurer. With the approaching retirement of Wright, Arthur Willey was appointed to replace him on the executive committee. These were yearly appointments, and by yearly appointment Prince, Knight, and Canon Huard served on the executive committee throughout the period 1912 to 1924, also Macallum until his enforced retirement in 1919. Upon Wright's retirement and departure for England, later in 1912, J.P. McMurrich was appointed to represent the University of Toronto on the Board.

The Act of 1912 gave the Board the right to determine and carry out its own policies, not only in operating the stations but also for the 'conduct and control of investigations of practical and economic problems associated with marine and freshwater fisheries, flora and fauna.' The scope of the Board's activities rapidly widened, as the yearly reports of the stations soon revealed, and the nature of those activities were more and more related to the practical problems that confronted the fisheries of Canada.

St Andrews work,
1912

A.G. Huntsman was again curator at St Andrews in 1912, and he retained that position almost continuously until 1922, when he was named director. The first year did not immediately reflect the new direction, as certain basic studies from previous years were continued and others were initiated. A.B. Klugh was first man on the scene, arriving prior to the official opening and staying to the end of June. He initiated a three-year study of the algae of the region. L.W. Bailey continued his investigation of the diatoms, which continued through the entire period under review. Similarly, continuing studies were made of the amphipod and cumacean crustaceans by Huntsman, of the plankton of Passamaquoddy Bay by Willey, of parasitic copepods by V. Stock of the University of Toronto, of 'fish lice' or Argulidae and the Isopoda by N.A. Wallace of the same university, of the Sporozoa by J.W. Mavor of Harvard College, of the polychaetes by W.H.T. Baillie and the bryozoans and hydroids by C.A. Miller of Queen's.

Absent-minded
Professor Mavor

(Calder has reminisced briefly about Professor Mavor, whom he knew as the classical absent-minded professor. One morning he went out to collect frogs in a paper bag, but got them mixed up with another bag containing his lunch. He had a lively lunch that day. On another occasion, Mavor was making his thoughtful way along St Andrews' main street, with one foot in the gutter. An acquaintance stopped to enquire of his health. 'Oh, I'm very well, thank you,' Mavor said graciously, 'though I'm a little lame today.')

A start was made in the preparation of museum specimens in 1912. Three sets of faunal specimens were prepared, one set being sent to the Fisheries Museum in Ottawa, one to the Victoria Memorial Museum there, and the third to the Imperial Institute in London. Huntsman made a collection of 45 species of fish, assisted by Douglas Jeffs of the University of Toronto.

In the direction of practical research, N.A. Wallace of Toronto studied the food of fishes as revealed by their stomach contents and J.D. Detweiler of Queen's began a two-year study of the distribution of molluscs. Experiments with lobster traps, arising from an Order-in-Council requiring a wide lath space, got under way, and bait experiments begun the previous year were continued, confirming results previously obtained.

More practical
flavor of 1913
studies

The following season, 1913, saw an extension of activities when a new 50-foot motorboat, named the *Edward E. Prince* and popularly referred to as the *Prince*, was delivered to the station on 10 June. Since it could sleep six, longer trips became possible. One trip to St Mary Bay in Nova Scotia for a survey and collection of material lasted a week.

Dr Philip Cox of the University of New Brunswick was named assistant director for the season and Huntsman was again curator. Although basic studies continued,

The *Edward E. Prince*, 1913–32, rebuilt 1916

there was a more practical flavor to the research in the second year. Klugh made a collection of seaweeds and forwarded them to the Central Experimental Farm at Ottawa for chemical analysis to determine their value as fertilizer. A.R. Cooper of Toronto began a study of parasitic flatworms, Prince studied the larval fishes, and Cox undertook the identification of fishes obtained and the preparation of a list of the fishes of the region, including those in Chamcook Lake. He also experimented with the influence of light on the movement of adult eels and constructed a net-box to test the possibility of keeping herring and squid alive for bait purposes. The study of food of fishes continued and W.H. Martin of Toronto conducted experiments on the survival of fish after freezing. He also collected data on the temperature and density of the waters of the region, and spent two weeks investigating the physical and chemical conditions of a pool at Long Beach, Digby County, which had been purchased by the government for a lobster hatchery.

Experiments with lobster traps, with laths spaced at various widths, continued, but a scarcity of lobsters of the proper size made it impossible to complete them. A jar of soft-shelled boiled lobsters was sent to Macallum to determine their food value.

Prince's idea for the perfect fish pass was tried out by Knight during the season and was found to be experimentally successful. Martin prepared a number of museum specimens, including one of the largest lobsters on record. The length of the body was 20-3/4 inches, and the total length to the tips of the claws was 34 inches.

J.W. Mavor was in charge of the station as curator in 1914 while Huntsman was in Europe, and Philip Cox again assisted with the administration. A major activity in the early part of the season was a study of the life history of several important food fishes, under Mavor's supervision. Cox, Dorothy Duff of McGill, R.P. Wodehouse, A.B. Macallum, and E. Horne Craigie of Toronto, and J.D. Detweiler and C.B.

Food fish
investigations, 1914

79

Waite of Queen's were all involved. Winter flounders were obtained in quantity in Brandy Cove near the station; these were measured and scales taken, and a number were tagged and released to study their movements. Other supplies were obtained around Passamaquoddy Bay and at Wilson's Beach, Campobello. Cod, pollock, haddock, and hake were examined at the fish market at St Andrews, and at Wilson's Beach and Grand Manan. Salmon and shad from Saint John were also examined. Investigations included recording of weights, lengths, weight of gonads, determination of sex, and samples of scales. On one occasion 600 pollock were involved, and on another day, 575 hake and pollock. Material was also brought back to the station for further study. Duff made a special study of the haddock, with reference to rate of growth, age at first spawning, and size most abundant in the catches. Craigie determined the rate of growth of the hake from a study of the scales. Wodehouse tested the method of determining the age of the cod from the scales and studied the rate of growth, and Macallum studied the rate of growth of the pollock from three to six or more years of age and determined the year-class found most frequently in the catches. Macallum also studied the physiology of dogfish and skate tissues and W.T. MacClement, who spent the greater part of the summer at Long Beach, took part in the studies when he was at the station.

The *Prince* was equipped that year with a hoisting engine, a Pettersen-Nansen water bottle, and a Richter reversing thermometer. With the new equipment a survey of the bays along the southern coast of New Brunswick between the St Croix River and Saint John was made to determine their suitability for oyster culture. Mavor directed the operation, assisted by Craigie, who was responsible for the hydrographic observations, and Detweiler, responsible for the dredging. Other hydrographic surveys were made in the St Croix River and the Bay of Fundy, and in the late summer on a trip to Nova Scotia Macallum, Cox, and Cooper made a series of hydrographic observations at four stations between East Quoddy Head, Campobello, and Petit Passage, Nova Scotia.

Knight makes
a point

Knight was active that season undertaking an investigation into the suitability of the department's lobster pond at Long Beach. He was assisted by Professor H.G. Perry and A.B. Dawson of Acadia University, Dr W.E. Sullivan of the University of Milwaukee, and W.A. Mersereau of the University of New Brunswick. MacClement was also there at Knight's invitation to study the effect of diatom growth on lobster larvae. This marked the beginning of a long and intensive study of the lobster on the part of Knight, sufficiently complete to enable him a few years later to tell a group of skeptical fishermen: 'You'd better listen; I know more about lobsters than any man alive.'

At Malpeque Bay, Prince Edward Island, researches on the oyster went on all summer under the direction of A.D. Robertson of the University of Toronto. Investigations were made of areas where the production had failed, to determine the cause. Marine algae were identified by Klugh, and the diatomaceous oyster food by A.H. MacKay. Water samples were examined for chlorine and total solids content at Macallum's laboratory in Toronto.

Cox conducted two special investigations. One involved conditions at Buctouche, where the high death rate of stored quahaugs (hard-shell clams) was causing concern. Cox inspected the storage and other conditions there and brought back water and clams for further study at the station. He made several recommendations regarding storage and shipment of quahaugs. Another investigation concerned a disease of herring which had occurred in two successive years along the Straits of Northumberland. However, before Cox reached the area, the schools had disappeared. He was able to obtain only two specimens, and after an examination he forwarded a report on the character of the disease and the conditions prevailing to the department.

In 1915, Canada became involved in an oceanographic research venture with other countries. During the last decade of the 19th century and the first of the 20th rapid strides had been made in obtaining knowledge of the oceans and fish resources of northern Europe, and as early as 1902 the International Council for the Exploration of the Sea was set up to coordinate the research efforts of the different countries. Methods of telling the age of fish were discovered, first using length frequencies and later the marks on their scales and bones. Some work that attracted wide attention was that which culminated in a famous paper by Johan Hjort of Norway: 'Fluctuations in the Great Fisheries of Northern Europe.' Hjort related changes in the catches of herring, cod, and other species to changing climatic conditions over several centuries. A fact previously unsuspected was that, even in the short term, broods of fish hatched in successive years can vary greatly in abundance. The year 1904 was particularly favorable for the survival of larval cod and herring, and the 1904 year-class dominated the Norwegian spring herring fishery for many years.

These were exciting discoveries. It was natural to inquire whether a similar situation existed in the western Atlantic and, if so, whether events there were synchronized with those in the old world. Accordingly, in the fall of 1914 the Biological Board invited Dr Hjort to come to Canada and study samples of herring that had been collected at points from northeastern Newfoundland to Massachusetts. Results were so encouraging that in 1915 the Fisheries Branch was prevailed upon to support a major survey under Hjort's direction. Thus the Canadian Fisheries Expedition was mounted, with participation of Norwegian, Canadian, and United States workers. The main objectives of the expedition were: to determine whether our herring were all of one race or whether several different races existed; to look for variations in growth rate in different waters; to obtain information on year-class strengths and past fluctuations in abundance; and to study the regime of ocean temperatures, salinities, currents, and plankton contents in east-coast waters.

In 1915 Paul Bjerkan of Norway accompanied Hjort to Canada, and the expedition's headquarters was set up at Souris, Prince Edward Island. Others who assisted with the field work included A.G. Huntsman and Arthur Willey from Canada, J.W. Mason of Union College in Schenectady, New York, and a Mr Nightingale from the United States Bureau of Fisheries. Three ships were used, the government cruisers *Princess* and *Acadia* and a Scottish steam herring drifter known only as *No. 33*. The work covered the whole Gulf of St Lawrence, Cabot Strait, and the offshore regions from Newfoundland to western Nova Scotia. Studies of the herring were prepared by Hjort and Einar Lea of Norway. It turned out that the 1904 year-class was, in fact, a very strong one in the Gulf of St Lawrence, although not quite as predominant there as among the Norwegian spring herring, but off Nova Scotia and in the Bay of Fundy this situation did not prevail. Other reports by Norwegians included H.H. Gran's study of the phytoplankton and Protozoa, Alf Dannevig's identifications of fish eggs and larvae from the plankton, and Bjerkan's hydrographic observations. Hjort also prevailed on W.J. Sandström of Sweden to reduce the expedition's oceanographic observations to the first general account of the hydrodynamics for any part of our east-coast waters. Willey reported on the copepods, the principal food of herring, and Huntsman wrote papers on the arrow-worms and on the growth of the small sardine-sized herring of the Bay of Fundy.

Huntsman had been appointed permanent curator of the St Andrews station in 1915, but during his absence on the expedition the work at the station was under the direction of Macallum and McMurrich. Cox was in charge of two cruises of the *Prince* investigating the fauna of the Bay of Fundy. The first, to St Mary Bay, involved dredging at 24 stations, and the second, in the Annapolis Basin, at 9 stations. Craigie made hydrographic observations at the 4 stations established in 1914 and at 11 new stations in the Bay of Fundy, in addition to those at St Mary Bay and Annapolis Basin.

McMurrich began a qualitative study of the plankton collected the previous winter near the station by A.E. Calder, and Prince studied variations in the structure of sardine herring in an effort to identify local schools and their migrations. Mavor concentrated on a two-year study of the pollock, assisted by Craigie who, before Mavor's arrival, had been making measurements and collecting scales of pollock, including a lot of 652 fish caught in Casco Bay of Campobello Island.

A practical problem of fish processing was tackled: an attempt to improve the quality of smoked 'finnan haddies' was begun by Olive Patterson of the University of Toronto, assisted by the staff of the station. Principal F.C. Harrison of Macdonald College examined for bacterial content haddock that had been salted and smoked immediately after landing. He also sent a box of spoiled haddies to the college for laboratory examination.

Dr Clara C. Benson of the University of Toronto began a seven-year study of the biochemical content of the blood and tissues of marine animals, especially lobsters and food fishes, and Dr J.B. Collip of the University of Alberta studied the composition of herring ova, working between St Andrews and Nanaimo.

Field work that season included Knight and MacClement's ongoing lobster study at Long Beach, assisted by Andrew Halkett. For his continuing oyster study at Malpeque Bay, A.D. Robertson was fortunate enough to obtain the collaboration of Dr Julius Nelson of New Jersey, one of the best-known oyster scientists in the United States.

In October, Huntsman investigated an outbreak of disease among the salmon of the Northwest Miramichi River. He visited the hatchery at South Esk, New Brunswick, examined conditions in the rearing pond, and sent samples of the diseased fish to Harrison at Macdonald College for bacteriological examination. At the pond Huntsman examined affected fish for organisms other than bacteria which might have caused the fungus found in the salmon.

It had been hoped to continue the Fisheries Expedition in 1916, moving up the coast into eastern arctic waters, but the war brought these plans to naught. At the station, although fundamental research continued, there was emphasis on practical problems of the fisheries. Thus, while Willey resumed his study of plankton, assisted by Mary Currie of McGill, and Clara Fritz, also of McGill, began a two-year study of the seasonal distribution of the diatoms in the region, the finnan haddie investigation was continued, and samples of the product were sent to various places across Canada for testing and criticism. On the initiative of F.C. Harrison, Wilfred Sadler of Macdonald College came to the station to investigate 'swelled' canned fish. He visited several canneries in New Brunswick and Maine to study processes and to obtain material, which included sardines and other herring, haddock, lobster, and shrimp. Since sardines represented the main industry, bacteria of swelled sardine cans was the subject of the chief study, which continued at Macdonald College at the end of the season. Sadler also studied the bacteria responsible for the destruction of the copepods which form the food of many fishes.

Alexandre Vachon of Laval University made an analysis of starfish as a possible manure and also studied the hydrography of Passamaquoddy Bay and points in the Bay of Fundy. Huntsman concentrated on the growth of herring and the distribution of macroplankton. The collection of plankton was continued on a systematic basis during the winter months and extended to include weekly or monthly collections of plankton and hydrographic samples at certain stations established in the St Croix River, Passamaquoddy Bay, and the Bay of Fundy.

At Willey's suggestion a series of popular lectures was inaugurated at the station; these included 'Fish Eggs and Larvae' by Prince, 'The Origin of the Sea' by Macallum, 'Plankton' by McMurrich, and 'Evolution' by Willey. These informal gatherings proved extremely successful and were continued through subsequent seasons.

Field work included two surveys of St Mary Bay, Annapolis Basin, and Yarmouth harbor by Huntsman with the *Prince* and crew. Collections were made and information secured from fishermen concerning the fishes, their spawning times, and methods of capture, and plankton and hydrographic samples were taken. On the second trip the *Prince* made a short survey of the Kennebecasis and lower St John rivers.

Robertson pursued his oyster investigation, examining New Brunswick oyster beds at Cocagne River and Bay, Buctouche River and Bay, Richibucto River, and Baie du Vin, accompanied by Ernest Kemp on the *Ostrea*. Bad weather prevented completion of the program. D.A. MacKay of Ottawa, assisted by A.B. Dawson of Acadia Univeristy, continued with the lobster work at Pictou. Knight and MacClement undertook extensive mating experiments with lobsters. Berried lobsters were kept at Harbour de Loutre on Campobello Island during the winter, in a car constructed for that purpose. Halkett forwarded eggs taken from berried lobsters to Knight and made observations at the close of the lobster season.

During 1917 the war dragged on and world shortages of food became more and more serious. An attempt was being made by the United States Bureau of Fisheries to popularize dogfish (a small shark) as food, under the slightly more appetizing name of grayfish. This prompted a study by Mavor of the life history of this and other sharks, their distribution and present uses, and the nutritive value of their flesh. He was assisted by Dr. Emil J. Baumann of the University of Toronto, who made analyses of samples of the canned product. Mavor also assisted Dr Robert Chambers of Cornell University in a study of the winter flounder by investigating its distribution. Chambers made microdissections of the egg cells of this species as well as of the ganglia cells of the lobster and the starfishes *Asterias* and *Solaster*; and in collaboration with Bessie Mossop of Toronto he experimented with fertilizing *Solaster* sperm with *Asterias* eggs. Miss Mossop also began a three-year study of the life history of the sea mussel, examining mussel beds in the region. Gerald Coote of Laval University made a biological study of the soft-shelled clam.

Along with assisting Sadler in his continued investigation of swelled sardine cans, Eleanor Shanly of McGill worked on the intestinal flora of the herring. Vachon continued with his hydrographic sampling, analysing the samples at Laval after the season ended.

The major field effort in 1917 was an expedition into the Gulf of St Lawrence, headed by Huntsman, to carry on further research, particularly on the undeveloped fishery resources of the region and the fate of the vast quantities of herring eggs spawned on the Magdalen Shallows. Frits Johansen of the Victoria Museum in Ottawa studied the cunner, and Philip Cox went along to take part in the scientific work. Mavor filled in for Huntsman at the station until the arrival of McMurrich on 29 June.

St Andrews tackles wartime problems, 1917

83

The *Prince* left St Andrews in May and put in at Eastern Harbour, Cape Breton, where Huntsman, Johansen, and Cox joined it. There they were able to obtain a vacant house to be used as a laboratory, making it a base from which to conduct operations. The survey was carried on from May to September, for the most part at stations between Eastern Harbour and the Magdalen Islands. To establish the relation between physical conditions and the distribution of various species, water samples and temperatures were regularly taken at each station during the period and collections were made by use of plankton nets, drag seines, gill nets, young fish trawls, dredge, hoop traps, and hand lines. The vast amount of material and information obtained provided the basis for reports and papers over many years. Most of the material was brought to the station by the *Prince* at the close of the survey.

Upon his return from the expedition, Huntsman made an investigation of the herring spawning area at the southern end of Grand Manan at the request of the Fisheries Department. He also studied the American plaice, a food fish not yet exploited at that time, using material obtained on the 1915 expedition and some from Passamaquoddy Bay and the Bay of Fundy. Cox also continued a study of the lumpfish which he had begun on the expedition, and Johansen worked up his data on the cunner.

In the ongoing lobster investigation under Knight, MacClement conducted further lobster mating experiments at Long Beach and D.A. MacKay studied the output of young lobsters at Georgetown, Prince Edward Island, and Shemogue, New Brunswick. At Bay View, Nova Scotia, experiments in hatching and rearing lobsters were conducted by A.B. Dawson. Arrangements were made to send an experimental shipment of lobsters to British Columbia for planting in Pacific waters, but owing to hot weather in transit the lobsters all died.

During the season a 'blight' broke out among the oysters at Malpeque Bay and the Board was asked to determine its nature and cause. It was suspected that the causative agent might be a tubellarian flatworm, similar to one which had appeared along the coasts of Florida and Connecticut. Such outbreaks were usually of short duration, disappearing after running their course, but meanwhile they wiped out all susceptible oysters. It proved, however, that the disease had nothing to do with any flatworm, and it became endemic at Malpeque when the bay's oysters acquired a tolerance to it. In later years it spread to New Brunswick and temporarily reduced stocks there.

Further Gulf of St Lawrence studies, 1918

Investigations of unutilized food fishes continued in 1918 with further study of the lumpfish by Cox, while C.J. Connolly made a study of the angler or monkfish and W.A. Clemens studied the muttonfish or eelpout, assisted by his wife Lucy Smith Clemens.

Clemens was in charge at the station in 1918 as most of Huntsman's time was taken up with further investigations of the Gulf of St Lawrence fisheries. For the latter project the *Prince* left the station on 20 May for the Miramichi, with Loggieville as a base. There a building was obtained as a laboratory. Huntsman had already initiated the survey before the arrival of the *Prince*, using local boats. In addition to the crew of the *Prince*, he was joined by Cox, who concentrated on the tomcod and striped bass; Vachon, who worked on the hydrography of the region; Klugh, who studied the algae of the Miramichi River; and Captain P.A. Larkin, who assisted in collecting plankton. Huntsman also continued his investigation of the spawning and growth of the herring and the biology of the smelt and the alewife, as well as the distribution of the fishes and larger invertebrates. Experiments were also made on the effectiveness of various kinds of bait.

Lobsters shared the spotlight. The discoloration of canned lobsters was resulting in yearly losses of thousands of dollars to the industry. Concerned about the problem, the Council for Scientific and Industrial Research (later called the National Research Council) provided a grant for the study of the cause of the discoloration. Jennie McFarlane of Toronto began a two-year study at the station of the bacterial flora of both blackened and unblackened canned lobster meat as well as of fresh lobster. She continued this at the university, and found certain bacteria present in all the affected cans but absent from the unaffected cans.

Knight, meanwhile, was at the station for two weeks, carrying on mating experiments with lobsters obtained from the great Conley lobster pound at Deer Island. He then supervised investigations on the lobsters of Northumberland Strait between Prince Edward Island and Nova Scotia. Dawson was in charge of the work along the south shore of the island, examining the catches at the traps and at the canneries as well as those made by fishing on special permit from the Fisheries Department. Observations were made on hatching, sizes, mating, and fishing regulations. D.A. MacKay carried on similar work off Cape Tormentine, New Brunswick, and Halkett recorded the catches at the Magdalen Islands, along the north shore of Prince Edward Island, and along the southeast coast of Nova Scotia.

The possible exhaustion of the lobster industry by the taking of egg-bearing lobsters was a matter of serious concern, and an educational program to secure the cooperation of fishermen and canners to prevent this calamity seemed essential. Knight drew up a program, which was agreed upon by the Board and by the department. It included demonstrating means of conserving berried lobsters, increasing lobster mating facilities by using pens near the canneries, and giving talks to fishermen and canners to enlist their cooperation. The area was divided up. Alexandre Vachon took the gulf coast of New Brunswick; Professor H.G. Perry of Acadia University took the south shore of Prince Edward Island; MacClement took the southern portion of Northumberland Strait; Dr M. MacGillvray of Queen's, the southern portion of New Brunswick and western Nova Scotia; and Professor J.T. Herbert of the University of New Brunswick, Cape Breton Island. Halkett also assisted in the campaign in Prince Edward Island and at the Magdalen Islands.

Addresses were given to the fishermen in halls and canneries, on wharves, and wherever a handful of listeners could be gathered. The imperative need of returning every egg-bearing lobster to the water was strongly emphasized. Opinions were obtained from the canners as to possible limits to the canning season as further protection to the industry. A conference was held on 8 August, to which all persons interested in the lobster industry were invited. This resulted in recommendations to the government which, in turn, produced certain amendments to the lobster fishing regulations.

Other activity that season at that station included Eleanor Shanly's continued studies of the bacteria of the gills and viscera of sardine herring and also bacteriological work on canned sardines which Sadler, because of illness, was unable to continue. Dr Louis Gross of McGill investigated the rate of spoilage, and the effect of retention of the gills on this rate, in some of the common food fishes, and studied the question of immunity in lower organisms.

Work on the suitability of certain fishes and marine animals as food continued in the 1919 season. Cox, working with the tomcod and lumpfish, was assisted by Marian Anderson of the University of New Brunswick, who also made a study of the seasonal growth of plaice and flounders, with Mavor. Huntsman continued his work on the growth of fishes and the spawning of herring. Bacteriological investigations were continued by Miss Shanly on canned clams and finnan haddie, as well as

Educational
program launched

Major programs
continue after the
war

85

on the salmon. Further studies of fish as food were carried on at outside laboratories with material supplied by the station for the purpose. Willey at McGill studied the food of plaice and other flatfishes as shown by stomach contents; Johansen studied the life history of the cunner at Victoria Memorial Museum; Dr Clara C. Benson at the University of Toronto made a chemical study of the flesh of elasmobranchs; Miss Neff studied the flesh of the hake to determine reasons for unsatisfactory refrigeration, a histological examination of the flesh being made by Professor W.H. Piersol at the University of Toronto; and Clemens continued his study of the muttonfish. At Halifax a chemical investigation of the spoiling of canned lobsters and clams was made by Professor E. MacKay under the auspices of the Council for Scientific and Industrial Research.

A special committee to consider methods of preparing fish for food and comparative values of various species was formed, with representatives from the household science departments of the University of Toronto (Dr Benson as convenor and Professor A.L. Laird), Macdonald College, Ste Anne de Bellevue (Miss A.E. Hill), and Macdonald Institute, Guelph (Miss M.U. Watson). The program included study of the best way to handle and prepare familiar as well as new food fishes, with special attention to economy in time and money, undertaken by Misses Laird and Hill, and the chemical changes taking place during these processes, directed by Dr Benson. A report on the cooking experiments was submitted by Doris McHendry, and one on the skate and grayfish as food by Dr Benson.

Knight was at the station for some time studying age determination in lobsters, but the main lobster program continued at Malpeque Bay, with MacKay and Klugh exploring areas where berried lobsters could hatch their eggs to best advantage and where the young fry would have the best chance for survival. This required finding young lobsters in abundance. The search was conducted with special traps designed by Klugh, and with hoop nets and a beam trawl, using a variety of baits. Guy E. Johnson assisted. The educational program was resumed among lobster fishermen and canners during the spring lobster season. Halkett, MacGillvray, Connolly, MacClement, Perry, and Vachon took part, and in the winter the campaign was followed up by Halkett, who gave lantern slide lectures.

New programs initiated, 1919

A two-year hydrographic investigation, which concentrated particularly on the circulation by Bay of Fundy waters, was begun by Mavor during the 1919 season. To determine what general movement might exist, apart from tides, 396 drift bottles were set out during the summer by the *Prince* on lines crossing the bay, and hydrographic observations were made in six sections. Vachon analysed the water samples. During the previous winter systematic fishing operations had been carried out by the *Prince* to determine the seasonal movement of fishes, and information was obtained on the spawning and migration of a number of species, as well as temperature and salinity observations to make a complete yearly record.

A continuing and extensive study of the life history of the shad was begun by A.H. Leim of the University of Toronto, in an effort to find the factors responsible for the decline in the fishery over the previous 45 years. Leim joined Huntsman at Lawrencetown in June, and collections of shad were made in the Annapolis River, where the effects of the dam at Lawrencetown on the spawning of shad in the river were studied. After an abandoned shad hatchery at Cody's on the Washademoak River, New Brunswick, was visited, some lakes in Rockwood Park, Saint John, were examined for their possible productiveness. Leim then went to the Shubenacadie River in Nova Scotia to study conditions there. He also made general collections and a special study of the life history of the shrimps of the genus *Spirontocaris*.

Summer staff at St Andrews, 1919. Front row (l. to r.): C.J. Connolly, J.W. Mavor, A.G. Huntsman. Middle row: Miss E. Shanly, E.E. Prince, A. Vachon. Back row: L. Gross, Miss B.K. Mossop, Miss Rose Prince, Miss J. MacFarlane

One of the major field activities of the season was a systematic faunistic study of St Mary Bay, with the *Prince*, from June until September. Little River was chosen as a base, and space was rented for a laboratory. In addition to periodic collections of plankton and hydrographic material, investigations with dredge, beam trawl, shrimp trawl, seine, and traps were made, covering the entire bay. Cox spent the summer there examining the fishes and other fauna; Connolly was there for a short period studying the cunner and investigating the bottom fauna, and Huntsman also took part in the survey. It had been hoped that the bay would prove suitable as a spawning ground for lobsters, but investigation extinguished this hope.

Other scientists at St Andrews that season included F.S. Jackson of McGill, who made a comparative histological study of the muscles of fish; Prince, who did general work; and L.W. Bailey, who continued preparation of diatom slides. Huntsman visited Campbellton, New Brunswick, to investigate a disease which had broken out among the salmon of the Restigouche River.

Most of the continuing studies were resumed in 1920, and expansion of the physical plant was begun to accommodate the steady increase in work dealing with pressing problems. The facilities of the station had become sorely taxed, leading Huntsman to recommend to the Board a number of changes: the installation of a gasoline plant to replace the inadequate blue-gas plant then in use in the laboratory and for hot-water heating in the residences and cottages; the installation of electricity for use in the laboratory and for lighting; an addition to the residence to provide bedroom space for the increasing number of workers; an addition to the laboratory

Expansion marks
1920 season

87

building to provide biochemical and bacteriological laboratories; the construction of a library and cold-storage rooms; and the appointment of a collector to provide material for the investigators, arrange local explorations, and assume other duties.

In addition, Jackson had been conducting a series of experiments on histological changes taking place in fish muscle as a result of freezing under various conditions. This work emphasized the need for an experimental cold storage plant or freezer for temperature control in these and other experiments on the freezing and cold storage of fish, and the need was brought to the attention of the Board, which applied to the Department of Fisheries for funds to carry out the project. These were made available, and in November a refrigeration plant was constructed and an ice machine installed.

F.C. Harrison was at the station briefly, superintending the work of Jennie L. Symons of McGill on the bacteria which produce blackening in canned clams; that of Jane Williamson of Columbia on the normal intestinal flora of the sardine herring, a continuation of the work of Eleanor Shanly; and Jennie McFarlane's work on the bacteriology of discolored canned lobster meat. Investigations were also made on the shrinking of lobster meat during canning operations.

Margaret E. Reid of Toronto studied reproduction and development of the arrow-worm *Sagitta elegans* in the Bay of Fundy, using material taken in plankton tows. C.J. Connolly made observations on the young stages of decapods, and Arthur Willey continued his investigation of the free-living copepods and their economic value as food for fishes, including material obtained by Huntsman and himself from the Quill Lakes, Saskatchewan, as well as from the Miramichi estuary. Philip Cox completed his work on the tomcod and Miss E.H. Chant of Toronto investigated the smelt, particularly the eggs and young. Several more series of drift bottles were put out in the bay in 1920, hydrographic and plankton sections being made at the same time. Alexandre Vachon again analyzed these water samples as well as those from the Quill Lakes.

Activities broaden

In 1920 experiments were begun on the effects of physical factors on marine animals. Huntsman observed the effects of light on growth of mussels, and Knight studied the influence of temperature on newly hatched lobsters. A method was developed by Knight for rearing lobsters to an advanced stage, and experiments were conducted by him until the end of July, when he left to superintend the lobster work in Prince Edward Island. Prince continued the experiments, and a number of young lobsters reared through the early stages were preserved for later examination of the changes which young lobsters undergo in their life history. At Malpeque Bay, MacKay continued a search for yearling and two-year old lobsters, while Dawson studied the distribution of small lobsters at Egmont Bay.

Robertson resumed investigations on the growth of the oyster, which had been interrupted the previous year. He was assisted by Mrs Robertson and a small staff. In September the *Prince* made a survey of Minas Basin and Channel, and Scotsman Bay. In October a series of tow-net operations was conducted off the southern end of Grand Manan to ascertain the distribution and movement of herring from the spawning ground.

The field work took Huntsman and Willey to distant parts. In May, while en route to the Pacific coast for a meeting of the Canadian Fisheries Association, they spent two days at the Quill Lakes of Saskatchewan, making collections of plants and animals and taking water samples. The object was to study the fishery possibilities of these saline lakes. Huntsman suggested that they might be suitable for the growth of eels, and later arranged for an experimental shipment of elvers from New Brunswick.

The 1921 season opened with Huntsman in charge as director. The proposed Fruitful 1921 season extension to the laboratory building had been approved and was carried out that year. Biochemical studies of fish as food were advanced and made more accurate by the use of the new refrigeration machinery. Under Dr Benson's direction studies were made of rigor mortis of fish with Jean R. Panton of Toronto making an examination of stroma proteins and Miss A.E. Dempsey of the same university examining the proteins of the muscle juice. Jackson studied the histology of frozen fish tissues. Collip continued studies previously initiated at the Pacific Biological Station on respiratory processes in the clam *Mya arenaria*.

The Council for Scientific and Industrial Research had commissioned Harrison to investigate the cause of pink or red discoloration which had been frequently appearing in green-salted, cured, and dried codfish, and which had seriously affected the market value of the product. During August Harrison and Margaret Kennedy of Macdonald College worked out of the station, and Miss Kennedy visited a number of places in the surrounding district where fish were cured. She obtained samples of the fish, scrapings from containers and wharves, and samples of discolored salt from which cultures were made and incubated. Conditions under which the fish were cured and marketed were also studied. The reddening was found to occur when sea (or solar) salt was used, but not mined salt.

Dorothy E. Newton of Macdonald College isolated spore-forming bacteria from various sources such as sea water and the alimentary canal of many of the common food fishes and shellfishes with the object of determining which forms are not killed by sterilization and hence may produce hydrogen sulphide, causing blackening when canned.

Researches on the effects of certain physical factors, particularly as affecting reproduction and growth, were continued by Huntsman and a group of workers. Freida Fraser of Toronto studied the effect of light on growth of intertidal animals; Miss E.M. Taylor, of the same university, studied the effect of hydrogen-ion concentration on marine animals, and A.H. Leim, hydrogen-ion concentration in relation to copepod life. Prince continued his larval lobster rearing studies, observing the effects of temperature and light. Miss M.E. Reid of Toronto studied the distribution and development of the cunner and, from collections of plankton tows taken in several areas along the coast showing considerable variation in physical conditions, attempted to establish some correlation between the conditions and the development of the species.

Continuing the investigation of the life history of the smelt, in the spring of the year the newly appointed collector and former captain of the *Prince*, E.G. Rigby, started enquiries concerning their spawning and established a number of stations in the St Croix River, Passamaquoddy Bay, and Lake Utopia, searching for eggs and taking tows for the young fish.

Knight continued his lobster rearing and canning studies at the station, assisted by MacKay, who experimented with various foods and varying degrees of light and temperature, using hatchery boxes designed by Knight for retaining berried females alive and for rearing and feeding fry.

Klugh began a series of experiments on the culture of certain Entomostraca (ostracods and copepods). From tows taken at various points along the coast Connolly made a study of decapod larvae.

The major field work of 1921 was a three-month survey of the southwest coast of Major field work Nova Scotia, made with the *Prince* using Harrington Passage as a base. The importance of this region derives from its proximity to the Gulf Stream, and its serving as a mixing point for the various waters that enter the Gulf of Maine and the

Bay of Fundy. Conditions in the region were studied by collecting fishes, plankton, and bottom material, and taking hydrographic sections each month over the continental shelf and each fortnight in inshore waters, as well as special sections east and west of Cape Sable. Cox accompanied the expedition, studying biological conditions, assisted by M.I. Sparks of Toronto. Huntsman spent two short periods with the survey and Connolly made a 10-day study of the decapod larvae.

In the spring and in the fall, conditions in the Kennebecasis and in the Saint John River at the mouth of the Kennebecasis were observed at a series of stations which had been established in 1916 and at two newly established stations. Trawling was done and hydrographic and bottom samples taken.

Complaints had been received about the color and quality of canned lobster meat from some of the canneries on the Atlantic coast, and the department arranged with the Board for an educational campaign among the canners concerning the cause of deterioration. Dr G.B. Reed of Queen's University and C.J. Tidmarsh of McGill were appointed and under Knight's direction visited the canneries in Prince Edward Island, showing with a miniature laboratory the growth of bacteria under unsanitary conditions and the resulting discolored meat of inferior quality. A noticeable improvement took place in the quality of the fall pack in the area covered by the demonstrations. Knight continued his study of the post-larval life of the lobster during the latter part of the summer at Summerside, Prince Edward Island.

Robertson shifted his investigations on the oyster to the east branch of the Hillsboro River, about 10 mines from Charlottetown, assisted by Misses H.I. Battle and M.A. McIntosh of the University of Western Ontario. The work concentrated on the effects of extreme salinities, temperatures, and types of bottom on the growth of the oyster as well as on its food.

A technical course in fish culture for hatchery officers was arranged by Knight and given in Truro, Nova Scotia, during August. Lectures in physics and chemistry were concluded and examinations given at the end of the course. Addresses on the migration of the fishes and the growth of bacteria were given by Knight and Prince to the inspectors and fishermen at their conference in Charlottetown.

Huntsman had been named to represent Canada on an international committee on marine fishery investigations, consisting of members from Canada, Newfoundland, and the United States, which had been formed to determine what measures of international cooperation might be desirable. Two meetings were held in 1921. The committee later became known as the North American Committee on Fishery Investigations.

HUDSON AND JAMES BAY EXPEDITIONS

The years just prior to the First World War marked the height of the railway boom in Canada. Among the routes that seemed scheduled for early construction were a line through northern Manitoba to Port Nelson on Hudson Bay and an extension of the Temiskaming and Northern Ontario Railway to Moose Factory on James Bay. Anticipating that these regions would soon be opened to commerce, in 1914 the Department of Fisheries decided to send three survey parties to investigate their fishery potential. Using the National Transcontinental Railway, then under construction, as a point of departure, the first party under C.D. Melvill went down the Missinabi and Moose rivers by canoe. They then explored the east side of James Bay as far as Cape Jones, using canoes and small vessels belonging to the two local trading companies. The second party, under A.R.M. Lower, went by canoe down the Nagogami, Kenogami, and Albany River to Fort Albany, and made a similar survey of the west coast of James Bay. The third party, led by the legendary Napoléon Comeau of Godbout on the St Lawrence, sailed from Halifax in the small schooner *Burleigh*, which was fitted with an auxiliary motor that would make 'about 2 knots.' It entered Hudson Bay from the north and proceeded to Port Nelson. From mid-August to the end of September Comeau's party fished in the rivers and estuaries of that region with rod, seine, and gillnet, and gathered information from residents.

The three parties all discovered that large quantities of brook trout, whitefish, and tullibee could be taken and were being taken by the local people in the rivers and nearby brackish waters. Suckers too were numerous and were used for dog food. All these fish moved out into the bay during summer for a period of rapid growth, and then returned to spend the winter in the rivers because the salt water became too cold for ordinary fish to survive. Reports were heard of capelin and Greenland cod in the bay, but they lacked vessels and equipment to search for these species. The small white whales were numerous on the western side of the bay, but Comeau sighted only one bowhead, a lone survivor of the great stocks of former years.

When the three parties returned they found the war in full swing and the railway boom about to collapse, so that the lines to Hudson Bay were built only many years later. However, in 1920 the Biological Board sponsored a supplementary trip by Frits Johansen, a naturalist employed by the Department of Naval Service, to collect aquatic life in Hudson and James bays. He, too, traveled north by canoe, and went up the east coast as far as Richmond Gulf, collecting invertebrates mainly. From the specimens that he brought back, reports were published on the Foraminifera, echinoderms, ascidians, copepods, amphipods, and sticklebacks – all of which faunas reflected the severe arctic nature of their habitat.

9   Pacific work by the Biological Board, 1912–21

Activities at the Pacific Biological Station during the Biological Board's first 10 years were not as wide-ranging as at St Andrews, and the number of investigators was not as great, but, in general, there was the same trend toward somewhat greater emphasis on applied research.

In 1912 the continued illness of the Reverend Taylor made it necessary for the Board to relieve him of all responsibilities, although retaining him as curator. Greatly weakened by his illness, Taylor maintained a keen interest in the work of the station and clung to the hope of returning to health to resume his own activities. Almost to the end, in spite of efforts of the staff to dissuade him, he daily made the effort of a quarter-mile trip from his home to the station to watch the progress of the work. Professor J.P. McMurrich acted as director during the summer months, and C. McLean Fraser was engaged to fill in for Taylor as curator. Taylor died on August 22 and Fraser was appointed to replace him, taking over formally on 1 October.

Charles McLean Fraser

Born in Huron County, Ontario, in 1872, Fraser graduated from the University of Toronto in 1898 and received his master's degree in 1903. He took a teaching post that same year as a science master in the high school at Nelson, British Columbia, and became principal of the school the following year.

As has been noted Fraser worked at Canso in the summers of 1901 and 1902, spent some time at the Minnesota Seaside Station during the summer of 1903, and, in 1908, with the establishment of the Board's biological station at Nanaimo, appeared on the scene to begin studying various marine forms, particularly hydroids, work which would eventually establish him as a world authority on those animals. He received his doctorate from the University of Iowa in 1912, the year he was appointed curator of the Pacific Biological Station at Nanaimo, a position which he was to hold until 1924. In 1920 he became professor and head of the

Department of Zoology at the University of British Columbia. His study of hydroids extended over a period of 40 years, until almost his death in 1946, and covered life-histories, systematics, ecology, and distribution.

At the Pacific Biological Station Fraser carried out studies of parasitic copepods, marine borers, and commercial clams, as well as the life histories of herring and Pacific salmon. He began the daily recording of sea-water temperatures and specific gravities at Departure Bay and William Head, and served on a commission that studied the sea lions of the British Columbia coast.

During the summer of 1912 McMurrich resumed his studies on the young stages of the Pacific salmon and made a trip in September to the longlining banks to study the life history of the halibut, whose catches from traditional areas had already greatly decreased. Plankton was collected at various points around the Queen Charlotte Islands, the west coast of Vancouver Island, and other locations. The anemones taken became the subject of a special study by McMurrich. The 1912 season

Dr W.F. Thompson of Stanford University had been retained by the British Columbia Commissioner of Fisheries to investigate the stocks of shellfish of the province. He made the station his headquarters for two weeks, collecting clams and oysters, and as a side-line the gephyreans and annelids as well.

McLean Fraser began investigations on the life history of the herring and studied the eggs of the lingcod, often called green or blue cod from the color of the live flesh of some specimens. Other studies included the possible destruction of young salmon by seals in the estuary of the Fraser River and the alleged injurious effects of seines at Point Grey. Fraser continued his herring study through the winter of 1912–13 at Departure Bay, where they abounded, and in March obtained ova which he hatched out as young fry towards the end of the month.

In 1913 a 40-foot gasoline launch with 20 horsepower engine, the *Ordonez*, well-equipped for marine researches, was purchased for the station and berthed at a dock completed the previous year. Running water was supplied for the first time from a nearby creek. Taylor's valuable library was purchased by the Board and became the nucleus of the station's library. Launch available in 1913

Professor E.M. Walker spent the summer at the station, continuing his studies of the life history and ecology of dragonflies and damselflies, and making a collection of isopod and decapod crustaceans. Mr T.B. Kurata of the Biological Museum of the University of Toronto made collections of marine, freshwater, and terrestrial insects and other invertebrates for that museum. Mr and Mrs Dayton Stoner of the University of Iowa also collected the local fauna for three weeks. Fraser, assisted by his wife, made extensive observations on the various species of salmon, and he continued his study of hydroids.

During August and the first part of September a man who was to loom larger in a later period of Board history, Professor A.T. Cameron of the University of Manitoba, collected material at Nanaimo for a systematic study of the distribution of iodine in marine plants and animals. He also made observations on the kelps available for economic purposes in the locality and on the salinity of the water in Departure Bay. He came for three summers, and then, after a three-year break, spent another two summers at Nanaimo.

Early in the 1914 season Arthur Willey, who had resigned from the Board in order to work for an honorarium (a practice not permitted a Board member), came to the station and used it as a base to pursue investigations on the Pacific halibut. Their natural history was studied, particularly their spawning and migrations in relation to maintenance of the stock. These studies took him to the Queen Charlotte Islands, Hecate Strait, the Gulf of Alaska, and the west coast of Vancouver Islands. He and Halibut and crabs studied in 1914

Fraser also studied scales of halibut as a possible indicator of their age.

In addition to his other duties, Fraser made a thorough examination of the shores along both sides of the Strait of Georgia, seeking a possible favorable lobster planting area. He found no survivors of government plantings made in 1896, 1905, and 1908. Bottom conditions, available foods, temperature and salinity of the water, predators, and other factors were studied. A.B. Klugh of Queen's assisted in this survey. Klugh also made a systematic study of the crabs of the region and the marine algae, and took photographs of fish scales.

Continuous records were kept of the temperature and density of the water as well as air temperatures from 12 May to 10 September, and Fraser and Cameron made a study of the relationship between these factors and the distribution of certain marine animals and plants.

Professor F.W. Weymouth of Stanford University, commissioned by the provincial government to investigate the life history of the large Dungeness crab, used the station's facilities for three weeks.

Chinook salmon and sea lions, 1915 Continuing as year-round curator at the station, in 1915 Fraser undertook a study of the 'winter check' used in age determination of fish by the scales. The spring or chinook salmon, available in the strait the year round, was chosen as the species for investigation. Material taken during the fall, winter, and spring of 1914–15 was examined. He also studied the rate of growth and migration of this species, and their possible relation to the temperature of the water.

Fraser also took part in the work of the Sea Lion Commission, which had been established to ascertain the number of sea lions killed for bounty, to estimate the number remaining, to investigate the extent of damage done by the sea lions to the fisheries and to nets, and to study their food. Owing to the lateness of its start, 30 August, the commission did not complete its work until the following year.

Dr Th. Mortensen of Copenhagen, on a two-year collecting trip around the world, spent two months at the station collecting specimens of echinoderms. A.B. Macallum was also there. But possibly the most significant event of 1915 was the opening of the University of British Columbia, which soon became closely associated with the station, which, in turn, provided a training ground for qualified workers in marine research and for key personnel for the university.

In 1914 the Pacific Fisheries Society had been founded. At that time David Starr Jordan and his students had made Stanford University the focus of fish and fishery research on the Pacific coast, so the society had a strong California orientation. However, Canada was represented in 1915 when Fraser attended its second meeting. His paper on the growth of chinook salmon was an important one, because it introduced the technique of subtracting a constant before making a proportional back-calculation of fish size from scale size. But after the publication of its *Transactions* in 1916 the society apparently became a war casualty, for nothing was heard of it subsequently.

One-man team in 1916 In 1916 Fraser was practically a one-man team at the station. He examined and reported on sockeye salmon hatched in the spring of 1913 from eggs taken from Harrison Lake in the fall of 1912, in an experiment to determine differences of growth, spawning, and other factors between retained pond-fed salmon and those which normally returned each year to salt water. A study of the coho salmon, which had recently become commercially valuable, was begun. Growth determinations from the scales of 10,000 salmon, including five different species, were also made by Fraser. In addition, he made a study and report on the habits and peculiarities of the lingcod and the red cod or rockfish, in response to requests for the protection of these species on the Pacific coast.

The only other scientist at the station that season was J.B. Collip, who prepared material for biochemical analysis, chiefly muscle, ova, and milt of three species of salmon (coho, sockeye, and pink), and also investigated the pigmentation of salmon.

In 1917 Fraser continued his studies of the life history of the salmon, obtaining much of his material from the canneries at Quathiaski Cove, Lasqueti Island, and Nanaimo. He paid particular attention to the rate of growth of various species. In March 1,000 spring salmon fry were marked at Cowichan Lake Hatchery by removing a fin, the first of many such experiments. Fraser also completed his study on the hydroids of eastern Canada, using much material forwarded from the collection at St Andrews. In addition he made a study of fish parasites.

Fraser and the remarkable Berkeleys, 1917

The report of the Sea Lion Commission was considered by the Board, but since the evidence concerning food was not considered sufficient, Fraser was asked to try to secure additional stomachs of sea lions for examination. And it was in 1917 that the remarkable Berkeleys first appeared at the station.

Cyril J. Berkeley was born in London, England, in 1878. He was educated at St Paul's School and the Realschule of Nürnberg, Germany, proceeding to University College, London, where he specialized in chemistry under Sir William Ramsay. Later he studied agricultural bacteriology at the Agricultural College of Wye. Edith Dunington Berkeley was born in South Africa in 1875, and was educated at Wimbledon High School and the University of London, where she studied zoology. The Berkeleys were married in 1902, and went to India almost immediately. Cyril had been appointed government bacteriologist to work in the state of Bihar, where his job was to improve the cultivation of the indigo plant and the technique of recovery of its dye. There he introduced measures that eased somewhat the last years of the indigo industry, which already realized it could not compete indefinitely with the new aniline dyes.

Cyril and Edith Berkeley

Although Cyril could have remained in India doing other work, in 1914 the Berkeleys moved to British Columbia. They farmed for three years near Vernon, and then both of them taught for two years at the new University of British Columbia. In 1919 they moved to Departure Bay, where they purchased the house near the biological station that had belonged to the Reverend Taylor and furthered the contact made with the station in the summer of 1917. From that time onward much of their energy was devoted to an extensive garden, planting rare species and cultivars of barberries, irises, rhododendrons, etc., and even developing a few new ones.

At the station Cyril Berkeley was to serve as assistant curator during 1920 and 1921, and thereafter to work as a volunteer. For a time he made biochemical studies of various invertebrates and kelps, being particularly interested in the oxidase found in the peculiar crystalline style of clams and oysters. Later he assisted his wife with the classification of marine annelid worms. In undertaking studies at the station, Edith Berkeley had intended to carry on the anatomical and physiological studies on polychaetes that she had begun more than 20 years previously under Professor Weldon at University College, but she discovered that the classification of the species found on the coast was in chaos and so undertook a fundamental reorganization. This led to extended taxonomic studies, including work on materials sent from elsewhere, and ultimately taxonomy filled her time. For the first 12 years she worked alone; then Cyril came to her aid and they worked and published jointly thereafter. Between 1923 and 1962 some 40 publications appeared. Many new species were discovered and described, and many new distributions were recorded. As world authorities on polychaetes, the Berkeleys received collections

95

for identification from all parts of the world.

Edith Berkeley died in 1963. In 1968 Cyril Berkeley was awarded an honorary LL D by the University of Victoria, an honor which he insisted pertained equally to his deceased wife. When he died in 1973 at the age of 94, it ended an association of more than 50 years with the Nanaimo station.

The amazing sockeye salmon, 1918

The year after the Berkeleys first came to the station, 1918, saw the Board's first direct involvement in a situation that had quite abruptly become an economic disaster for British Columbia salmon fishermen. Much the largest sockeye salmon runs in the province were those to the Fraser River. Scale reading by McLean Fraser and others had shown that these sockeye had predominantly a four-year life cycle, and like all Pacific salmon they died after one spawning. What made them unique was that every fourth year they appeared in phenomenal numbers, their abundance then being many times greater than in the intervening years. Yet so it had been for at least a century, as shown by old records of the Northwest Company in New Caledonia.

The disaster resulted from construction of the Canadian Northern Railway in 1912–13, which had dumped rock into the Fraser and created obstructions that prevented a great majority of salmon from passing through its canyon in two successive years, 1913 and 1914. The obstructions were alleviated by 1915, but unfortunately one of the runs blocked had been the 'big' year of 1913. In some quarters there were hopes that the damage had not been really severe, but these hopes were shattered in 1917 when unusually intensive fishing produced only a quarter as many sockeye as in 1913, and very few escaped the nets to reach the spawning grounds. United States fishermen were affected as much as Canadians were, because their gear took more than half of the catch of Fraser-bound salmon while they were migrating through Puget Sound.

The collapse triggered efforts to restore the damaged runs. Indeed it was expected that the Fraser's overall production could be greatly increased, for at that time it was assumed that, whatever might have been the original cause of the big-year phenomenon, it should be possible to bring all four lines of sockeye up to the abundance of the most productive one. A preliminary to effective action was to ascertain the exact routes and rates of movement of sockeye during their migration, what percentage of them were being caught, and what tributaries the survivors went to for spawning. So in spite of the increasing stresses of the war, in 1918 a cooperative tagging program was arranged between the United States Bureau of Fisheries and the Canadian Department of Fisheries. As part of the Board's contribution, Fraser spent a month at the salmon traps off Sooke, near Victoria, tagging and releasing sockeye caught there. Although the tag used was soon lost from most fish, enough returns were obtained to indicate rates of migration and the fact that the earlier migrants mostly went to tributaries above the Fraser canyon while later arrivals spawned in downriver tributaries.

At the station in 1918, Collip investigated the biochemistry of salmon flesh, particularly its red pigment, and Edith Berkeley again collected materials for her university work. When not tagging, Fraser continued his herring and salmon life history studies. He found, for example, that the so-called blueback of the Strait of Georgia were coho salmon starting their final season of growth.

Increased activity in 1919

The war being over, there was a marked increase of activity at the station in 1919. Cyril Berkeley was back, on a year-round basis, to make arrangements for biochemical investigations and to collect Bryozoa, and Edith Berkeley started her work on the local polychaete worms. Professor J.J.R. MacLeod of the University of Toronto collected material for research on the glycogen content of invertebrates and fishes. Professor C.H. O'Donoghue of the University of Manitoba and his wife Elsie collected and studied the sea slugs of nudibranchiate molluscs – which are fascinatingly beautiful animals. Irene Mounce, now at the University of British Columbia, made twice-daily collections of diatoms in the bay for Dr A.H. Hutchinson. The diatoms were measured, and records were made of the wind direction and general weather conditions.

The daily records of water density and temperatures, which had been kept over a five-year period, were studied by Fraser in an attempt to determine the effect of variations in these conditions on the migration, spawning, and other activities of marine animals. He also continued his investigations of the life history of Pacific salmon and other faunal studies. During the early part of the year he obtained leave for a few weeks to lecture at the University of British Columbia.

In the following year, 1920, the station was faced with the possible loss of Fraser's services as curator, as he had been offered the post of professor of zoology at the University of British Columbia. However, an arrangement was made whereby he would remain as curator at the station during the summer and move to the university for the balance of the year. Cyril Berkeley was appointed assistant curator for the year, taking charge in Fraser's absence.

Cameron was back that season to examine the waters of Departure Bay and adjacent areas for chemical content. Irene Mounce continued her collection of planktonic diatoms, and studied the relation of various physical factors, such as temperature and specific gravity of the surface water and direction of wind, to the abundance of phytoplankton. Collip studied the biochemistry of certain fishes and other marine animals, particularly the carbon dioxide content of blood and body fluids and the effect of changes in environment on carbon dioxide content of the coelomic fluid in the clam, *Mya arenaria*. Edith Berkeley continued her studies on the Polychaeta.

During the summer Fraser attended the Pan-Pacific Science Congress at Honolulu as a representative of the Board. He chaired discussions on ocean currents and their relation to the abundance of marine organisms, mapping of the Pacific, and planning of joint biological research.

In 1921 Fraser spent four months at the station. In addition to continuing his systematic study of the hydroids and other faunal investigations, he worked up material and data gathered over several years on the life history of the herring.

The general work of collecting biological material occupied all workers at the station that season. Professor O'Donoghue and his wife resumed their work with sea slugs. In addition they reported on the local bryozoans, and also some that had been collected during a short trip to Friday Harbor in Puget Sound.

Cameron made physicochemical analyses of water samples to ascertain causes for variations in composition and also analyzed certain worm tubes for iodine content. In collaboration with O'Donoghue he studied the effects on narcotized animals of sudden changes in light intensity. Experiments were conducted on coelenterates, echinoderms, crustaceans, molluscs, and fishes taken near the station.

Irene Mounce continued her study of the disappearance of diatoms in Departure Bay during the summer and the possible causes. Loss of nets reduced the number of observations for the season. H.A. Dunlop of the University of British Columbia studied the distribution of free-swimming copepods. R.E. Foerster from the same university made a study of the Hydromedusae, including live specimens collected during the summer and specimens in the station's museum which had been collected over a period of nine years.

Fraser's new post, 1920

The 1921 season at Nanaimo

Cyril Berkeley started a series of studies on the occurrence of manganese in the tube and tissues of certain marine annelids, and Edith Berkeley continued her study of the polychaete worms.

IMPORTANT NEW TRENDS IN RESEARCH

During the first 10 years of the Biological Board, as the reports of both stations clearly show, the trend towards applied science continued strongly. Two other trends were equally clear: the greater participation of biologists and scientists from other disciplines in the work at the stations, and an increased flow of scientific papers resulting from this increased and widened participation. The first year of the Biological Board saw the fewest number of workers at the stations: 16. The last year of the 10-year span, 1921, saw the largest number: 40. During that period a total of 230 papers were published. This figure does not include papers published later than 1923, but which may have resulted from work done between 1912 and 1921. And a fairly large percentage of these papers, in contrast to the 1898–1911 period, dealt with the practical problems of the fisheries. There was no doubt that the Board was in the process of making itself valuable to the government and fishing industry alike. It had already established its credentials with the scientific community.

10 The Board and the Assistant Deputy Minister: a confrontation

The relationship between science and government is a subject of more than academic interest; it vitally concerns and involves the fate of every human being on earth. Science, in the service of government, has already created the power to destroy all living things. The relationship needs to be studied, analyzed, understood.

Science and government

Is the function of the scientist supported by government simply to seek out scientific truth and record it as he finds it, pursue it as the logic of the search dictates? Or is it to place his scientific knowledge at the service of his government and apply his best effort to solve those problems which the government considers most urgent, irrespective of the conflicts which may develop with his search for objective scientific truth? Or is there a possible compromise: can he at the same time pursue pure scientific goals and meet the practical demands of applied science as conveyed to him through the government agency with which he is associated?

As the first honorary scientific body to be created by the Canadian government, for the specific purpose of developing a scientific understanding of Canadian fisheries, the Board became involved in the interaction of science with government right from the start. Its members were established figures in the universities, scientists of stature. They did not take happily to pressures and interferences from government officials, and although these were undoubtedly softened and mitigated by the diplomacy of the genial Professor Prince, the Act of 1912 made their point with emphasis.

In spite of the increasing emphasis on 'practical' work after 1912, both at St Andrews and at Nanaimo, W.A. Found, the assistant deputy minister of fisheries, was not satisfied. In fact he came to regard the Board as an obstacle to plans of his own. The Biological Board Act of 1912 said that the Board 'shall have the conduct and control of investigations of practical and economic problems associated with

Dr Found is not satisfied

99

marine and freshwater fisheries, flora and fauna.' Whether this wording implied an exclusive franchise for such activities is perhaps debatable: Board members certainly did not think it did, but Dr Found apparently assumed that he was required to pass on to the Board all scientific problems relating to the fishing industry. He seemed to consider that he had no recourse but to accept the Board's decision as to the seriousness of any problem, when it would be able to deal with it, and indeed whether its resources would permit it to undertake it at all. There is of course no instance where the Board turned down a request for an investigation that was accompanied by adequate financing.

Sometimes the Board may have been slow in replying to Found's requests; at any rate friction did develop. Dr Huntsman has explained the situation as follows:

Dr Found was not satisfied with what the Board was doing, and that was to be seen as perhaps natural because the Board could not be said to be very effective. There was a question, of course, as to whether or not this was to be expected at that stage in our knowledge. Dr Found was undoubtedly a very able man and, in my opinion, one who was prepared to put things through. He wanted to get things ahead as well as to merely carry on. He was more anxious to make definite progress, in my opinion, than any other deputy minister with whom I had familiarity, and I am familiar with quite a large number. He, being dissatisfied with the Board, had his minister bring in a bill for a scientific division in the department.

An alarming development

A bill for a new scientific division within the department would not have alarmed Board members. Indeed they would have welcomed it, as J.P. McMurrich pointed out, so that work could be done that was not suited to the volunteer approach, or not within the scope of their modest budgets. But the bill that was actually introduced on 14 May 1919 was different: its effect would have been to transform the Biological Board itself into a division of the department. It took control of the stations out of the Board's hands, and restricted its work to 'such investigations ... as may be assigned to the Board by the Minister.'

The move apparently caught Board members by surprise. Yet they had been warned. At a meeting of the Board Executive on 12 February 1918, Prince reported that the Deputy Minister of Fisheries desired some change in the relations with the Board, so that its work would come more directly within the department and the staff would be departmental officers to whom work could be assigned and from whom the solution of fishery problems could be secured with certainty and expedition. The Deputy Minister thought that the system then in use did not ensure these certain and expeditious results because the workers were mainly volunteers and not under official orders. The executive committee strongly opposed this suggestion, stating that academic conditions and not the rules of the Civil Service were more suitable for the Board's work.

The 1919 bill would in fact have effectively robbed the Board of all initiative in the choice and perhaps even the conduct of scientific investigations. It was introduced without any prior consultation with the Board or any of its members, even Prince. The master diplomat was caught napping. The bill passed the House of Commons without a ripple of dissent.

The Board fights back

The Board did not submit tamely. Though there is little concrete evidence to support the assumption, it is clear that it mounted an active lobby in the Senate with the hope of having the bill revised there. Macallum is credited with making the most active fight, but even the gentle McMurrich is on record in a letter addressed to Senator Lynch-Stanton, which explained the situation in detail:

The Bill concerning which I spoke to you the other evening in the House is *No. 106*. It has passed the House and will no doubt shortly appear in the Senate.

It repeals and gives a substitute for Section 5 of the Act of 1912 which created the Biological Board of Canada (2 George v, chap. 6) and as I understand the situation, is due to an idea in the mind of the Minister that the Act of 1912 took away from the Department of Fisheries all right to carry on scientific investigations on Fishery questions, this becoming the exclusive prerogative of the Biological Board. The idea is, I think, quite erroneous and is not sustained by the wording of the Act.

Section 3 of the Act establishes the Board, 'which shall be under the control of the Minister.'

Section 4 determines the constitution of the Board – two members appointed by the Minister and one member appointed from each of such Universities as shall be named by the Minister.

Section 5 reads as follows (1) 'The Board shall have charge of all biological stations in Canada, and (2) shall have the conduct and control of investigations of practical and economic problems connected with marine and freshwater fisheries, flora and fauna, and (3) such other work as may be assigned to it by the Minister.' The numbers in brackets are not in the original but I have introduced them to facilitate reference.

Sections 3 and 4 give the Minister full control of the Board, all the members of which, I may say, give their services without remuneration, as is provided for in Section 8.

In Section 5, the Board is given (2) 'the conduct and control of investigations, etc.' If it had been the intention of the Act that the Board should have a monopoly of the control of investigations, it would have read 'all investigations'. Instead, the indefiniteness of the wording clearly implies that the Board should conduct such investigations as may seem to it desirable and should have control of the methods to be adopted in attacking the chosen problems. There is no desire on the part of the Board to monopolize scientific investigations of the fisheries. Far from it, indeed; for the raison d'être of the Board is the encouragement of scientific investigation and the more that is accomplished the better pleased the members of the Board will be. But we feel that we should be just as unhampered in our choice of problems and in the choice of methods of attack as we would be in the case of problems being worked out in our own laboratories. We are not employees of the Department in the ordinary sense of that word but rather experts in biological investigation called in by the Minister to initiate and carry out scientific investigations of problems connected with our Fisheries.

In the substitute for this paragraph (2) the words 'and control' are omitted and the word 'such' is introduced. 'The Board shall conduct such investigations ... as may be assigned to the Board by the Minister.' This seems very objectionable to the members of the Board. It takes away from us all initiative and we must sit down and wait for the Minister to assign problems to us. It takes away all reason for our existence as a Board. If the Minister wishes to have a certain problem investigated the Board is bound by paragraph (3) of the Act to take it up and has always been glad to do so when it is possible. But that is a very different matter from having all our problems assigned by the Minister, and all initiative suppressed. Scientific investigation can flourish only when free and unhampered. Indeed, scientists are in many ways just as 'kittle cattle' as are poets and artists. If the Board is to do what it was intended to do, that word 'such' should be deleted.

As to the omission of the word 'control' I personally do not feel so strongly. To an investigator the word as used here merely means that he shall have the determination of the methods to be used and the lines of attack to be followed in carrying out a research. It is practically a repetition of the word 'conduct', since an investigation cannot be conducted without control. But to others, the word might imply something more.

Paragraph (1) of Section 5 of the Act of 1912 placed all biological stations in Canada in charge of the Board. That is now repealed and thereby the Board is deprived of the opportunities to carry on investigations which the Act requires of it. The general control of such stations has always been in the hands of the Minister but they were placed in charge of the Board that the purposes for which they were established – the scientific investigation of Fishery problems – might be carried out. The Stations and their equipment constitute the armementarium of the Board and unless the Board has charge of these, its endeavours will be seriously curtailed and even rendered futile.

The members of the Board feel very strongly on these matters. We have given a great deal of time and energy to the work, finding our sole remuneration in the satisfaction that comes from participation in the solution of problems of scientific and economic interest. We seek no financial reward for our work.

The Bill No. 106 was introduced without consultation with any member of the Board. We learned of its existence only after it had received its first reading and no opportunity was given us for any expression of opinion until after it had passed the House. As it stands it allows of a situation which will deprive the Board of all freedom of action and diminish its utility almost to the vanishing point.

We have suggested a way out of the difficulty that the Bill be amended in the Senate by the omission of the first 'such' in Section 5, and that we be allowed to add to our By-Laws a section placing the Biological Stations in charge of the Board. The Deputy Minister when we saw him on Thursday last expressed himself as agreeable to these propositions. They practically restore things to the *status quo ante* even though the Act may, in the face of it, be amended, and the Ministerial control of the Board made more definite.

If you can throw your influence on the side of these suggested modifications, I, personally would be greatly obliged and I am sure that you would thereby be acting in the best interests of both the Fishing Industry and the development of scientific research in Canada ... [See Rigby and Huntsman 1958: 116–18]

Bill No. 106 was referred to a special committee of the Senate, which reported on it on 23 June 1919 as follows: 'Your Committee finds that the preamble of this Bill has not been proven to their satisfaction. The ground on which they have arrived at their decision is that the passage of the amendment proposed by the said Bill would not be in the public interest.'

Macallum's departure

Thus passage of the bill was successfully blocked in the Senate, but Macallum paid dearly for his role in the fight. Shortly afterward the minister advised him that he was no longer secretary-treasurer or even a member of the Board. He retired as chairman of the National Research Council at the same time and returned to full-time academic life at McGill. Thus the Board lost a forceful leader.

Soon afterward there was an attempt apparently to establish a tighter control on the Board by changing its by-laws. A draft of proposed amendments, probably framed by Found, was approved by the Minister, who submitted it to the Board for adoption. However, these did no more than establish clearly what had always been recognized *de facto*: the need to obtain the approval of the Minister for the broad lines of the Board's activities. With some modification of the more restrictive clauses, the Board adopted the new by-laws on 18 May 1920.

Following the forced retirement of Macallum, Knight was appointed as a representative of the department on the Board, and for a brief period Prince acted both as chairman and secretary-treasurer. This situation was rectified at the Board's annual meeting on 18 May 1920, when Knight was appointed chairman and Prince secretary-treasurer. Professor W.T. MacClement was nominated by Queen's University to replace Knight as its representative on the Board. Professor R.F. Ruttan became McGill's representative, replacing Willey, who had retired from the Board in 1914 when he undertook paid work for the Board. The University of Manitoba had been invited to nominate a representative and they selected Professor A.R. Buller, who served from 1914 until his retirement in 1923, when he was replaced by Professor C.H. O'Donoghue. In the latter year St Francis Xavier University of Antigonish, Nova Scotia, was given representation on the Board, selecting Professor C.J. Connolly. Bailey also retired in 1923 and was replaced by Professor Philip Cox. The following year Professor A.H. Hutchinson was appointed to represent the University of British Columbia.

11 Practical biological and technological research, 1921–5

Knight takes over

Although A.P. Knight's tenure as chairman was one of the shortest in the Board's history, 1921 to 1925, it was marked by significant developments: a definite orientation towards practical studies, and the establishment of technological stations at Halifax, Nova Scotia, and at Prince Rupert, British Columbia. Knight had been an active researcher and a dominant figure during Prince's long chairmanship and he had declared his intention to direct the Board's efforts towards more practical studies. His personal research covered many such fields, ranging from the effect of sawmill pollution on fish in the early years of the century to lobster studies that led to the enactment of a government grading system during his regime. But his most significant accomplishment as chairman was to engineer a mutually agreeable rapprochement between the Board and the department, which permitted a rapid increase and diversification of fishery research activities.

Board membership broadened

At the annual meeting of the Board held on 23 May 1923 the deputy minister of marine and fisheries, A. Johnston, and the assistant deputy, W.A. Found, attended by invitation to discuss with the Board the possibility of achieving closer relationship with the department by a change in organization. Two suggestions were considered. One was to increase the membership of the Board by additional representation from the department and by representation for the fishing industry. The second was to employ permanent investigators and to establish stations with commercial canning and curing equipment so that more problems of fish processing could be investigated.

Knight was appointed as a representative of the Board to cooperate with the department in drawing up an amendment to Section 4 of the 1912 Act. An agreement was reached for a reorganization of the Board by the appointment of three additional members to represent the department. An Act amending the Biological Board Act, to provide for this, was brought before Parliament and came into law on 30

June 1923. Repealing Section 4 of the 1912 Act, it substituted: '4. The Board shall consist of five members appointed by the Minister, and one additional member appointed by such Universities (named by the Minister) as may engage in the work of Biological research.'

John Dybhavn of Prince Rupert, British Columbia, and A.H. Whitman of Canso, Nova Scotia, were appointed by the Minister to represent the fishery interests of the Pacific and Atlantic coasts respectively, and J.J. Cowie of the Fisheries Branch was appointed as the additional third member to represent the department.

It is clear that relations between the Board and the department improved notably after Knight became chairman. In Dr Huntsman's words:

Dr Knight was the member of the Board more than any other who did establish very good associations with the department and with Dr Found. He did his well-known work on trying to rear lobsters, then on investigating the lobster hatcheries so that they were discontinued on his recommendation because they weren't accomplishing anything. Then he did work on the canning of lobsters, improving the canneries ... Professor Knight was the one to press for practical things and he greatly influenced me. In his time ... the Board gave me a very free hand as director of the station at St Andrews to deal with the department and deal with the industry ... Professor Knight wanted to get these things done and working with the department.

Later, according to Huntsman, Found told him personally that he was perfectly satisfied with the Board when it gave its attention to the practical problems of the fishing industry. With the new emphasis on applied research, relations were happier with the department, and the Board entered upon a new period of expansion.

ST ANDREWS

The installation of a new bacteriological and biochemical laboratory with new and improved equipment in 1922 at St Andrews facilitated continuing studies of canned lobster, the bacteriology of frozen fish, the effects of various types of common salt on the metabolism of bacteria, the chemistry of food fishes and the proteins of their muscle juices, as well as the source of insulin in fishes, their body temperatures, and the effects of physical factors on marine animals. The life history of the smelt and the spawning of mackerel were also studied, and A.H. Leim completed his study of the natural history of the shad and the factors limiting its abundance in the Shubenacadie River. He concluded that the unfavorable conditions of temperature and salinity were largely responsible and made suggestions for improving the fishery.

Practical studies

Knight spearheaded the lobster investigation, examining the question of an earlier opening of the season and recommending that it would be against the interests of the fishery to do so. He also made an examination of lobster canneries and came up with recommendations for increased sanitation and grading, delivering lectures to fishery officers on 'Sanitation in Lobster Canneries.'

Huntsman had been made permanent secretary of the International Committee on Deep Sea Fisheries Investigations and in 1922 the committee developed a plan for drift bottle experiments to determine the currents along the Atlantic coast from Newfoundland to New York. That year the station set out 1,736 bottles in four sections, and bottles were released in three additional sections along the United States coast by that country. An additional 2,325 bottles were put out the following year in cooperation with the Newfoundland and French governments. The third

Ocean currents charted

The Atlantic Biological Station complex, St Andrews, 1924. Left to right: cottage on hillside, water tower, residence, laboratory (with 1921 addition), and workshop

year saw 1,526 bottles put out by the station, 500 of which were distributed by the United States Coast Guard cutter *Tampa*. By October of that year 1,245 cards had been returned to the station from the bottles.

The International Committee on Deep Sea Fisheries Investigations also launched a plankton and hydrographic survey of the waters contiguous to the Strait of Belle Isle, with an expedition planned and organized by Huntsman on behalf of the Board and supported in part by the Fisheries Branch of the Newfoundland government. Two ships, the *Prince* and the CGS *Arleux*, took part. Sections were studied across the Labrador Current, and its course was followed down the Newfoundland coast by means of drift bottles. A general survey was made of the region from Cabot Strait and Anticosti Island through the Strait of Belle Isle and around the eastern and southern coast of Newfoundland. The movement of cod and their abundance or scarcity at the important fishing grounds at Blanc Sablon on the north shore and in the Quirpon region of Newfoundland were found to be definitely related to the temperature produced by the mixing of gulf water with the cold Labrador current in the Strait of Belle Isle.

The appointment of Dr A.H. Leim as ichthyologist and assistant director at St Andrews in 1924 marked the beginning of the shift from voluntary investigators to full-time scientists in the Board's working staff. The same year saw the first appearance of A.W.H. Needler, who was then a student at Toronto. Both were to be future directors of the Atlantic Biological Station.

Needler has provided some recollections of his first summer at St Andrews:

I think I was being groomed to be a paleontologist with the Geological Survey. I had been accepted for field work in the summer of 1924, working with the Geological Survey in Red River, Alberta, digging up dinosaurs, and I was hurrying to a singing lesson, believe it or not, when I was stopped in the street by Professor Turner, who told me that I should see Dr Huntsman ... I gave up my summer opportunity of dinosaur digging and came to St Andrews. In 1924 there were, I suppose, 25 seasonal workers at the St Andrews station. I would think that about half of them were university staff and about half were students or graduate students ... I came down on a very odd sort of job. I was asked to be a statistician and of

The residence, built 1908, as it appeared in 1922

course I didn't know anything about the theory of probability. It wasn't that kind of statistics. I was asked by Dr Huntsman to see what information I could get out about the public fisheries system and I did this rather than some other alternative ... Whenever there was an expedition or a boat going to sea, I usually went along. The marine life greatly impressed me and I made a collection to satisfy one of the requirements of the University of Toronto honor biology course.

That golden summer of 1924, when some of the giants of Canadian scientific research worked as volunteer investigators at St Andrews, has also been vividly recalled by F. Ronald Hayes, who was to become chairman of the Fisheries Research Board in 1964. Professor J.N. Gowanloch of Dalhousie University had taken Hayes to St Andrews as his assistant. Hayes had intended to become a doctor:

The golden summer of 1924

I gave up the idea of medicine because of advice which I received in St Andrews. The people in St Andrews were a remarkable lot. One of them was J.J.R. MacLeod of the University of Toronto who had just won the Nobel Prize with Banting for the insulin treatment of diabetes.

107

MacLeod had some of his students, younger people from Toronto, with him on the staff there, and the atmosphere was very informal. So I went to MacLeod and said I was enrolled in a medical course and would he advise me to finish it with the idea of going into science. I said I knew I would never practice. MacLeod said: 'If you are sure you are not going to practice, a medical doctorate is of no more concern to you than a doctorate of divinity. You would be wasting your time to take it.' So I asked him why he had taken a medical course and he told me that at that time the medical course was the only way in Scotland in which a man could get a general training in biology ...

Canada's first
Nobel Prize

It was a very mixed group at St Andrews that summer. With MacLeod there, it had very much the atmosphere of Canada's first Nobel Prize. MacLeod was a professor of physiology, a life-time researcher in carbohydrate metabolism and a world figure in this field. Dr Banting, who, I think, had been in London, Ontario, came up to Toronto to work in MacLeod's laboratory and was provided with MacLeod's orientation in carbohydrate metabolism. That's why he went there and was working in MacLeod's laboratory, helped by MacLeod. Banting was assigned a student who was more or less chosen at random, and this was Best. So Banting happened to hit on one particular technique which was a key one, namely that if you ligate the pancreatic duct, there will be an accumulation of insulin in the pancreas. You can then remove the pancreas with this large amount of insulin in it, and this was the first source of insulin.

So the prize was divided between MacLeod who was the professor – the carbohydrate metabolism expert – and Banting, the student, who had hit on a particular technique for accumulating insulin in quantity. Banting divided his half of the prize with his student, Best, and MacLeod divided his half of the prize with Collip, who was also a very young man at that time and who had done some biochemical work which was essential to the whole problem.

Now at that time in St Andrews, Banting's name was held in very low esteem as a man who had attempted to claim all the credit for something away from MacLeod; there were public debates between Banting and MacLeod which appeared in the press, about stealing his work, and Banting, I thought, behaved very badly. We, as young students, were all 100 percent MacLeod men.

Scientists at work
and play

MacLeod had brought some other of his students with him to St Andrews. They were looking for a source of insulin from the Islets of Langerhans ... The Islets of Langerhans in a fish are not part of the pancreas; in fact a fish has a very diffuse pancreas. The islet is a discrete body that can be picked off the end of the gall bladder, and it was thought at that time that it would be possible to pick off the islets from large cod and other fish, and mash them up and get a source of insulin. They did in fact obtain insulin from this source. But gradually, over the years, the mammalian sources came to predominate and they developed better techniques, and fish never became a commerical source of insulin. But at that time MacLeod was there with his assistant, busy on the project. He was a very charming Scotsman.

It was an extremely exciting period. Just to be able to talk to these people in a very informal way. There used to be tea on the verandah every afternoon, and we would all assemble. There was [J.P.] Logan, the biochemist from Queen's, and A.P. Knight came down. Knight was getting very old by this time, and he didn't do much research in St Andrews, but he came down and spent a couple of weeks there. He had a particular suit that he wore every summer, a kind of knickerbocker outfit with a light khaki sort of look ... He was very charming ...

There was Gowanloch, to whom I was attached as a student, and Father Connolly from St Francis Xavier, who was a member of the Board ... G.B. Reed, of course, was working there at that time on the abolition of black smut in canned lobsters ... and he cleaned up that bacteriological problem and got the canning of lobsters on a firm basis.

Incidently, this was another trouble. The National Research Council claimed that they had solved the problem, and not Reed ... Jealousies, as you know, are still quite large in all sections of government service.

We had tea, and we saw these people, and Huntsman, who was an extraordinarily Huntsman's games unhumorous man in general, undertook to organize a social evening once a week, and put on games which, I suppose, had been played in his youth – various kinds of parlour games.

I remember one time they cut a hole in a sheet – and that in itself was quite a thing considering the finances of the Board at that time – and different people put their noses throught the sheet, and you tried to identify the people by their nose. There was one woman at that time who had a very large nose; there was some pause and puzzlement when any nose came through, but in this case there was an immediate shout: 'Miss So-and-so.' It was a little bit embarrassing. Then we would have little slips of paper to fill out, and guessing games and other things which Huntsman thought up. On other evenings there would be a scientific session in which somebody would give a paper on his work. The first paper that I ever presented in my life was at St Andrews on the periwinkle work ...

The atmosphere as far as people of my age were concerned was rather like a boy's Other socializing boarding school. We used to play tricks on each other, put jellyfish in people's beds, and shave people's heads forcibly, raid the pantry, and all this kind of business. There were some girls around of the same general age, so there was a certain amount of social life between the sexes.

Dr Tate was another celebrity; he was professor of physiology at McGill at that time. He was a senior man, a Scotsman, and he used to jump off the end of the wharf like a walrus and swim in the Bay of Fundy waters, which run in the 40s usually. It is very stirred up from the bottom there and damned cold. I could plunge in for a minute or two, but Tate used to wallow around as though it was the tropics, and all his family were swimmers, aquatic people ...

E.E. Prince used to come to St Andrews every summer ... He was gray with a gray walrus Prince vignette moustache, and he was a laughing man whom we all liked. He used to give a speech which I heard him give a couple of summers, with lantern slides ... of a trip across Canada. A slide would come up and Prince would look at it for a while. Then he would say: 'This is one of our grand Canadian rivers. We have some of the most magnificent rivers in the world. I don't exactly remember which one this is, but there are jolly fine rivers in Canada.' So he took us across Canada and told some yarns of his experiences and so on, and the final slide came up and it was a picture of King George V, and old E.E. would say: 'Our King, and a good King he is, too.' And this was the end of the lecture.

NANAIMO

At Nanaimo an important new practical study that developed during Knight's Practical studies regime was R.E. Foerster's work on the salmon. Initially it was conducted at Cultus and Harrison lakes, where he investigated the food and the enemies of young salmon, while at the station H.A. Dunlop of the University of British Columbia studied the growth of young salmon as determined from scales provided by Foerster. Another practical study conducted there, which took place over the same period that a similar study was being conducted at St Andrews, was the investigation of damage done by wood borers, using test blocks supplied by the Department of Public Works. The importance of this work came home to the station a year or two later, when its wharf collapsed.

There was a considerable increase in the number of investigators, but their activities were largely along purely scientific lines and had little bearing upon the practical problems of the fisheries until the advent of Dr W.A. Clemens as a full-time director there. The need for a full-time director at Nanaimo had been

W.A. Clemens, director of the Pacific
Biological Station, Nanaimo, 1924–40

recognized by the Board in May 1923, when it authorized its executive committee to
make such an appointment. They selected Clemens, who was then assistant professor of biology at the University of Toronto and director of the Ontario Fisheries
Research Laboratory. It was a happy choice.

Wilbert Amie
Clemens

Clemens was born at Millbank, Ontario, in 1887. He graduated from the University of Toronto in 1912 and later received his MA there and his PH D from Cornell. He
taught zoology at the University of Maine and made a survey of the trout streams of
Oneida county for the State of New York Conservation Commission before joining
the staff of the University of Toronto as a lecturer in biology in 1918. There he
became involved in B.A. Bensley's efforts to develop the Ontario Fisheries Research Laboratory in 1920, and was director of that laboratory when he was
approached for the Nanaimo post.

Clemens was reluctant to accept the appointment, stating that he was almost
totally ignorant of the fauna of the Pacific coast, especially of the fishes. He had
been at Go Home Bay Station in 1912 to study mayflies, and the ecology of the
mayfly genus *Chirotenetes* was the subject of his PH D thesis at Cornell. In 1918 he
was at St Andrews investigating the muttonfish as a possible food source. In 1920 he
studied the ciscoes of Lake Erie, and later the fishes of Lake Nipigon. But he
quickly became familiar with the western fauna. *Fishes of the Pacific Coast of
Canada*, which he co-authored with G. Van Wilby, was published in 1946 and
became the standard reference work from California to Alaska.

At Nanaimo Clemens turned out to be the perfect combination of scientist and
administrator, and the station grew rapidly in stature under his guidance. In 1940 he
resigned as director to become professor of zoology and head of that department at
the University of British Columbia, succeeding Fraser. From 1943 to 1956 he was a
member of the Fisheries Research Board, and in 1946 he was chairman of the Royal
Commission on Fisheries of Saskatchewan. On the Pacific coast he contributed
largely to the development of fishery research, by teaching and by his laboratory
and field work. He successively sponsored and brought into being the Institute of
Fisheries and the Institute of Oceanography at the University of British Columbia,

110

and made an important contribution to establishing the Vancouver Public Aquarium. After his retirement as head of the Department of Zoology he served as a director of the Institute of Fisheries (1953–5) and of the Institute of Oceanography (1955–9). He also served as a consultant to the staff of the International Pacific Salmon Fisheries Commission and as a member of the board of management of the British Columbia Research Council. He died in 1964.

W.S. Hoar and J.L. Hart, in their preface to Clemens' autobiography, *Education and Fish* (1968), wrote: 'Wilbert Amie Clemens was regarded with affection by all who knew him. He possessed a singular ability to use discussion, compromise and persuasion in promoting united action by people with divergent interests. His success was based on a friendly and tolerant attitude and a steadfast willingness to treat all associates as worthy individuals deserving a chance to be heard or to try out their ideas. This attitude is implicit throughout the autobiography but it is never cited as a formula for success, if indeed Dr Clemens so regarded it.'

When Clemens assumed the directorship at Nanaimo in 1923, he was faced at the start with three major projects: to examine the efficiency of the salmon hatcheries in British Columbia; to get a technological station started at Prince Rupert; and to develop a general program of fishery research including the accumulation of basic information on the fishes and the fisheries.

Clemens takes charge

Although the greater part of his time during the first season was taken up with administrative duties, Clemens was able to carry on studies of the rate of growth of fishes. He also arranged a series of conferences on British Columbia fishery problems, under the chairmanship of John Dybhavn, the newly appointed representative of Pacific fishery industries on the Board. A number of recommendations from these conferences were forwarded to the Board for consideration.

Clemens attended two meetings of the International Halibut Commission, in Seattle and Vancouver, and, with Fraser, he was appointed to represent the Fisheries Branch at an international conference on salmon problems held in Seattle. There a continuing organization was formed; a salmon-tagging program was outlined for the Pacific coast from Alaska to California; and coordination of salmon investigations in the United States and Canada was planned, including a basis for collecting statistics. Recommendations were made for the establishment of a commission to investigate and control the salmon fisheries in extraterritorial waters off the coast.

While basic studies from previous years were continued at the station, G. Van Wilby of the University of British Columbia began a study of the life history of the lingcod, an important food fish on which the Board desired information. C.R. Elsey, a teacher at the Point Grey High School in Vancouver, began a study of the Japanese oyster which had been introduced to British Columbia waters, and R.E. Foerster became the Board's second continuing employee, apart from the curators and directors.

Earle Foerster was born in 1899 in Neepawa, Manitoba. He graduated from the University of British Columbia with a BA degree in 1921 and an MA degree in 1922, proceeding to the University of Toronto where he obtained the PHD degree in 1924. Later he was elected a Fellow of the Royal Society of Canada, and attended its meetings as often as possible.

Russell Earle Foerster

Foerster's professional life was to be spent with the Board, except for three years with the International Pacific Salmon Fisheries Commission in 1938–40. He was director of the Nanaimo station from 1940 to 1949, later in charge of salmon research, then special consultant until his retirement in 1962. His 422-page compendium, *The Sockeye Salmon*, published in 1968, describes the research done on

111

that species by Canadian, United States, and Russian scientists from the earliest beginnings.

Foerster had originally set out at the University of British Columbia to become a doctor, but came under the influence of McLean Fraser who, he recalled in 1972,

> got me so keenly interested in biology that I dropped my idea of medicine and decided to go into marine biology. In 1921, when I graduated, Dr Fraser gave me a summer job at the station. I went over there and he assigned me to a study of the jellyfish, a taxonomic study, which I enjoyed very much and carried on with at the university the next year and completed as an MA thesis. That is why I stayed with the Biological Board, as it was called then, and I was contemplating going back to marine biology in 1922. Through the instrumentality of Dr Fraser I was awarded a studentship from the National Research Council which was worth $1,000. But when I got back to the station, Dr Fraser thought maybe I would be more interested in the study of sockeye salmon behavior and development. Therefore I got introduced to the sockeye salmon problems and have carried on from there ever since.

Foerster began his salmon studies at Harrison Lake in 1922, studying the early development and the source of food supply. In 1923 he shifted to Cultus Lake, where sockeye were more numerous and more accessible. That year the Board indicated that they wished Foerster to join the staff of the station on a permanent basis, but he decided that in the winter of 1923 and spring of 1924 he would work on his PHD. He applied to Dr Clemens at the University of Toronto asking to be a candidate for the PHD degree, and Clemens accepted him. Foerster continued:

> Just at that time, it's interesting to note, Dr Clemens had been offered a position as director of the biological station at Nanaimo, to succeed Dr McLean Fraser; and knowing that I had come from there, he was of course interested and quite keen on finding out what I thought of the job, the place, and so forth. He thought, of course, that the coastal area was populated mainly by Indians, and that the conditions were rather unsatisfactory. But I quickly assured him, quite the contrary, that I thought it was a wonderful opportunity for a young biologist to come out and start in a new field. As a result of that I sort of feel that I was responsible for Dr Clemens' final decision to accept the position and come to Nanaimo. I did not realize at that time that I would be working for him for many, many years. But so it happened, and I've been very grateful ever since to Dr Clemens for all he did for me.

TECHNOLOGICAL STATIONS

A key development
The decision to develop two new technological stations, at Halifax and Prince Rupert, was a key decision which arose directly out of the reorganization of the Board. This involved an expansion of the work to include investigation of technological problems of the fishing industry. A sum of $70,000 had been voted by Parliament in 1923 for this expansion, with the expressed understanding that the Board take over responsibility for educational work in connection with the fisheries.

A special meeting of the Board was held in Ottawa on 21 December 1923 to consider the policy for the conduct of the two new stations. J.J. Cowie, who was inspector of fish curing in the department, prepared a memo which outlined the work which the department wished carried on at the new stations. There was objection to the plan at first, as it seemed to some members that it involved another attempt to take the conduct and control of the work away from the Board. It was also suggested in the memo that the staff of the stations would probably have to be appointed under the rules of the Civil Service Commission since the money voted for the stations was a departmental vote for the fisheries of Canada and not part of the Biological Board vote. There was objection to this also, but the Board finally adopted the resolution 'that this Board accepts responsibility for carrying on educational fisheries work, as suggested in the communication to the Chairman of the Biological Board from the Deputy Minister, and agrees to draw up a suitable plan, embodying education, demonstration and investigation, in accordance with the requirements of the fishing industry, in addition to extending the researches hitherto conducted by the Board.'

A request, rather naive under the circumstances, that the votes for the work of the Board and for the establishment of the new stations be combined, was predictably refused. Found had no intention of relinquishing this leverage on the Board.

Meanwhile a committee to choose a location for the new Atlantic station had reported through Huntsman that Canso, Halifax, and Lockeport had all been considered. Both Canso and Lockeport had advantages, but Halifax appeared a poor location, as it was far from the fishing grounds and the amount of shore fishing there was small. The department, however, considered Halifax as the best location because of its advantage for educational purposes which, in the department's view, was the main consideration. And it was the department's money. A site on King's Wharf was finally chosen where there were some vacant federal buildings that could be obtained. Atlantic Technological Station

A subcommittee on education appointed by the Board met on 6 March 1924 to consider the educational work in fisheries which had already been carried out, as outlined by Knight. It made the following recommendations for the proposed station:

1. That a Director be appointed to the Halifax Station at the earliest possible date to initiate and carry out educational work.
2. That the said Director procure information on the technology processes of the fisheries as a preliminary step to conducting a scheme for educational work.
3. That there be prepared as regularly as possible pamphlets for issue on matters of educational interest and in connection directly with the work of the Biological Board.
4. That the said Director plan the equipment of the Station so that it may be used for demonstrating fishery methods and processes.

At a meeting of the executive committee on 19 May 1924 an estimate was drawn up covering costs of fitting up the buildings at King's Wharf, Halifax, proposed for the new station, and salaries and other expenses for a year. At the annual meeting held the same day, qualifications necessary for a head of the station were considered and several persons were mentioned for the position. A committee was appointed with Knight as convenor, and given the power to appoint a director, to proceed with the equipment, and to initiate the whole work. Knight was appointed acting chief instructor until a regular chief instructor could be secured.

At a meeting held on 4 July at Halifax, Knight reported on the educational work he had carried on as acting chief instructor in connection with the fall fishing season for lobsters, courses for fishery officers, and other activities, and made proposals for continuing the work.

After considering all applicants for the directorship, it was decided to request Huntsman to assume the post while retaining the directorship of St Andrews. Huntsman wears two hats

113

The Atlantic Experimental Station, Halifax, 1925–6

Huntsman agreed, and the Deputy Minister was advised of the appointment. Found was likely delighted; he knew that Huntsman was a glutton for work and was very much influenced by Knight's emphasis on applied science.

Arrangements were made to assist Huntsman and to continue the classes under Knight temporarily. The remodeling of the building on King's Wharf was authorized by the Board and the work was initiated at once. The building, which was ready for occupancy in the spring of 1925, was named the Atlantic Experimental Station for Fisheries. The first problem to be selected for experimental investigation was the smoking of fish.

Pacific Technological Station

The same special meeting of 23 December 1923 which initiated the Atlantic Technological Station also dealt with a letter from John Dybhavn urging action on a site for the new Pacific station. Dybhavn was authorized to visit the fishing centers of British Columbia for views on the most pressing problems facing the industry.

The Department of Public Works of British Columbia offered to provide a suitable building for the use of the Board at Prince Rupert for $1 per year. At its meeting of 30 December 1924 the Board approved building plans and informed the British Columbia government that work on the construction should proceed at once, on the understanding that all expenditures in excess of $5,000 would be met by the Board. Clemens was appointed temporary director of the new station and was requested to take steps to initiate the work. In his 'Reminiscenses of a Director' (1958), he wrote:

The day after I arrived at the Nanaimo Station, Mr John Dybhavn of Prince Rupert, the Board's representative for the fishing industry on the Pacific coast, and Mr John McHugh, resident engineer of the Dominion Department of Fisheries at Vancouver, arrived and

114

informed me that I was leaving that evening for Prince Rupert to look over possible sites for the technological station. And so that night I found myself on board the steamer headed for the northern city.

It had been suggested that accommodation for the station might be obtained in the Trans-Pacific freight shed constructed by the Grand Trunk Pacific Railway. The building was a huge barn-like affair and the expense of fitting up laboratories in one section would have been great. Moreover, it was a very considerable distance from the fish wharves and cold storage plant. Several other possibilities were considered but none was remotely satisfactory. I made a somewhat discouraging report to Dr Knight, who replied that some of the best scientific work had been done in attics and shacks and that it should be possible to find some temporary quarters in which investigational work could be started. During the next several months of search and consideration, it became evident that the most desirable place for the laboratory was on the Provincial Government wharf, where large quantities of fish were landed. Negotiations were entered into with the Provincial Government resulting in the construction of a two-story building which also provided space for the Halibut Commission.

Meanwhile Dr C.H. O'Donoghue, a member of the Board, found a young man recently graduated in Chemistry at the University of Manitoba, Mr D.B. Finn, who was interested in refrigeration problems and in the possibility of an appointment to the directorship of the Station. In due course the appointment was made. Mr Finn arranged to secure temporary quarters in the basement of the old unoccupied hotel near the fish storage plant. In a short time a laboratory was fitted up, and here Mr Finn and an assistant, Mr Roger Reid, started their investigations. A couple of rooms were fitted up as living quarters but a plague of rats almost drove the men to distraction. They waged an unceasing battle and were able to endure the situation until the construction of the Station was completed and the move made up-town.

Finn was to prove another distinguished recruit to the Board. Born in London, England, in 1900, he obtained his B SC from the University of Manitoba in 1924, his M SC in 1928, and got a PH D from Cambridge in 1932. He was an associate chemist with American Cyanamid in 1924–5, and then joined the Board's Halifax Technological Station. His move to Prince Rupert came a year later, in 1926, and over the course of the next four years he rose successively from assistant scientist (in 1927) to assistant director (in 1928) and finally director (from 1929 to 1934). In 1935 he returned to Halifax to become director of the station there until 1939, when he became chairman of the Salt Fish Board. In 1940 he became deputy minister of fisheries, and in 1946 he was appointed director of fisheries for the Food and Agriculture Organization of the United Nations, a position he held until his retirement to an Italian castello.

Donovan Bartley Finn

PRINCE RETIRES

An event that concluded the period of expansion of the early 1920s was the retirement of the genial professor. When he retired as commissioner of fisheries in 1924, Prince notified the Chairman that he desired also to be relieved of his duties as secretary-treasurer of the Board. J.J. Cowie was temporarily appointed to the position at a meeting of the executive on 25 October 1924 and confirmed in the appointment at the annual meeting of the Board on 20–22 May 1925. At this later meeting the Board placed on record its appreciation of Prince's long service, and he remained an ex officio member of the Board until his death in 1936.

12 Widening horizons, 1925–30

McMurrich as
chairman
The early years of J.P. McMurrich's chairmanship of the Board were marked by a substantial increase in technological work, brought about by the opening of the two new stations and the acquiring of permanent scientific staff to operate them. At the same time several more scientists were employed year-round for biology and oceanography. The budget in the 1925–30 period increased correspondingly, going from $70,000 in the 1925–6 fiscal year to $360,000 in 1930–1, a figure that was to prove an all-time high for the following 15 years. It provided a vivid contrast with the early Board of Management days, of which Macallum had remarked in 1918: 'We had great difficulty getting money that was allotted, we had great difficulty in getting any money at all. The amount was increased, $5,000, $6,000, $7,000, $8,000, $11,000, $12,000, but it was only in 1911 that it was $15,000 or $17,000 ... But it did not matter what was voted, we were not allowed to spend it. The maximum amount we could spend in any one year was $7,000–$8,000 out of that $15,000, and all because of the machinery that blocked the way.' (See Hachey 1965:131)

Knight had initiated the program of permanent staff recruitment, and McMurrich carried it forward. In 1925 the meeting of the Board's executive in March 'agreed to recommend to the Board that all permanent appointees to the scientific staff should begin at a minimum salary of $2,000 per annum and should receive yearly advances of $200 per annum to a lower maximum of $3,500.'

Salaries, pensions,
problems
In 1928 a subcommittee was appointed to look into the matter of classifying positions and salaries, and by the time they had reported back, in 1929, the total year-round staff consisted of 28 scientists and 21 administrative and technical personnel. For the summer season a number of scientific and technical assistants were employed on a temporary basis and the stations also had the usual volunteer workers.

Although the salary scales and classifications recommended by the Board were in line with conditions at that time, the proposed scheme contained the provision that the existing status of employees should remain unchanged but for salary purposes each could be placed under the recommended classifications as the Board might decide. This scheme, though adopted, was never applied. The result was that scientific workers, although classed as director, assistant director, senior scientist, and so forth, in most cases did not draw the salary that went with the classification.

Pensions became another source of future discontent when at the annual meeting of May 1925 it was reported that 'the Finance Department had decided against employees of the Board as eligible for superannuation.' At the executive meeting of January 1926 correspondence between the Board and the departments of Finance and Justice with regard to superannuation was submitted. Both departments gave their opinion that the Board's employees were not eligible for superannuation under the Civil Service Act. In May of that same year the executive committee recommended to the Board that the Carnegie Teachers' Insurance Scheme be adopted. As explained to the Board, this scheme was 'one that the universities have; and it is the best one in the Committee's knowledge. The employer pays five percent of the salary and the employee, five percent.' Since coverage was voluntary, a number of staff members did not join the scheme.

ATLANTIC WORK

Meanwhile the programs of the new stations were formulated. At the Atlantic Fisheries Experimental Station the proposed plan of work was to take one special industry or line, investigate it thoroughly, record in detail how it was being conducted at each of the principal points along the coast, and then to:

Program pushed at Halifax

1. prepare an account of the best method of carrying on the industry;
2. plan an educational campaign according to the needs and deficiencies that were found; and
3. undertake experimental work for solving difficulties and for providing a rational basis for technical procedure.

Huntsman arranged to have an advisory committee for the station, composed of persons from the fishing industry along the coast. Smoking of fish was selected as the first subject of investigation, and the basic study of the process of fish smoking began.

It was found that to obtain the best finnan haddies the fish should be treated when past the stage of rigor. If lightly brined, and the surface quickly dried and smoked, the fish provided an excellent product of a golden color. Formaldehyde proved the main constituent of the smoke to provide best results. Some of the earlier work of Dr S.A. Beatty at McGill was on studies of haddock flesh treated with formaldehyde. Bulletins on the preparation of dried fish and smoked fish were issued in the first year of operation.

In the 1926–7 fiscal year, the educational program was four-pronged, covering: (1) instructions in processing salted cod fish; (2) courses in fish curing for executives and foremen; (3) courses for fishery officers; and (4) preparation of bulletins on fish freezing and smoking. Experimental work was planned on problems of the smoking, drying, and refrigeration of fishery products. This combination of educational and experimental work was to become the established policy of the Fisheries Experimental Station.

In 1926 it was decided to erect a building for demonstrating particularly the brine

The Atlantic Experimental Station (later called the Technological Laboratory) at Halifax, showing the main building built about 1930, and the Demonstration Building at left rear, built about 1927. The wooden structure at right rear was demolished in 1961 in order to erect a building to house live fish and shellfish facilities. In 1950 the main building was renovated, a fourth floor added, and a four-story structure erected between the existing building and the Demonstration Building

Smoking fish at Halifax, 1926–7

freezing of fish, and this was completed and a refrigeration plant was in operation by the end of 1927. Major extensions were required by 1930. Finally a new laboratory building was erected in 1930–1, and further expansion took place over the following years.

Huntsman pioneers quick-frozen fillets

In 1926 the Board also decided to demonstrate the possibilities of producing frozen fish of high quality, by developing improved methods of production and handling, and by marketing the product on a small commercial scale. The project resulted in the introduction of 'Ice Fillets' to the Toronto market in January 1929. For over two years the Board maintained a limited supply of Ice Fillets in the Toronto market, and although they sold at a premium, demand always exceeded the supply. Despite this fact, the experiment was finally terminated. Huntsman has commented: 'We did it, we showed them how, and in the opinion of the industry on the Atlantic coast, the continuation of what we started in Toronto was killed deliberately by the wholesale fish dealer who distributed the fish. He was against it, although he handled it, because it was in competition with the stale fish that he sold otherwise.'

The Ice Fillets had been prepared to meet the most exacting demands and thus had to be of the highest quality. Into the process came the need for holding the fish at

118

proper temperature after capture, for the prevention of drying and surface changes, for immediate and rapid freezing, for storing at proper temperatures without fluctuations and with drying reduced to a minimum, and for proper handling until the product reached the consumer.

In 1929 more than 50 tons were sold at a standard high price through the very few channels properly equipped to handle the product. These sales were chiefly in Toronto, but small quantities were sold in Halifax, Montreal, Ottawa, and even as far west as Winnipeg. By this time, too, in addition to the Board's product, Ice Fillets were being produced by Lunenburg Sea Products Co. and the Lockeport Co.

However successful the demonstrations, interest on the part of the industry and the trade finally dried up. Whether this was, as Huntsman claimed, because of the opposition of the wholesaler, or whether it was the general effects of the depression of the 1930s or lack of effective promotion that doomed this early effort, we know that today quick-frozen fillets make up a major portion of the retail sales of fish. The Board was some 25 years ahead of the times.

Another brilliant innovation was the 'jacketed cold storage room,' which was also ahead of the times. Devised by Huntsman, it demonstrated the effectiveness of a jacket of circulating air surrounding a cold storage room in cutting down the 'dry-out' effects on food products. The first such room was constructed at the Halifax station in 1927, and a second and improved version was built at St Andrews in 1929. A third was built at Prince Rupert, but the refrigeration industry showed little interest in the new development for several years. (The Prince Rupert station was to bring this and another Huntsman inspiration – refrigerated seawater – to successful fruition in later years.) Another Huntsman first

In 1928 the federal government obtained jurisdiction over the oyster areas of Prince Edward Island, for the purpose of developing oyster farming. First action was taken in 1930 with the establishment of a substation of the Atlantic Biological Station at Ellerslie, Prince Edward Island. A.W.H. Needler was engaged on a permanent basis to head it up. New substation at Ellerslie

Alfred Needler was born at Huntsville, Ontario, in 1906, the son of a professor of German at the University of Toronto. He graduated from the same university in 1926, and received his MA in 1927 and PH D in 1930. A later academic honor was an honorary D SC awarded by the University of New Brunswick in 1954, and he became a Fellow of the Royal Society of Canada. Alfred Walker Hollingshead Needler

Needler first worked for the Board in the summer of 1924 and had been directing oyster studies at Prince Edward Island as an assistant biologist for a year before he was named director of the substation, in charge of Atlantic oyster investigations. Later, from 1941 to 1954, he was to serve as director of the St Andrews station, including a period between 1948 and 1950 when he was also assistant deputy minister of fisheries. In 1954 he transferred to Nanaimo as director, and in 1963 he moved to Ottawa as deputy minister of fisheries, a post he held until 1971, when he retired to become director of the new Huntsman Marine Laboratory and special adviser to the minister of the environment.

Much of Needler's career has revolved about international commissions and negotiations. In the early 1930s he became active with the North American Committee on Fishery Investigations, a predecessor of the International Commission for the Northwest Atlantic Fisheries (ICNAF). He was senior adviser to the Canadian government in the preliminary meetings that led to the formation of ICNAF in the 1940s, and also served in that capacity with the International Passamaquoddy Fisheries Commission between 1931 and 1934, and with ICNAF between 1951 and 1954. When he went to the west coast, the International North Pacific Fisheries

The biological substation at Ellerslie, as it appeared in 1967

Commission had just been formed, and its meetings and research programs occupied him for the whole of his time there. He was a member of FAO's Advisory Committee on Marine Resources Research, and later of its Committee on Fisheries. In recent years he had been one of the Canadian commissioners at ICNAF and head of the fishery delegation to the Law of the Sea Conference. Not surprisingly, Dr Needler is regarded as a skilled negotiator and specialist in international fishery matters.

Recalling his time at the Ellerslie substation, Needler has said:

I think I was the major factor in deciding where the little station would be built. It was a building 25 feet by 40 feet which was about as solid as the Rock of Gibraltar, because we used specifications that were set out by the maintenance man at the St Andrews station, Elmer Rigby. I remember that it had a $7^1/_2$ foot basement with walls that were a foot thick at the bottom and 10 inches at the top, and the floor joists rested on stringers which were 2 by 12 inches and 18 feet long. The whole building was built very solidly. I think [Huntsman] was impressed by the fact that we had a concrete tank in the attic for our salt-water system.

I started working on oyster farming and I remember I had never seen an oyster alive before I went over there, and all the local people considered me an expert. The only instructions Dr Huntsman gave me were that if I was to learn about the problems of oyster farming, the only thing I could do was to try growing them. So, along with a student assistant, I carried out some experiments and made spat collections and so forth.

A year later, when the depression hit, the department, which had an appropriation for oyster culture, didn't want to employ someone to look after it. In 1930 I took over the responsibility for the department's oyster culture program, the administration of an oyster farming program. So these two jobs, the Board's investigation research and the department's oyster culture program, remained together, and I remained in charge until 1941.

A.P. Knight, who had discredited the lobster hatcheries, had reservations also about trout and salmon hatcheries. F. Ronald Hayes has reported:

White and Hayes test hatchery programs

In the years 1926 and 1927 I would check in at St Andrews, get my gear, and set off for the Miramichi, where I did an experiment in the survival of young salmon in a brook in the Miramichi comparable to the work that Harley White was doing on trout ...

White was an extremely interesting individual. He was a very expert field man. I suppose if you were a gardener you would say he had a green thumb, for he could keep animals alive and many theoreticians in biology weren't able to do this. I think that Huntsman relied a great deal on Harley's skill as a field observer ... Harley White was from Queen's University and a protégé of Dr Knight. Knight was a grand fighter and had concluded on the basis of some work which he had done himself, and other observations, that the hatchery program of the department, which was led by a man named [J.A.] Rodd in the Fish Culture Branch, was a waste of time; that these hatchery trout – and hatchery salmon – were not surviving ... So Knight sent Harley White into Prince Edward Island's streams to check the actual survival of hatchery-planted trout and, as Knight expected, White found that the hatchery program was worthless. The department people, of course, simply didn't believe this. They thought that White was faking his results, and there was a great deal of bitterness.

So, I suppose, I was sent up to look into the same question in salmon. As a kind of neutral or third observer, I perhaps wasn't as well known, therefore wasn't assumed to be as crooked as they thought White was. So I went up and took a branch of a brook with the full collaboration of the local head of the Fish Culture Branch, and set up the nets at the top and bottom of this stretch of brook, and planted some young salmon ... After a length of time I took them all out again to see how many were left ... There was quite a heavy loss, something like 90 per cent or more. So my work on the whole supported White's, although it was subject to observation at all times by the Fish Culture Branch, and I don't think anyone said I was faking my results. Well, this duly came out, and the bad relationships continued between the Board and the Fish Culture Branch until, ultimately, Rodd retired at the usual age and then [A.L.] Pritchard came in and more or less accepted that the hatchery program, as then conducted, was worthless.

From the first the Board had recognized the need for information on the physical and chemical properties of the waters in which fish stocks and other animals live. Observations had been made sporadically since 1915 in the Pacific and since 1912 in the Atlantic. The St Andrews station had reams of data from Passamaquoddy Bay, and from the Canadian Fisheries Expedition and various other expeditions along the coast. The need for further analysis of these data and for additional work led to the appointment of a resident oceanographer there in 1928. The man chosen was H.B. Hachey, who eventually became dean of Canadian oceanographers, serving as the Board's chief oceanographer from 1946 and as secretary of the Canadian Committee on Oceanography in 1959–66, working out of Ottawa.

Oceanography is recognized

Harry Hachey was born in West Bathurst, New Brunswick, in 1901. He received his bachelor's degree from St Francis Xavier University and an MSc from McGill in 1925. His move to the St Andrews station came after two years as professor of physics at the University of New Brunswick. Later he carried out wartime oceanographic work for both the army and the navy, for which he was awarded the Order of the British Empire. His scientific contributions were recognized when he was elected a Fellow of the Royal Society of Canada and made an honorary Doctor of Laws at St Francis Xavier University in 1950.

Henry Benedict Hachey

Hachey's first important mission for the Board was as head of the 1930 Hudson

Bay Expedition, described below. For the rest, he has reported on and explained the ocean weather and climate in all our east-coast waters, north to Labrador and east to the Gulf Stream, and made one excursion inland to the Great Lakes.

Another Hudson
Bay Expedition
A perennial dream of Canadian Atlantic fishermen had been to establish a major fishery in Hudson and James bays. Here was a huge body of water that extended south to 52°, the same latitude where there was excellent cod fishing off Labrador. Furthermore it was all Canadian, so that there would not be any complications from foreign competition. The bay had been navigated for 250 years by vessels of the Hudson's Bay Company, and was regularly visited by whalers during the 19th century, without producing reports of any abundance of marine fishes. However, no one had yet made any serious effort to search for them. The 1914 expeditions (see chapter 8) were not equipped for ocean fishing, but they had brought back reports of Greenland cod and capelin. Furthermore, if there were really no important marine fish stocks in the bay, a big question was, why not?

Accordingly in 1930 a new Hudson Bay Fisheries Expedition was organized, with Harry Hachey in charge. It made a survey of the bay in the steam trawler *Loubyrne*. Not a single marine commercial fish was taken in trawl hauls scattered throughout the bay, although gear set near shore took fish of river origin, mostly trout and whitefish. A considerable collection of marine specimens was made, and the general oceanography of the area was studied. It turned out that, in spite of its southern location, Hudson Bay was a truly arctic environment, in which winter temperatures below zero Celsius were fatal to most fishes. Only a few species can survive such high-arctic marine conditions; they stay quiet, and their blood remains unfrozen even when cooled below the freezing point. The Greenland cod reported earlier are evidently quite local in distribution, and to this day the bay supports no salt-water fishery.

Atlantic biological
studies
Investigation of the biology of fishes and invertebrates continued to occupy both volunteers and staff members at St Andrews during the late 1920s. For example, the growth of mussels was studied by H.S. Coulthard, of periwinkles by F.R. Hayes, of clams by Margaret and Helen Wetton, of cod by G.C. Duff, and of haddock by A.W.H. Needler. Sex ratios in marine fishes were reported by E. Horne Craigie. A number of experimental studies were begun: for example, the effects of different temperatures and salinities on development of flounder eggs by W.C.M. Scott, and the effect of light on the copepod *Calanus*. Huntsman gave close attention to herring and the herring fishery, tracing its historical change from a fishery for adults to one based almost wholly on 'sardines.' Research on scallops was begun in 1927 by J.A. Stevenson, eventually including studies on growth rates, gonads, maturity, spawning, and the requirements for successful egg development, drift of larvae, and the varying success of year-classes.

PACIFIC WORK

Prince Rupert
program initiated
On the Pacific scene, D.B. Finn was appointed director of the Pacific Fisheries Experimental Station in 1926, and on 6 November of that year the station in the building on the provincial government wharf was officially opened. The first studies initiated by this station were directed to by-products, including such items as fish meal and fish glue. These were shortly followed by studies of fish oils and problems of the refrigeration and freezing of fish. Problems relating to the discoloration of halibut opened up fundamental bacteriological investigations. The researches on fish oils produced early and continued information of value to industry. In particu-

122

lar, much of the early work on vitamins in fish oils helped to develop an active and valuable vitamin oil industry on the west coast.

It was a starry and talented team that was assembled under Finn's leadership at Prince Rupert. The original nucleus was provided by Dr H.C. Williamson as zoologist and Dr H.N. Brocklesby as chemist. Then Otto Young joined the team. Finn has provided a colorful account of the early days at Prince Rupert; it begins with his first arrival there:

The site had been chosen and according to the rumors that had reached me, the building was supposed to have been built. When we got up there, we found nothing, absolutely nothing. The site was chosen, but there wasn't even a wharf. So we had to do some pretty hustling to get a wharf built and get a building. They had a set of plans. I looked at these plans and the man who drew them had never seen a lab before; totally unsuitable. So we drafted new plans and sent them down to Vancouver, where they put them into working drawings. This is the way it started ...

We had to do something while this building was being built, so we went to Seal Cove and found the old abandoned mill boarding house, to which we had our material sent. Well, we unloaded it there and we set up a lab. In the meantime, I knew what I was interested in because I had already come in contact with the pioneer work of the low temperature research station in Cambridge under Sir William Hardy ... In the Halifax station they had the reports of the research of the low temperature lab, and I read all those reports and I knew the nature of the problem.

He knew what he wanted

We started to examine halibut, which was the big thing with the frozen trade in Prince Rupert. It was really to find out what happened when a fish was cooled below the freezing point. We knew something about what happened up to the point of death, although very little. We hardly knew anything about the muscle proteins ... What we did was really to embark on a study of post mortem physiology. What happened when the ice started to freeze and salts came out of solution, and so forth, and, of course, denaturation wasn't known in those days, and we didn't discover the course of events at all until I went to Cambridge. I had the pioneering glass electrodes and we set up this little lab ...

We had to make our own apparatus. We had some glass tubing and a blow torch. We couldn't do much because we didn't have proper gas, so one of the things which we had to do in the new lab was to put in a gas-making machine. That made it possible for us to do glass blowing and everything else ...

The first man that I recruited was an organic chemist by the name of H.N. Brocklesby ... He was trained in biochemistry and just before I graduated I had been very interested in vitamin A and B, which were then being discovered really. I knew, of course, the source of the one which I was interested in was cod liver oil, and I wondered what was the concentration of any vitamin, if at all, in any other fish oils. So I got this organic chemist, an oil chemist as a matter of fact, and together we got a colony of rats and housed them, and we started to feed them rachitic diets, so as to induce rickets. Then gradually we fed them all the fish oils ... It was a well-controlled experiment, and halibut oil seemed to be extremely promising. At last Brocklesby was just ready to publish when we had a notice from Germany that somebody else had just got there first. A man called Schultz, I think. Anyway, we worked on pilchard oil, we worked on cod liver oil, salmon oil, and so on, and we catalogued the potency of these oils as far as we knew them in respect to vitamin A and B ...

Finn hires Brocklesby

We found that pilchard oil contained something more than A and B; some vitamin that we couldn't explain. We called it growth factor, and that was discovered, not with rats but with chicks ... I think it turned out to be vitamin B-12 ... In addition to that, of course, we found that pilchard oil, being an unsaturated oil, could be used for a lot of things which it wasn't

123

The old mill boarding house at Seal Cove, Prince Rupert, the first accommodation of the Pacific Fisheries Experimental Station

being used for at all ... By proper conditioning it made a very flexible film, much more flexible than linseed oil, and was extremely good for outdoor use on steel girders and bridges and things of that sort, because it withstood the weather so much better. As a matter of fact, we erected a panel for testing these various oils, and the uses to which we could put them. We wanted to bring about thorough use of the by-products ... We had a fellow called Louis Smith, a UBC graduate, working on the production of fish glues, and he developed a glue that was stronger than the material itself ... There were so many opportunities. They only used halibut and salmon, because that is where the money was, but, heavens above, on the British Columbia coast, the herring oil, the pilchard oil, crabs, shrimp, and so forth, there was an abundance of things to do ...

Broad terms of reference

When Mr Cowie gave me my terms of reference, he said that it was my job to apply science to the industry. That was all ... So we kept coming to this new scene with young, enthusiastic fellows. All they had to do was to get wind of a problem and they were on top of it ... They worked too hard, really. The mayor of the city ... he took me to the Rotary Club and he said: 'You know, I've been watching the Fisheries Experimental Station, and you're wasting a lot of current.' And I said, 'What do you mean?' He said, 'Well, you leave your lights on all night.' But the point was that they were working there, and it very often came to a point that I was afraid their health was going to break down.

In one case, the case of Orville Denstedt, who became professor of biochemistry at McGill, I had to say: 'Now, look, Orville. You're working too hard; you've got to go to Vancouver, take a break.' However, I don't think he liked it at all, but nevertheless he was doing some very fine work, he and Brocklesby, both oil chemists.

124

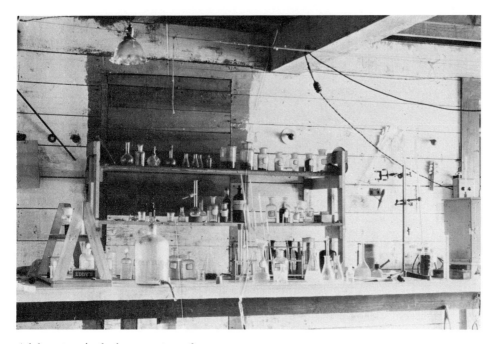

A laboratory in the basement, 1926

In 1929 work on refrigeration was initiated under O.C. Young, who was to build a reputation over the years as an outstanding refrigeration engineer. A low temperature research laboratory was under construction, and the earlier experiments on the storage of frozen fish were carried forward with the new facilities. A series of engineering studies upon cold storage practices was initiated, which represented the start of one of the outstanding and continuing contributions of the Experimental Station over the years. Basic to these experiments was the 'jacketed cold storage room' that Huntsman had pioneered.

Finn later said of Young:

Another chap who was an extremely valuable man was Otto Young ... He was a research engineer, which was a rare, rare thing ... We were concerned with freezing, and not only freezing goods but also transporting them in a frozen state ... Otto came along and the first job Otto ever had had nothing to do with freezing at all. At that time I had persuaded the Board to give us another grant to build another building, as the one we had was simply overcrowded. By that time I had quite a lot of people, summer workers and so forth there. So we built this building, not on the wharf but above the wharf. The first job Otto had to do was to make the drawings and supervise the construction of the building, which he did, and he put in a cold storage, a controlled lab, incubators, freezing apparatus, and so forth.

Incidentally he was the man who developed the idea which I got from Huntsman, and that was a jacketed cold storage. Now Dr Huntsman tried to put one up in Halifax when I first went down there, but they didn't have an engineer on the thing ... The principle was absolutely marvellous, so we turned Otto onto this thing and he did the pioneering work. He got hold of the railways and got their cooperation, which is very hard to do. They built freight cars and we equipped them. Eventually we sent people from the Pacific coast right down to the Atlantic coast. They had other carriages put on and they were running sort of a rolling laboratory, as it were, and this was the thing that revolutionized transportation of frozen fish.

125

Board investigators on the west coast, 1927. From the left, sitting, are W.A. Clemens and D.B. Finn, and, standing, H.C. Williamson, G.H. Wailes, L.F. Smith, H.N. Brocklesby, and R.E. Foerster

Of course they quickly adapted this to fruits and everything else, but it was Young who really pioneered the whole thing. So very often this happens that these people who work in small places make initial discoveries ...

And Brocklesby, too: his bulletin on chemistry of oils. For a long time that was the standard work on the subject, and it all developed in this tiny little place.

Life at Prince Rupert
After I had been there for about six weeks, I wrote to Mr Cowie and I said: 'I think you sent me up here under false pretenses. It is true that you told me to expect rain every day, but you didn't say a damn thing about the nights!' As a matter of fact it was pretty hard to take until you got used to the uniform, which was a slicker, a sweater and gum shoes, and you put these on. Once you got used to that, it was very nice ...

About that time we had [R.H.] Bedford, Brocklesby, Smith, Denstedt, summer workers, too, but a very close unit, because Brock was a very gifted pianist and an organist. I had a tremendous lot of music which I brought with me from Winnipeg, because I used to belong to the Philharmonic Society for choral music. Well, Brock started a choir, and they did some fine stuff and really electrified those people because they had never heard anything like it before. We didn't have any shows, we didn't have any movies, and we didn't have any means of amusing ourselves except by manufacturing our own amusements.

The choir was very much in demand. It went outside the staff at the station, of course. Anyone who felt they could sing, Brock used to test them and take them into the choir if they could sing, and painfully teach them and give them all this stuff ... There were amateur theatricals, too, and we had to make our own amusements. We had hoedowns and usually there was a sort of coalition between the hospital nurses and the training nurses, and on Sunday we all used to go over to the other side of the harbor for picnics.

Major study of hatcheries launched
At Nanaimo over this period of six years intensive long-term studies were developed under Clemens' leadership. These included salmon propagation and

126

migration, herring and pilchards, oyster and other shellfish, trout propagation, and oceanography. From this period onward the work of the station became more and more directly concerned with pressing fishery problems.

A.P. Knight's suspicions that the release of artificially hatched fry was doing little or nothing to increase fish stocks extended also to the Pacific salmon hatcheries, of which 11 were operating during the early 1920s. The possibility of estimating their actual contribution to the fishery was discussed with Board members and with Dr Clemens, who took appropriate action. In Clemens' words (1958):

After the departure of the summer people in 1924 I arranged for the formation of a Fisheries Advisory Committee consisting of Chief Supervisor of Fisheries Major J.A. Motherwell; Mr J.P. Babcock, Assistant to the British Columbia Commissioner of Fisheries; Dr A.H. Hutchinson and Mr John Dybhavn, members of the Fisheries Research Board; Dr C. McLean Fraser; and Dr C.H. O'Donoghue. In the meantime Dr Foerster worked assiduously on an outline of a program of hatchery investigation. After a number of conferences with him I arranged a meeting of the above Committee to which Dr Foerster presented a brief on the proposed program. It recommended that a direct attack be made on the problem by actually counting each year the number of seaward-going yearly sockeye resulting from the actual number of adult females present 18 months previously. The comparison of the relative efficiencies of natural reproduction and the fish cultural methods of fry and egg planting [was] to be based upon the number of eggs involved in each procedure ...

I believe that the members of the Committee were somewhat overwhelmed with the novelty and magnitude of the proposal but after a lengthy and critical discussion, they unanimously endorsed it and recommended it to the Board. As I recall, the reactions of the Board members were much the same as those of the Committee members, but the investigation was authorized ... The basic program proceeded according to schedule and when three tests of each of two methods of propagation and two tests of the third had been made the Board decided to bring this phase of the investigation to an end, since no significant differences appeared among the methods – each showing approximately a three-percent production of yearlings on the basis of the number of eggs available.

The following statement was adopted by the Board and, in view of this, the Department of Fisheries closed all the salmon hatcheries in British Columbia: 'On the basis, therefore, of the above results, it may be concluded that in an area such as Cultus Lake, where a natural run of sockeye occurs with a reasonable expectancy of successful spawning, artificial propagation for purpose of continuing the run in that area is unnecessary, and if producing any additional results over natural spawning, these would not appear to be in any way commensurate with the cost.'

This investigation was an original method for the study of fish production and will always remain as a classic. It is greatly to Dr Foerster's credit that through his vision, careful planning and careful attention to details, the project was carried through without any change in or disruption of the basic program.

Clemens had attended the Seattle conference on salmon in the autumn of 1924, and out of this came the International Pacific Salmon Investigation Federation, which sponsored a program to tag coho and chinook salmon. It was carried out under the direction of Dr H.C. Williamson, an investigator attached to the station and described by Clemens as 'a wiry, bearded Scotsman.' The investigation produced the first firm information on the coastwise movements of the two species concerned. Clemens said: 'At the second meeting of the Federation I was able to report a real contribution to the knowledge of salmon movement along the Pacific coast. Unfortunately the American agencies to the south were unable to carry out

Pacific salmon tagging

any work, partly, it was said, because of the refusal of cooperation on the part of the fishermen. The result of the 1925 tagging program was so interesting and valuable that further programs were undertaken in subsequent years.'

Actually the program was continued for six years. The fish to be tagged were caught by trolling off Vancouver Island and the Queen Charlotte Islands. Most recaptures of chinooks were made in or near the Columbia River, which had not yet been ruined by dams and had much the largest chinook stocks along the coast. Smaller numbers of tags were returned from the Fraser and various other rivers, south as far as the Sacramento. Cohoes exhibited less coastal movement than chinooks, but nevertheless they distributed themselves quite widely from any given tagging site.

Later this program proved to be of value in quite a different and unforeseen manner. Scales were taken from the salmon caught, and their ages determined by C. McC. Mottley. These were available for comparison with fish caught in the same manner during the 1950s and 1960s, when D.J. Milne found that the average age and size of the chinook salmon had decreased dramatically.

Annual sampling of sockeye salmon

In 1924 Clemens had assumed responsibility for a continuing study that had been begun for the provincial government in 1912 by Professor C.H. Gilbert of Stanford University. This was an annual compilation of the size, age, and duration of freshwater residence of sockeye salmon in the four or five major watersheds along the coast, using samples obtained in a systematic fashion by fishery officers. Lucy Clemens assisted with this. That the busy director of a station should occupy himself and his wife with this apparently humdrum project illustrates his keen awareness of the importance of continuing observations of variable natural phenomena. According to W.E. Ricker:

The information from this work by Gilbert, the two Clemenses, and their successors has shown up sustained trends in fish size that can only be a result of the selective action of the fishery in modifying the gene pools of the various stocks, as shown by H.J. Godfrey. It has also made possible an explanation of the mystifying decline in sockeye catches on the Skeena and elsewhere: it turns out that the quantitiative relation between catch and parental spawning stock is different, comparing a period of increasing fishing effort with one of stable or decreasing effort. As a result, some decrease in yield was inevitable even without any decline in the biological productivity of the system. These data are still [1973] being used as part of the input for computer programs that seek to identify optimum fishing strategies.

An unusual experiment

When Dr E. Horne Craigie of the University of Toronto came to the Nanaimo station as a volunteer worker in 1925, Clemens discussed with him the possibility of carrying out a study of the relation of the olfactory organs of adult salmon to their ability to find their way into the spawning stream. Craigie experimented with cutting the olfactory tracts of fish and then applied the technique to sockeye salmon. At Deepwater Bay near Campbell River, working with a tagging crew, he cut the olfactory tract of 259 tagged salmon; and they also tagged 254 normal sockeye. The normal sockeye were subsequently caught in the Fraser River three times as frequently as the operated ones.

Clemens concluded: 'Aside from what conclusions may be drawn from these results, the experiment was of interest in indicating possible future experimental work. Unfortunately, for several reasons this line of endeavour was not continued, nor were other researches on sense organs ... There appeared to be a lack of men in Canada who were interested in and trained for this kind of research, and it was not until such men as Drs K.C. Fisher and F.E.J. Fry at the University of Toronto developed the field of experimental zoology that young men became available for such researches.'

Progress in salmon rehabilitation

After the spectacular decrease of the Fraser River pink and sockeye salmon stocks in 1915 and 1917, respectively, the major fishery problem on the west coast was how best to bring these great runs back to their former level of productivity. The prevailing view was that intensity of fishing should be curtailed, so as to get more fish up to the spawning grounds. However, to reduce fishing required the cooperation of the state of Washington, and at first the negotiations to arrange this showed little sign of being successful. During the 1920s the Fraser sockeye catch came mainly from downriver stocks, whose spawners had not been affected by canyon obstructions. These stocks had an additional advantage: they were fished less heavily because they ran late in the season, so that many of them got through during the annual closed period. In 1926 a late run numbering about 300,000 reappeared above the canyon in Adams River, where it was inspected with amazement by Major Motherwell, Dr Clemens, and many others. This run continued to increase until it included about two million spawners toward the end of the 1930s, the level at which it is still maintained today. And just as before 1913, it occurred abundantly in only one year out of four, although in an even-numbered year rather than the former odd-numbered one.

Another hypothesis to explain the continued low abundance of early sockeye runs upriver was that they were again being held up at Hell's Gate. John McHugh, Alexander Robertson, and others of the Department of Fisheries made a careful study of this toward the end of the 1920s and charted the water levels that presented difficulty. Small numbers of sockeye were observed to be permanently blocked, while pink salmon had scarcely been seen upriver at all since 1913. Accordingly plans were prepared for removing additional rock and for constructing a fishway that would operate at the most difficult level, but the depression arrived before these could be implemented.

Meantime efforts were being made to rehabilitate upriver runs by transplanting eggs taken from several downriver stocks. Lack of obvious success prompted a request that the Board examine the transplantation picture. One possibility was that downriver stocks lacked the physical stamina to make the difficult journey far upriver, and a test of this was arranged by R.E. Foerster from the Cultus Lake substation. Ponds were built on a tributary of Shuswap Lake where sockeye could be reared to a size large enough to mark by removal of fins, and a comparison was made between fish of downriver origin and those obtained from the nearby Adams River stock. It turned out that none at all of the downriver sockeye returned, whereas a few of the upriver stock did so, in spite of the handicap of their missing fins.

Trout study

Research on trout production, then a responsibility of the federal government in British Columbia, was started in a preliminary way in 1928 and expanded in 1930 under C. McC. Mottley. It included studies on hatchery rearing of rainbow trout and their survival in Paul Lake near Kamloops. Professor J.R. Dymond of the University of Toronto spent a summer surveying the situation, assisted by Mottley, and produced an illustrated bulletin *The Trout and Other Game Fishes of British Columbia*. John Hart has said of Mottley: 'He had the happy knack of getting the right answers with insufficient data.' Bill Ricker remembers him as 'a lively person, a great joker. His goal was to put trout management on a sound factual basis by making comparisons based on statistical techniques. After leaving the station he pursued this objective for a while at Cornell, but during the war he got into operations analysis for the United States Navy, and he has stayed in that field.'

Japanese oysters had been introduced to British Columbia in 1912 or 1913 to supplement the rapidly vanishing small native oyster, but large numbers were brought in for the first time in 1926. They eventually became the dominant oyster in the Strait of Georgia. Atlantic oysters too had been introduced into the Boundary Bay area south of Vancouver. C.R. Elsey, first as a volunteer worker and later on the staff, undertook a study that lasted for several seasons. The detailed basic knowledge of the biology of the three species of oysters – native, Japanese, and Atlantic – in British Columbia waters, and of their economics, derived from his studies. A school teacher and biological investigator who got his PH D the hard way, Roy Elsey ultimately went to work for British Columbia Packers and soon became a director of that company.

The systematics and life histories of the shrimps of the genus *Pandalus* in British Columbia waters were studied by Alfreda Berkeley, daughter of the remarkable Berkeleys and a talented scientist in her own right. Hart has observed: 'She was the one who first established on the Pacific coast the fact that shrimps change sex in the course of their life history. They start off as males, and after functioning as males for one year, they change sex and become females. This is a spectacular enough finding to establish her reputation.' She later married Dr Alfred Needler and moved to the Atlantic coast, where she continued to study various invertebrates and also produced a bulletin on the haddock. She died, comparatively early in her career, in 1951.

Next to salmon and halibut, the two clupeid fishes pilchards and herring were the major commercial species in British Columbia waters. Herring had been sampled and studied sporadically by Fraser and others, partly because they were caught in the Strait of Georgia in large numbers and could readily be sampled and experimented with. Pilchards, however, were a pelagic fish that migrated north from California each year, sometimes invading the inlets and even wintering in them, but abundant only offshore. Canning of pilchards started in 1917, but the fishery boomed only after reduction plants began operating in 1925, and at one time there were 23 such plants scattered from Barkley to Quatsino sounds along the west coast of Vancouver Island. This major industry called for knowledge about its resource base, and the man employed to obtain it was John Hart, a recent graduate of the University of Toronto.

Hart was born in Toronto in 1904, and studied at the university there, receiving the PH D degree in 1930. As undergraduate and graduate student he worked with the Ontario Fisheries Research Laboratory under Professors Clemens and Dymond, first at Lake Nipigon and then on Lake Ontario. His PH D thesis concerned the whitefish of these two lakes. He was a lecturer at the University of Alberta when Clemens offered him the position on the staff at Nanaimo, and since it 'was considerably more remunerative' than the one he had at Alberta, he accepted at once. Hart has recalled his early years at the station:

I arrived in Nanaimo on the first of July 1929. This by accident happened to correspond with the day that they opened the new residence building. I can just imagine how glad Mrs Clemens, who was acting as housekeeper, was to see a new and unexpected face show up at that time. But, at any rate, I was hired for that day, so that was when I showed up ...Mrs Clemens acted as supervisor of staff and this sort of thing ... That was an arrangement that started then and continued through for several years, as long as the residence stayed open, I guess, or as long as Mrs Clemens was there ... She was a very considerate and thoughtful woman. Her actions were all very well planned ... Lucy Smith Clemens was a PH D in her own right and worked to some extent with her husband on some of the papers. But she was particularly dutiful as a wife and mother, and she didn't have much time left for research, and consequently this side of her life went into an eclipse.

At Nanaimo, actually there were two dining rooms. Most of the year only one of them was used, and the Clemenses sat as a family at one end of the table and the rest of us sat around. Sometimes when there were many present, smaller tables were used ... It was run as a family affair, with many of the advantages and some of the disadvantages that can go with that sort of establishment ... Quite a few of us stayed in the residence after marriage, mainly because we were doing a lot of field work and it seemed economically impossible to maintain decent homes in Nanaimo and to do the field work that had to be done ... Most of us were away from Nanaimo for four to eight months of the year, but the residence at Nanaimo was always very flexible indeed. As long as you gave warning at all, you were not charged for meals that you did not eat, and of course this was better than you would have done at any commercial boarding house. The meals were good and the service was good.

Later Hart initiated groundfish studies, and he had side interests in fur seals, capelin, dogfish, eulachons, albacore, amphipods, smelts, and pompano. He became director of the Nanaimo station in 1950, and in 1954 went to St Andrews with the same position. He retired in 1967, but continued for four years to work on a new *Pacific Fishes*. This great work appeared in 1973, not long before his untimely death.

John Hart was brought to Nanaimo initially to assist H.C. Williamson with studies on pilchards and herring. However Williamson returned to Scotland soon afterward, and Hart took over. He has recalled the beginning of that investigation:

Hart tackles pilchard investigation

I knew what my assignment was going to be before I arrived in Nanaimo. I had done some reading of the papers by W.F. Thompson and his group on the California sardine – their name for the pilchard – and I started to follow their rationale and to some extent their methods. I was greatly impressed by a statement in one of Thompson's papers in which he said there was little point in trying to explain a biological phenomenon until you knew what the phenomenon was. So I started on a system of sampling pilchards for size – and this seemed like a futile kind of thing to do – but ultimately I think that we did determine some key factors in the population ... While there was a pilchard fishery I continued to monitor it. But about the mid-forties I forecast that the fishery was going to collapse.

And collapse it did. The event was foreshadowed by a steady decrease in average age and size of the fish being caught, mainly because of an increasingly intensive fishery off California. It was principally the large old fish that migrated north in summer. When they were gone the British Columbia fishery lost its raw material base, and nothing could be done about it locally. The Californians did nothing either, continuing to fish until the stock had been reduced to a miserable remnant, at which level it persists to this day.

When Clemens first arrived at Nanaimo, he was surprised to find how little was known of the physicochemical conditions of the sea along the British Columbia coast. He wrote in 1958: 'Dr C. McLean Fraser, my predecessor, had instituted a daily record of temperatures and salinities at the Station wharf and also at the Quarantine Station at William Head near Victoria through the generous cooperation of Mr I.E. Cornwall. Dr A.T. Cameron and Dr Fraser had published some records of temperatures and salinities obtained in connection with a kelp survey. I was confident that the Board would approve of a sound program but I was so fully occupied with other matters that I postponed action.'

Oceanography program developed

Fortunately a group of volunteer investigators arrived to fill this gap. Headed by

The *A.P. Knight*

Professor A.H. Hutchinson of the University of British Columbia, the group included C.C. Lucas, Murchie McPhail, and student helpers. Starting in 1926, and continuing over the rest of the decade, they investigated the complex oceanographic conditions in the Strait of Georgia, including temperature, salinity and nutrient chemicals, and the plankton. Hutchinson was especially interested in the reaction of the plankton to seasonal changes in temperature and salinity related to the varying discharge of the Fraser River. These conditions, of course, differed greatly from those on the outer coast of Vancouver Island, where upwelling of deep water kept the summer temperatures in the range of 12 to 15°C, as compared with 20° or even 22° in the strait.

But although the Strait of Georgia is of major importance to British Columbia salmon fisheries, it is only a small part of the total marine environment. To go farther afield required a full-time oceanographer working at the station. In Clemens' words: 'I learned of a young chemist, graduate of the University of British Columbia, then completing a year of post-doctoral study in Germany. Thus it came about that (in 1930) Dr N.M. Carter became our chemist and oceanographer.'

Neal Marshall
Carter

Carter was what was at that time a rather rare thing in British Columbia, a native son. He was born at Vancouver in 1902, received his BA SC from the University of British Columbia in 1925, MA SC in 1926, and a PH D in organic chemistry from McGill in 1929. He was a lecturer in analytical chemistry at the University of British Columbia in 1926 and 1927 and was on a National Research Council traveling fellowship at the Kaiser Wilhelm Institute in Dresden, having just completed his year of postgraduate study there, when he received Dr Clemens' letter offering him the post. He said later:

132

I was certainly eager to look for an appointment, because I was married with two children at that time, so I said I would sure be interested. He recommended that on the way home from Germany I should visit the Halifax and St Andrews stations to get an idea of the type of work the Board was doing. I had an awful scramble to do it because I was practically broke by the time I landed in Canada after a year in Germany. As soon as I got back to Vancouver, I went over to Nanaimo and had a chat with Dr Clemens and I was very pleased to find that he thought he could use me, so I started work in Nanaimo as oceanographer and chemist around September 1930.

My first job was to put on a suit of old clothing and go aboard the station vessel *A.P. Knight*, and start taking water samples out in the Strait of Georgia. After getting enough samples to make it worthwhile, I brought them back to the laboratory and taught myself how to analyze them for what was needed.

Carter was to remain in Nanaimo until 1933, when he moved to Prince Rupert to become director of the Pacific Experimental Station. Both he and the station were transferred to Vancouver in 1942 when their building was requisitioned for military purposes. In 1955 he went to Ottawa as special assistant to the chairman and associate editor. He retired in 1962 and returned to the west coast, but immediately undertook one of his most useful pieces of work: the compilation of annual indexes of Board publications, and particularly two comprehensive indexes, one covering 1900–64 and the second 1965–72.

Outside of his profession Dr Carter is very well known for his extensive experience in climbing and mapping mountains, activities that made him a Fellow of the Royal Geographical Society.

INLAND LAKES INVESTIGATION

In addition to the activities on both coasts stemming from the four stations, the Board in 1925 initiated the Jasper Lakes investigation, carried out under Professor O'Donoghue by parties of investigators during 1925 and 1926. A concerted attempt to provide overall plans for investigations of Canadian lakes was drawn up by the Board's Research Committee on Fish Culture. At its second meeting in January 1926 it was decided to investigate specific lakes. These were Cultus Lake in British Columbia; Jasper Park Lake in Alberta; Quill, Bright Sand, and Moose Lakes in Saskatchewan; Clear Lake and Lakes Manitoba and Winnipeg in Manitoba; the Chamcook Lakes in New Brunswick; and Lake Ainslie in Nova Scotia. This program, along with those carried out by the provincial authorities in Ontario and Quebec, would constitute a basic national effort in lake investigations. Two key figures in the western operations were at the University of Manitoba, Ferris Neave and Alexander Bajkov.

The prairie investigation eventually included a wide selection of lakes in the Lake Winnipeg drainage system. Laboratory facilities were provided in part by the University of Manitoba. A summer station was established at Gimli on Lake Winnipeg in 1929. The Board supported it for only two years, but it continued to function under the auspices of the Manitoba government for several years longer. In addition to the work of Bajkov, Neave, and assistants, Professor R.A. Wardle of the University of Manitoba reported on the tapeworms found in local fishes, and M.V.B. Newton made a special study of *Triaenophorus*, a whitefish parasite of great concern commercially.

The prairie investigations had been triggered by a request from the Department of Fisheries for a biological survey to determine the potentialities of Lake Manitoba as

Prairie investigation short-lived

133

a fish-producing lake. For some years past, the lake had been producing quantities of fish far larger than any of comparable area in the west. Although there was still at this time no indication of depletion, it was of importance that the assessment of the possibilities of the lake should be based on fact and not merely opinion.

A year later (1930) it was decided that work on the Manitoba lakes consisted primarily of an investigation of the whitefish and pickerel. Requests were also received by the Board at this time for a comprehensive biological survey of Lake Athabaska, Lesser Slave Lake, and the connecting waters.

At the Board's meeting in May 1928 it was recorded that 'in view of the extent to which investigations are now being carried on in the Prairie Provinces, and of the advantages that would accrue from having more of the western universities represented on the Board, it was resolved to request the Minister to ask the University of Saskatchewan to name a representative for appointment.'

However, jurisdiction of the fisheries of the Prairie Provinces, along with other natural resources, soon passed out of the hands of the federal government. So, at the Board's executive committee meeting in January 1931 it was reported that 'the Department now has no jurisdiction over the fisheries of the Prairie Provinces, including fish culture there ... It was therefore resolved that no provision be made in the Board's budget for carrying on work in the Prairie Provinces after the thirty-first of March next.'

At the May meeting of the executive, it was noted 'that all work of the Prairie Provinces including Prince Albert National Park insofar as the Board is concerned, has ceased.' The Park reference alludes to work begun by Dr D.S. Rawson of the University of Saskatchewan, with Board support. Fortunately both Rawson and Bajkov were able to get additional support elsewhere, and over a period of years obtained excellent information concerning the lakes and the fishes of their respective provinces.

The ending of the prairie investigation seems an appropriate spot to provide an account of two of the key figures, Neave and Bajkov, both of whom were to work at the Board's stations again in later years.

Ferris Neave

Neave was born in 1901 at Macclesfield, England, and obtained his B SC and M SC degrees from the University of Manchester. In 1923 he made his first visit to Canada on one of the annual 'Harvest Excursions,' which were not tourist jaunts, but were designed to bring laborers to the prairie grain fields at the lowest cost possible. Ferris harvested his share of the wheat crop that year, and then had a look around. Among other things, he was powerfully attracted by the climbing opportunities in the west, for like Neal Carter he was an enthusiastic alpinist, so in 1924 he took a position as lecturer at the University of Manitoba, where he soon became professor. His formal education was completed in 1951, when he received a PH D from the University of British Columbia. Soon after he was elected a Fellow of the Royal Society of Canada.

Originally interested in aquatic insects, Ferris contributed to the Board's Prairie Lakes and Jasper Park investigations as noted above. On a trip to Churchill in 1930 with Bajkov he encountered Harry Hachey and the *Loubyrne* engaged in their Hudson Bay survey. Later he spent a summer at the Pacific Biological Station in 1934, and in 1938 joined its permanent staff. There he worked successively on the Cowichan River project, shellfish investigations, chum and pink salmon production, and the high-seas distribution of salmon.

Alexander Dmitrovich Bajkov

The emigrés who left Russia after the communist revolution included a larger-than-average proportion of individualists, to whom any form of regimentation was anathema. Among these, Alexander Bajkov was by no means the least original.

134

Born in Moscow in 1894, he took part in several of Professor Nikolai Knipovich's fisheries expeditions from 1914 to 1918: to the Caspian Sea, the White Sea, and far-eastern Siberia. He studied in the Fisheries Section of Moscow Agricultural College under the great geographer and ichthyologist Lev Berg, and later at the Engineering and Forestry College of Brno in Czechoslovakia. He came to the University of Manitoba in 1926, and received the MSC degree there in 1927 and the PHD in 1931.

During his work on the Board's Jasper Park and Prairie Lakes investigations, Bajkov introduced to the western world A.N. Derzhavin's method of computing minimum population size from the catch statistics and age composition of a fish stock. Later he made several expeditions to northeastern Manitoba and Hudson Bay; these were financed from various sources, including his own pocket. During 1933–5 he worked at the St Andrews station, and at another period he was employed by Ducks Unlimited. After leaving Canada, he worked first in Louisiana and later for the Oregon Fish Commission.

Bajkov always favored 'living off the land,' or the water, so that field work with him had its ups and downs, gastronomically. Ferris Neave recalls his supply of *balík* (dried sturgeon) that was kept curing at the top of the mast of their schooner the *Breeze* during studies on Lake Winnipeg.

Bajkov's final project, in 1955, was one that he had long cherished: to make use of marine plankton directly for human food, rather than following the more common method of letting shellfish accumulate the stuff and make it edible. In some way this was to be furthered by a drift on the ice from the North Pole southward, in the manner of Papanin. The drift party was equipped and got as far as Goose Bay, Labrador, in a chartered plane, but no arctic base could be persuaded to sell them gasoline for the hop to the Pole; evidently no one relished the possibility of an eventual all-out search for the party. Returning to Nanaimo, where his backers had already purchased some oyster leases as a base of operations, Bajkov died suddenly and the whole scheme collapsed. Neave's final characterization was that 'those who knew Bajkov well will long remember his versatility, originality and unflagging energy.'

13   The depression: contraction and consolidation

First restrictions The depression of the 1930s brought an end to the expansion of the Biological Board of Canada. It also brought to an end dependence on volunteer workers, whose achievements had done much to establish the credentials of the Board in the eyes of the scientific community. From the inception of the Board in 1898 and through the early 1930s facilities were provided at the Board's stations for teachers and students at universities, colleges, and even high schools, as well as a few non-academic amateurs. Thus close contact was maintained between many Canadian scientists and the Board.

The Royal Commission investigating the fisheries of the Maritime Provinces in 1928 noted:

The constitution of the Board provides a means for securing the voluntary service of the best scientific minds of the country for the planning and organization of the varied work. The fact that the universities are represented on the Board serves to stimulate interest in the universities, thereby providing volunteer investigators for the problems related to the fisheries ... From these volunteers permanent investigators for the Board's future work may be recruited. The varied connections with the fishery industry and with the Department serve to focus attention on the most important problems in the conduct of the fisheries, and to provide channels for the dissemination and application of results. [See Hachey 1965: 80–1]

However, as the Board's work increased in the late 1920s, and as a continuing staff was built up without any corresponding increase in laboratory space, facilities for volunteer workers began to be a problem. This was reflected in the policy toward volunteer workers expressed in the Board's Executive Minutes of January 1931, which made five points: preference would be given to those volunteer workers who undertook Board projects; if space was still available, it would be given to investigators who paid their own expenses; recent graduates and advanced students would either have to be able to work independently or serve as assistants to more experienced scientists; junior workers could be asked to leave the station if their work was deemed unsatisfactory; and the directors would be required to submit to the executive committee a list of the investigations approved by them, together with their reasons for considering the research of economic value.

Shortly after these regulations were promulgated in 1931, the full effects of the depression were felt, rapidly cutting down the funds for support of the year-round work of the stations, and soon few opportunities were available for volunteer workers. Chairman J.P. McMurrich was a strong believer in having a good mix of fundamental and applied research, and it must have been a bitter pill to swallow when he was forced to let the volunteer program go into eclipse. But the man who administered the coup de grace was Professor A.T. Cameron, who had himself been a volunteer investigator for five summers at Nanaimo.

Cameron succeeded McMurrich as chairman of the Board in 1934. He was strongly oriented towards practical research, and professed to hold a low opinion of volunteer workers, once referring to them as those who 'toil not neither do they spin.' He remained chairman during the balance of the Biological Board period (i.e. to 1937) and continued as chairman of the Fisheries Research Board until his death in 1947 (see the following chapter). Early in his regime he decided to stop paying transportation costs and providing food and lodging for volunteer workers. However, the other Board members would not have supported this move except for the Board's very straitened financial circumstances. The last four years of the McMurrich period had seen the budget cut in half. Although no great proportion of the Board's funds, not more than 10 percent, was being used to pay the expenses of volunteer workers, the demands of ongoing investigations and the desire to hold onto valuable permanent employees were sufficient to make this economy measure unavoidable.

The closest thing to an epitaph for the volunteer workers was given by Cameron in his annual report to the minister of fisheries for 1937, when he referred to changes of policy that had taken place during the previous few years, in particular concerning the training of young scientists in research work during the summer months:

In earlier years great facilities were given these young people, their travelling expenses and board were paid, and residences were constructed at St. Andrews and Nanaimo and largely used in housing them. Much of the time of the Directors at these Stations was occupied in summer in guiding them. Good results were undoubtedly obtained, and some of the brighter men were ultimately recruited to the permanent services of the Board. During the lean years, when our annual grant was steadily and considerably reduced, and we had to consider every cent, and what we could get out of it, and the many demands made on the Board by the Department and the Industry, we decided that the function of the Board did not include such training of young scientists in research, and that the time of the Directors was too valuable to be largely occupied in summer in such a way ... It is hoped that, before long, funds will again permit giving facilities to trained scientists, if they are willing to carry on research on the Board's problems. Of course the freedom of the Stations has always been available to scientists who would pay their own expenses, as far as accommodation permitted.

It was a rather churlish tribute, from one of its former members, to a group of people who contributed so greatly to the scientific status of the Board, and whose researches laid the foundation for marine science in Canada. Many of them became senior people on the staffs of the Board itself, the Department of Fisheries, the

National Research Council, and universities in Canada and abroad, as well as on the staffs of international fisheries organizations.

A different view Cameron's off-hand dismissal of the value of the volunteer workers was not shared by other senior people on the Board. In a memorandum to the Board in the early 1950s on 'Scientific Proficiency,' Dr J.R. Dymond proposed: 'As soon as adequate space and facilities can be provided, University scientists should again be encouraged to engage in research at the Stations ... The Board's research has profited in many far-reaching ways from the fact that a large number of the biologists of Canada have spent one or more summers at one or other of the Board's Stations. The work of the Board will lose much if closer contact between the universities and its work is not re-established.'

And W.A. Clemens, who as director at Nanaimo from 1925 to 1940 worked closely with the volunteer workers without suffering too much inconvenience, said in his 'Reminiscences of a Director' (1958):

Practically all of the investigations of the Station had been carried on by volunteer summer investigators from the Universities ... I am satisfied that I organized the volunteer worker system so that it made a real contribution to the Board's work. True, there were some disappointments in the research efforts, but with a few exceptions the investigators applied themselves steadily and faithfully to their problems not only during the summer months but after their return to the universities. Eventually they submitted detailed reports, the majority of which were subsequently published in *Contributions to Canadian Biology* or in other journals. Because of the limited time and the season, the amount which they contributed directly to fisheries research was limited but they did advance the knowledge of the systematics and life-histories of the marine plants, invertebrates (especially shellfish), fishes and oceanography, and all was accomplished with a relatively small outlay on the part of the Board.

The Depression of the early 1930s with its severe curtailment of funds brought the volunteer investigator arrangement to an end in 1934. That this or some similar system was not reinstated seems to me unfortunate, because the association of university and Station personnel was mutually stimulating and beneficial.

And Huntsman, who started as a volunteer worker, in his last annual report to the Board in 1953, wrote: 'I still believe, in spite of the very great material success of the Board and its phenomenal growth since that time, that its action twenty years ago in doing away with volunteer research was unwise and that organized research is decidedly expensive and unproductive scientifically in comparison with volunteer research.'

No protest from universities The phasing out of the volunteer workers in the 1930s was accomplished with little apparent protest on the part of the universities. Alfred Needler, in 1972, noted that the process coincided with 'a tendency within the universities to work that could be conducted in the laboratory, and the desire on the part of the universities to control the work that their graduate students would be doing, rather than to have graduate students working on field problems in cooperation with the Biological Board.' The universities became more and more reluctant to grant degrees for field work conducted with any but their own supervision, making it difficult for graduate students to work on fishery problems. They began producing more and more laboratory scientists. In 1972, he concluded, the trend is being reversed, away from the laboratory and back to field work.

Volunteer investigators did not disappear entirely. The Berkeleys at Nanaimo lived close to the station, and continued to divide their energies between their

138

garden and their polychaetes. A few university scientists still came to the biological stations 'at their own expense,' and some who might have been volunteers under the old dispensation were paid for making specific studies under the new. (It should be noted that after the Second World War research grants and sabbatical leaves with pay gradually became accepted adjuncts of the academic life, so that in time university scientists again became a common sight at most stations. For example, the Board's *Review* for 1971–2 lists 104 non-staff people who spent varying amounts of time at the stations; they included professors, post-doctoral fellows, graduate students, and a few amateurs, mainly from Canada and the United States but also from such countries as Britain, France, Japan, and Poland. However, it is true that today a much smaller proportion than formerly of the staffs of Canadian universities have had direct contacts with the stations. This is unfortunate, but the plain fact is that professors and students are now so numerous that no considerable part of them could be accommodated without major additions to Board facilities. Their needs are being met by three university-controlled marine stations that have been established since 1967.)

The volunteer workers were not the only ones to suffer the effects of the depression. In 1932 the Board was advised: 'The Treasury Board has had before it sundry submissions recommending changes in the organization of various departments. After giving the matter its consideration, the Board is of the opinion that at this time increases in the cost of personnel in the Public Service [are] not in the Public Interest, whether such increases are effected by the creation of new positions, reclassifications, or promotions.' It would be another four years before promotions and annual increases were once again even theoretically possible. Until then, with funds tightly controlled and overall temporary cuts made in all government salaries, many downward adjustments had to be made in Board salaries and operations.

Dr W. Templeman, who retired as director of the Newfoundland Biological Station in 1972, has recalled how the depression affected him. Upon graduating from Dalhousie University in biology in 1930, he said,

I was offered a scholarship by the Board to begin work on lobsters, working out of St Andrews ... I got a scholarship of about $1,000. A.A. Blair got one at the same time and he began work on salmon. I went to St Andrews and started work on lobsters with Professor [A.F.] Chaisson, and we went on from there. It was renewed again in 1931. Then one day Dr Huntsman called Blair and me in and said: 'These scholarships, you know, we can't count on their permanence. It would be better if you became members of our regular staff, as scientific assistants.' We grinned all over our faces and said that would be wonderful. Then we went out. The next day or so, Dr Huntsman called us in and said: 'The government of Canada has decreed that all permanent employees take a 10 percent cut. Your salary is now $900 instead of the scholarship of $1,000.' We didn't grin after that.

The 10 percent cut applied to government employees across the country, and in general it caused no resentment: everyone could compare his lot with that of his jobless friends. But additional sacrifices were soon required of the St Andrews staff. In 1932 a fire destroyed the main laboratory building, and Dr Huntsman took immediate steps to replace it. Construction costs were at rock bottom, yet *some* money was needed. Without waiting for action by the Board, he further reduced all the permanent staff salaries except his own, cut down on part-time employment, and instituted further economies that gave him enough capital to start reconstruction. As recently as 1960 one of the senior employees at St Andrews would

Salaries frozen, cut

The St Andrews fire

139

Later view of new yellow brick laboratory at St Andrews, rebuilt after the fire of 1932

facetiously point out to a visitor the section of the new building that he claimed as his personal property!

More serious than the loss of the building was the destruction of the library, and particularly data records and research materials. Morden Smith, for example, lost several years of observations on ecological succession in a series of small ponds, material that he had planned to use for a PHD thesis.

Cameron eliminates Huntsman

The main objectives in Cameron's first years as chairman of the Biological Board were to complete the phasing out of the volunteer workers and to organize activities both on the east and the west coast with full-time workers devoting their efforts to solving the problems of the fishing industry. One objective which wasn't spelled out but was certainly well understood between McMurrich and himself was that of regaining effective control of the Board's eastern activities from Huntsman. This meant, in short, removing him from the directorship at St Andrews, a move made the more desirable by the minor administrative chaos that existed there.

That a single director could have acquired such a dominant position in determining Board policies and activities as to require such drastic action is a mark both of Huntsman's remarkable and contradictory qualities and the rather casual approach to Board activities on the part of its presumed policy-makers.

Huntsman's idiosyncrasies

Templeman, who had started off his career in 1931 with a 10 percent cut at the hands of Huntsman, recalled in 1973: 'Well, I think he let us alone, mostly. He had to let us alone because he was doing so much. He was the director of the station, editor, chief adviser of the Board, I think, not only for St Andrews but also probably for Halifax. He was depended upon greatly, and besides, he was a great man for detail. I think he counted the laundry and looked after the residences ... There

140

wasn't much time, but he was an interesting man to talk to. Of course, he was so able, and he loved to argue ... The Board was Huntsman to us! We didn't see any further than that in those early thirties in our lowly positions.'

Ronald Hayes met Huntsman again in 1972:

He looked to me about the same as he did when I first knew him. It would be 50 years ago. He had iron gray hair. He had no flesh at all to spare. He was an intense worker. He used to have crackers and milk at his desk and continue working. He never recognized the necessity for relaxation or sleep or food, and he had ulcers and everything else because of his peculiar eating habits and overwork. He had the usual disposition of people with ulcers, certainly hardly a cheerful man.

He bought himself a car one time when I was around there, and turned up in it in St Andrews. He bought the car one morning in Halifax and he had never driven a car in his life before. He taught himself to drive and he just drove off. He took it away from Halifax and drove it to St Andrews in one day. This is the sort of thing he would do. He would undertake anything ... He attracted about as many anecdotes as Abraham Lincoln about himself. Some of them were true and some were not, but Huntsman was really not a good experimenter. Harley White was his assistant, and one of White's parlor tricks when the boys would get together was putting on his demonstration of Huntsman using his one-man seine in a little stream. Harley would get this out and it was really quite a show. Huntsman really didn't do field work well himself but he didn't know this ... He was, I suppose, a crusher of men. He took a man in the way a western cowboy would take a horse, and broke him, and to Huntsman a man was no good to him unless he was broken ... these people became simply the horses and this was one of his great difficulties in dealing with men. He could never deal with a man as an equal.

Another thing, of course, was that he had no idea in the world of the value of money in relation to time. He would quarrel for an afternoon about a fellow using 35 cents as bus fare to go up to St Stephen to buy something.

He was at one time director of both St Andrews and the new Halifax Technological Station when his insides gave out on him and he had to go down to Bermuda to recuperate. So he went down there and left word that for every item to be taken from the stock room, you had to send the chit down to Bermuda by boat, to be okayed by Huntsman and then sent back to Halifax. Then you could get the test tube or whatever it was. He was very peculiar in this way. He was unable to delegate authority and he was incapable of authorizing expenditures of funds.

These were very grave faults, against which you set his very high level of imagination. Perhaps his greatest characteristic was that he believed, as the Greeks did, that there was no distinction between science and philosophy. As you know, Aristotle provided the philosophy of the Catholic Church and also wrote the first textbook in zoology and so forth. Well, Huntsman, I think, was a follower in his mind of the Greeks. Most people today are either scientists or humanists, but it is difficult to bridge this gap. I think it could be bridged. It is the ideal which we are all struggling for at the present time in education. But Huntsman apparently believed that you could think a thing through without any experimental evidence, if you could think well enough. So he undertook to do this and he was, of course, frequently wrong.

A follower of the Greeks

D.B. Finn has also appraised Huntsman:

We owe a tremendous lot to Huntsman, a man of great imagination. I know a lot of people differ with him ... but it was his method of thinking that irritated people ... He and I had an argument over the meaning of the word 'good,' and it lasted for three years, and we didn't

The real founder of things

come to a conclusion. The nearest we came to it was 'ultimate order with infinite complexity' ... Huntsman turned his imaginative mind upon the problems of biology. He was a man who doubted everything. It didn't matter whether it was generally accepted or not, unless he could reason it out for himself, he had to reinvestigate. Which is a very healthy thing.

Up until when I was active, most fishery scientists of any significance in Canada went through his hands. He imparted something of himself to his students. Some of them, I know, don't regard him very highly because they couldn't get along with him. He was a one-man show, but he perpetuated himself in his students. Of course, he couldn't see that.

Now, as far as Huntsman went, of course, a man of his brilliance and his dynamic personality really dominated the Board, whose members honestly didn't think a lot about it during the year. They might have read the minutes and one thing and another before they came down to Ottawa for the annual meeting, but they depended upon his advice. He assumed, not his fault, that they had to get ideas, and he had them. He accumulated them from living with and thinking about the problem continuously. I think probably Cameron, when he first went there, saw that the real founder of things was Huntsman.

Alfred Needler said:

A gadfly He was an energetic man ... Very rugged, determined and persistent, and his scientific thinking, I think, was much influenced by the school that looks at individual factors ... the people that believe that animals depend on the operation of very simple factors. I don't think that he ever kept up with scientific thinking on such matters as animal behavior and migration. He still looked at it in a simple physical sort of way.

Dr Huntsman, when I first met him in 1924, was rather young looking, very dark ... He was shortly after that to begin to suffer in health a bit. Starting in 1935, I think, he had pretty bad ulcers but it didn't stop him from working. He actually would be feeling so ill that he would be lying down on a couch, but nevertheless he would be reading or dictating letters. He was in the Board for a number of years and he was a dominant factor. He dominated all the rest of us ... The turning point came when Dr Cameron was appointed chairman. Dr Cameron wasn't going to be told what the Board should do by an employee, and he really started deliberately to cut Huntsman down to size ...

Dr Huntsman tended ... to carry out a sort of one-man job usually, and a bit too much. I think he was too dominant for the people who were close to him ... He wouldn't let an oceanographer carry out his oceanography or a biologist his biology. Although he would be the first to deny this ... he would have to have things done his way. This didn't happen to me because I was just far enough away so that it wasn't possible. I was working in Ellerslie and his headquarters were in St Andrews. So I can't complain of this personally, but maybe it makes my opinion that much more effective. He was inspiring in some ways. One of the ways in which he inspired was because he argued always, and was very critical, with his PH D students. He taught them mainly by telling them all the arguments against their conclusions. As a matter of fact, you would go and discuss your problem with Dr Huntsman and come out with the impression that you were doing everything the wrong way and maybe you should go in some other direction; the next thing he would ask you was how you had progressed with the line that you had been carrying out. He was very stimulating. He sometimes termed himself a gadfly, too, as far as other people were concerned.

Tributes and an anecdote Morden Smith, a scientist on the St Andrews staff and a self-acknowledged disciple of Huntsman, said: 'He had a love for the flamboyant touch and for Greek names, but he put basic emphasis on the ecological study, on how fish react to the environment and to a changed environment, and in that as in so many things he was away ahead of his time.'

S.A. Beatty, who had a brilliant career as director of the Halifax station, recalled: 'Huntsman was a driver, with a legal mind and a good entrée into the universities. He persuaded people like J.J.R. MacLeod, G.B. Reed, MacClement, and many other professors to come to St Andrews in the summer to work for the Board ... Cameron and Huntsman were temperamentally too much alike. There could be only one boss, so Huntsman was deposed ... With him dropping out of the picture, the Board lost most of its good university contacts ... But he played a key role in developing fishery research in Canada.'

Captain Arthur Calder, whose connection with St Andrews goes back to the days of the movable station, said: 'I had more rows and fights with Dr Huntsman than with anyone else, but if it hadn't been for his bullheadedness in starting immediately to rebuild the St Andrews station after the 1932 fire, without waiting for the Board to make up its mind, we probably wouldn't have a station there today.'

The best story about Dr Huntsman, Bill Ricker avers, has to do with this same 'Cap' Calder. It seems that he had been detailed to go out dredging with a visiting scientist; call her Mrs X. While out in the bay the boat's engine caught fire. With customary presence of mind Cap took off his slicker and threw it over the engine, smothering the fire; but the slicker – his own – was ruined. Next day he went to see the director to try to get a replacement. Anticipating that this wouldn't be easy, he had rehearsed a rather dramatic account of the incident that ended something like this: 'So there we were, Dr Huntsman, a long way from shore, and Mrs X couldn't swim, so if I hadn't put the fire out she'd have had a terrible time. The fact of the matter is, I saved her life.' Dr Huntsman hesitated a moment, then replied with impeccable logic: 'Yes, you're right. You did save her life. *But you saved your own life too!*'

Meanwhile, during the dirty thirties, work at the stations continued, though badly hampered by lack of funds and cuts in staff. At St Andrews the staff was cut from 11 to 9, though in 1937 a full-time investigator was appointed for lobster research. Experimental oyster farms were established at two Nova Scotia localities – Gillis Cove of Bras d'Or Lake in 1936, and Malagash in 1937 – to make research results more readily available to the industry. Although work on clams was suspended for three years, it was resumed as soon as funds became available. Scallop research was continued in 1932 with studies of growth rates in various places around the Maritimes. By 1936, the studies covered food for adults, maturity, spawning, the physical requirements for egg development, drift of larvae, varying successes of year-classes, and the possible effects of the increasing commercial fishery. Most of these studies were suspended in 1936 following the drowning of investigator-in-charge J.A. Stevenson, while on duty at Digby, Nova Scotia.

Herring research continued until 1936, involving studies on parasites, the hydrographic and other factors controlling catches, and the relationship between food availability and feeding and fatness. Behavior studies showed that different sizes of herring acted differently.

A series of experiments was begun in 1932 to test the effects of increasing production of trout by flooding meadows so as to take advantage of the cycle produced by the decay of the drowned organic matter. However, it was found impractical to keep temperatures down and the oxygen level up in flooded areas that were fed only by surface drainage.

In 1931 a survey of the groundfish fisheries of the Fundy area was begun. This provided impetus for organizing additional research effort over the next 10 years in which the general life history of cod and haddock was explored, including growth,

143

distribution, migrations, temperature preferences, spawning and egg distribution, and the fisheries and their statistics.

In oceanography, from 1931 to 1939, the conditions off the Scotian Shelf were followed in a series of cruises and the results analyzed. They demonstrated three main movements: a southwest movement along the coast, an anticlockwise circulation in the southern part of the shelf, and a complicated vortex over the extreme southwestern portion of the shelf. For surface layers, the winter transport of water exceeds that of summer.

The investigation of wood borers was reopened in 1935 and continued through 1939, to assess the possibility of damage to pilings in Saint John harbor and also to learn more about methods of controlling the pests in oyster culture equipment.

INTERNATIONAL PASSAMAQUODDY FISHERIES INVESTIGATIONS

The possibility of transforming 'the ceaseless flow of the tides' into useful energy has a powerful emotional appeal. The fact is, of course, that there are few places in the world where the height of the tide and physiographic conditions could make tidal energy at all economic, but there are possibilities in the Bay of Fundy region. The idea of using Passamaquoddy Bay for tidal power was raised seriously during the 1920s, and one of the first questions concerned the effect on fisheries. In 1927 Dr Huntsman produced an analysis that convinced him that the sardine herring fishery inside of the proposed dams would be eliminated. Subsequently the matter was referred to a committee of the North American Committee on Fishery Investigations, which recommended a special study to determine effects on fisheries both inside and outside of the dams. The Canadian and United States governments funded investigations in 1931 and 1932, to be conducted, in part, by foreign experts in order to ensure impartiality. The scientists employed included Dr H.H. Gran and Trygve Braarud of Norway (phytoplankton), Drs C.J. Fish and M.W. Johnson of the United States (zooplankton), Mr Michael Graham of England (ichthyologist), and Dr E.E. Watson of Queen's University, Canada (hydrographer). The St Andrews station served as a base for their operations, and the *E.E. Prince* was made available for their work.

The report of these men confirmed Huntsman's forecast that the inner Passamaquoddy herring would be virtually eliminated, but anticipated little effect outside of the dams. This, however, did not satisfy Canada, who declined to participate in the project, much to the disappointment of the United States engineers. There the project was said to have the personal support of President Roosevelt – who had a summer home on Campobello Island – and it could have been financed as a Public Works Authority make-work project, so that the cost-benefit ratio, which was not particularly favorable, was not a worry. The Americans in fact started work on a similar project in the smaller Cobscook Bay on the Maine side, but abandoned it when a review showed it to be grossly uneconomic.

Bay of Fundy
fisheries program

In order to provide a better background for the Passamaquoddy study, the St Andrews station was asked to cooperate with departmental officers in making an intensive study of the Bay of Fundy fish and fisheries in 1931. All available personnel and a number of volunteers were pressed into service. A new research vessel, the *Zoarces*, was employed, as well as the patrol ship *Nova IV*. Oceanography was studied by H.B. Hachey and J.M. Morton; the herring by A.G. Huntsman; haddock and alewives by R.A. McKenzie; pollock, halibut, skates, and flounders by V.D. Vladykov; lobsters by W. Templeman; clams by C.L. Newcombe; scallops by J.A. Stevenson; cod, smelt, tuna, and swordfish by Dr C.M.

The *Zoarces*, 1931–40

Jeffers and shad by Anne M. Jeffers, both of State Teachers College, Virginia; salmon by Miss I. McCracken, and hake, cusk, and mackerel by Dr H.I. Battle, both of the University of Western Ontario; miscellaneous invertebrates and minor fishes by Professor P.M. Bayne of Acadia University; and seaweeds by Professors H.P. Bell of Dalhousie and A.B. Klugh of Queen's.

The above list includes the name of the Board's most distinguished female scientist, and that of another colorful Russian.

Helen Battle was born in 1903 at London, Ontario, and has spent a good part of her life in that city. Her bachelor's and master's degrees were from the University of Western Ontario, and after obtaining her doctorate from the University of Toronto in 1928, she returned to her Alma Mater as instructor and later professor of zoology, where she has guided a long series of students along the byways of biology. From the 1920s, however, she spent many of her summers at the Atlantic Biological Station, at first as a volunteer investigator and later as seasonal assistant.   *Helen Irene Battle*

Dr Battle's research has covered a number of aspects of biology. In physiology she studied the effects of high temperature on skate muscle and nerve tissue, digestive processes in oysters and in sardine herring, oyster maturation, and the effects of DDT and other organic chemicals on various organisms. Her work in anatomy includes the detailed structure of the oyster's digestive tract and cyclic changes in its gonads, and the comparative internal anatomy of freshwater and sea lampreys. In embryology she has illustrated the development of the goldeye and the Atlantic salmon. Finally, she has made general biological and fishery investigations of the hake and the bearded rockling in the Bay of Fundy. Altogether an extremely talented and versatile scientist.

The colorful Russian, Vadim Vladykov, was born in 1898 at Kharkov. After the revolution he went to Czechoslovakia, where he took a PH D degree in 1925 from Charles University, Prague. He became fishery expert and traveling lecturer in the Ukrainian-speaking Subcarpathian province, at that time part of Czechoslovakia,   *Vadim Dmitrovich Vladykov*

145

and later was director of the Ethnographic Museum of Uzhgorod. After two years in Versailles as an agricultural bacteriologist and phytopathologist, he came to Canada. He worked for the Board from 1930 to 1936 at the St Andrews station and at Toronto, where he cut a dashing figure with his long opera cloak and intriguing accent. After that he spent a few years at the Chesapeake Biological Laboratory of the University of Maryland, but soon returned to work for many years in the Province of Quebec's Biological Bureau, and eventually became professor of zoology at the University of Ottawa.

Through all his varied experiences, ichthyology has remained Vadim's overriding interest. He has produced numerous popular papers and scientific monographs on the morphology, systematics, and ecology of fishes, lampreys, and marine mammals, both in French and in English. His last major work for the Board was a bulletin containing an exhaustive account of the morphology of the cranium and caudal skeleton of five species of Pacific salmons, which had been commissioned by the Nanaimo station in the hope that it would reveal characters that might distinguish different stocks on the high seas.

ATLANTIC SALMON STUDIES

In 1931 Dr Huntsman published a comprehensive analysis of the catch statistics of Canadian Atlantic salmon. He found suggestions of a scarcity every 9.6 years and also of a 48-year fluctuation. For neither of these could he suggest a cause, and their reality has been seriously questioned. Better documented, and much more interesting, was a cycle related to the length of freshwater and ocean life of the salmon, which was different in different rivers. This he attributed to the pressure of a large brood in reducing the food supply or space available to subsequent broods, up to the time the large brood went to sea as smolts. In the Miramichi River, for example, most of the parr spent three growing seasons in fresh water, followed by part or all of three growing seasons in the sea to reach the large 'salmon' stage. Thus any 'line' that became larger than its neighbors continued to be dominant because it made things difficult for the two following lines. Two such dominant lines could exist within a six-year life history, and the catch data did in fact indicate a three-year 'cycle.' For other streams the cycle was two or four years, corresponding to different life spans.

Huntsman felt, however, that salmon production might be increased everywhere by increasing smolt survival, which meant reducing predation in the rivers. The most conspicuous predators were kingfishers and mergansers. Harley White studied the food consumption of these birds in the wild and in captivity, and concluded that the mergansers could account for most of the parr that disappeared from the rivers. Accordingly bird control campaigns were conducted on several rivers, and smolt production was in fact greatly increased. However, the mergansers proved to be extremely wary birds, and any large-scale application of the technique was considered impractical because it would be both costly and disturbing.

After 1934 Huntsman devoted most of his time to salmon. In addition to Harley White he recruited a small group of students and associates, including P.F. Elson, W.S. Hoar, and Viola Davidson. Research was carried out on several rivers, but particularly the Petitcodiac in southern New Brunswick. Natural stocks of salmon were compared with fingerlings planted from hatcheries, and the food of parr was compared with that of their competitors. Tests of the tendency of salmon to return to their native rivers were positive, but trials of the inheritability of their season of return to the river were inconclusive, so that possibility was rejected. This was must unfortunate, for Huntsman let it color his whole subsequent approach to salmon management, whereas later work indicated clearly that there is a large genetic component in the control of time of upstream migration.

Later the Conservation and Development Service provided Dr Huntsman with holding facilities for experimental purposes at the Grand Lake Hatchery near Halifax, the 'Biapoc Ponds.' This name was an abbreviation of the word 'biapo-crisis,' which he coined to mean 'the response of the organism as a whole to what it faces where it lives.'

SMOKEHOUSE DEVELOPMENT

The chief problem to concern the Halifax station during the 1930s was that of smoking fish. By 1931 there had been a gradual transfer of major attention from refrigeration to problems of smoking fish and fish products. The change-over came from the encouraging results being obtained from preliminary investigations of the problems. In the early 1930s little was known about the theory of smoking fish fillets, and the practice by present standards was an uncertain operation. Very little control over factors which affected color, sheen, and other characteristics of the product was possible; in fact, very little accurate information on the subject was available.

The smokehouses of those days were large cumbersome affairs, over two stories high, in which the fillets were hung on long racks at the second story level and wood fires were built on the floor at ground level. Constant attention was needed to fires and draft controls. Even under the best operating conditions, it was hard to produce a product of uniform appearance. There was also considerable loss of fish from fillets dropping off racks, being overheated or cooked, or being marked by support-ing racks. Ash contamination was also common.

In 1934 several papers were published on the effects on smoking efficiency of factors involved in the process: velocity of smoke, humidity and dew point, tem-perature, smoke density, type of smoke, need for cooling the smoke, and effect of drying.

In 1937 the prototype of the modern mechanical smokehouse was designed, much smaller in size for a given capacity than the old-style house, in which all the conditions could be rigidly controlled to give a uniformly high-quality product with no losses due to poor operating conditions. In addition to improving the quality of the product, it reduced the time required for smoking from 15 hours to 7 hours. This rapidly became the standard smokehouse for the Atlantic coast industry. Later the smoking time was reduced to 3 hours, and a single unit could produce 400 pounds of smoked cod fillet per hour. For greater production rates, multiple units could be assembled, two or three joined side by side.

GASPÉ STATION

Another important development in the midst of retrenchment was the opening of the Gaspé Fisheries Experimental Station in 1936, to promote the value of the fishing industry in the Gaspé region through scientific research and demonstration. Grande Rivière was selected so that the laboratory could serve not only the peninsula but the northern coast of New Brunswick and the Magdalen Islands as well. The station was opened in August under the direction of Dr Arthur Labrie, a very effective scientist who later became deputy minister of fisheries for Quebec.

New Station at
Grande Rivière

147

The Gaspé Fisheries Experimental Station at Grande Rivière, 1936

During the first few years of operation, the main activities of the station were directed to an evaluation of the fishing industry in the area served, followed by an educational program through circulars, conferences, and courses to fishermen. In research, the main activities involved problems of handling fresh fish, canning, smoking, production of fish meal, preparation of salt codfish, and cod liver oil extraction.

PACIFIC SALMON STUDIES

At the Nanaimo station, a major investigation of the 1930s was the pink salmon study. Clemens outlined the problem in 1958:

Large runs of pink salmon enter Masset Inlet of the Queen Charlotte Islands and then proceed to several excellent spawning streams tributary to the inlet. But the runs occur only in even-numbered years. Since the pink salmon matures invariably at two years of age, there can be no overlapping or inter-breeding between fish of the alternate years, hence the populations of alternate years are absolutely distinct from one another. Thus, one population cannot contribute to the other and one may decline and disappear as has apparently been the case in Masset Inlet. There were two canneries on the entrance to the Inlet and these could operate only every other year. It was natural therefore that the operators should request an investigation of the possibility of establishing a pink salmon run in the 'off' year. The Board approved of an investigation which would add to the knowledge of the biology of the pink salmon, discover the factors involved in the lack of salmon in the 'off' years, and include an experiment in transplantation to determine quickly the possibility of establishing a run in the 'off' year. Dr A.L. Pritchard was appointed to the staff for the project and later W.M. Cameron was appointed as assistant.

Andrew Lyle Pritchard

Andy Pritchard was born in 1904 at Alcove, Quebec, in the Gatineau valley above Ottawa. He received the BA degree from the University of Toronto in 1926, and the MA in 1927. Work with Lake Ontario ciscoes in the Ontario Fisheries Research

148

Laboratory provided material for a PH D thesis, which degree was granted in 1930. He began work at the Board's Nanaimo station in 1928, so was already a familiar figure there when he took charge of the pink salmon propagation and transplantation study. As will be noted later, Andy also led the Skeena River survey in the 1940s before his departure for Ottawa as head of the department's Fish Culture Development Service, and its successors Resource Development and the expanded Conservation and Development, where he was to play a key role as contact man for the Board and help to arrange many cooperative programs during the 1950s and 1960s. Although ill-health forced Pritchard to retire in 1966, he maintained some consulting work and other activities and remained a member of the Great Lakes Fishery Commission until his death in 1971.

For the pink salmon study a field station was established at McClinton Creek on Masset Inlet, where a weir was built across the stream so that adults migrating upstream and fry moving downstream could be enumerated. This was to determine the production of pink salmon as a percentage of eggs deposited and of fry produced, and also the causes of mortalities at the egg and fry stages. Six determinations showed an average survival from egg to downstream-migrating fry of 10 percent, with a range from 7 to 24 percent. The mortality was due to climatic and biological factors, including predation by trout.

Queen Charlotte field station

Transplantation experiments were carried out in 1931, 1933, and 1935. In 1933 an unexpected freshet washed the eggs out of the beds where they had been planted, but in 1931 and 1935, 878,000 and 506,000 fry, respectively, went to sea. Considerable numbers of these were marked by the removal of fins to provide proof of any return as adults. No pink salmon at all returned to the area from the 1931 experiment, and from the 1935 lot only four marked and two unmarked fish came into McClinton Creek. The three attempts to establish runs in the 'off' years were failures.

Thus the investigation refuted the 'accidental' theory of the origin of the blank years, and strongly suggested that there must be some kind of direct or indirect interaction between the two brood lines – something akin to what Dr Huntsman was postulating for Atlantic salmon. This is made possible by the fact that pink salmon practically all mature at the same age. Forty years later the exact mechanism of interaction has yet to be identified. One possibility is the 'depensatory predation' hypothesis of Ferris Neave, based on his observations that a small 'off-year' line is at a handicap because it encounters predators that are too numerous in relation to its own abundance, so that it remains permanently small or may even die out. On this view, if it can somehow be increased above a certain threshold level, an 'off' year line should be able to survive and increase to the size of the 'on' year. The other hypothesis, advanced by W.E. Ricker, suggests that the pinks are themselves their own predators: the returning adults may eat the outward-bound fingerlings in the ocean in proportion to the abundance of both, and computations show that under such circumstances a smaller line always tends to decrease and may be exterminated. Such a mechanism makes the establishment of a good run in the off years impossible. It may be that both mechanisms operate, reinforcing each other to different degrees in different regions; also, sometimes neither is very effective, for some rivers have more or less equal runs in both lines.

During the 1930s the Board was instructed not to make a general attack on problems of migration and production of Fraser sockeye salmon because an international agreement had been drawn up for that purpose and its ratification was expected momentarily (it actually did not happen until 1937). Meantime, however,

New ideas at Cultus Lake

a number of additional experiments and observations were carried out, most of which used the facilities at Cultus Lake; in fact a 1937 summary by Foerster and Ricker lists 8 major categories of work, including 32 individual projects.

One project was to cross the five species of Pacific salmon and determine the viability and fertility of the hybrids, with the idea that these might exhibit fast growth from 'hybrid vigor' or other characteristics favorable for artificial culture. Most of the crosses were in fact viable, and a few back-crosses were obtained.

Another experiment was to see if land-locked sockeye (kokanee) from interior lakes could be persuaded to go to sea and become anadromous sockeye if released as yearlings along with normal sockeye smolts. Some did, but not enough to make the scheme practical economically.

Of greater potential importance was an attempt to modify the lake's ecological balance so it would produce more sockeye. It was clear from the beginning that if the Cultus Lake experiment found the hatcheries to be ineffective, the negative result would be received with very little enthusiasm. What was needed was some alternative procedure that would be at once effective and inexpensive. Experimental netting had shown that many and probably most of the missing sockeye fry at Cultus Lake disappeared into the jaws of predaceous fishes, and the commonest one by far was the useless squawfish. Consequently an experiment was undertaken to reduce the abundance of this fish, which after three years had decreased the sockeye-consuming sizes to one-fifth of their former abundance. Concurrently the sockeye survival rate increased to four or five times its former value, and two very large returns of adult sockeye were obtained. The benefit-to-cost ratio was very favorable, and the lake's trout population had not been reduced. Unfortunately the Board's control of the Cultus Lake project ended before it could be determined whether sockeye production could be maintained at the high level. The squawfish had responded to control by increasing their recruitment of young, and these would have required intensified removal before they became large enough to be important predators.

No large-scale application of this principle has been made as yet, except that the Salmon Commission removed many of the large char from Chilko Lake. The major trouble is that in many British Columbia lakes the abundant predators are trout, which are as valuable as the sockeye they eat.

HERRING PROBLEMS

The British Columbia pilchard oil boom continued through the 1930s. At that time herring were mainly dry-salted and sold in the Orient, but this market did not absorb the large quantities that seemed to be available. Any greater utilization would have to be by reduction, but reduction plants were firmly under the control of the provincial government, as a result of a celebrated legal battle earlier in the century. Although the British North America Act specified that all fisheries were the responsibility of the federal government, the plants in which fish are processed are normally on the land or on piles over water and are subject to local regulation and taxation. An enterprising entrepreneur sought to bypass these by building a floating cannery, over which, he claimed, neither the province nor any junior government had any jurisdiction. The resulting court case was appealed right up to the Privy Council, which decided in favor of the province.

As far as herring were concerned, the provincial authorities preferred to see them used for human food, but were open to persuasion that if in a given area a large surplus was available, some reduction might be allowed. Thus it was time for a

150

thorough study of the herring populations, their abundance, discreteness, fluctuations, and response to the exploitation they had experienced to date. This job was turned over to a new Nanaimo recruit, A.L. Tester.

Al Tester was born at Toronto in 1908 and received his education in that city, ending with a University of Toronto PH.D. degree in 1936. While at university he worked on smallmouth bass at the Lake Nipissing field station of the Ontario Fisheries Research Laboratory before joining the staff of the Board's Nanaimo station in 1931 to carry out the herring studies, his major contributions being in delimiting the numerous stocks that occur along the coast and determining their abundance and rate of utilization from surveys of spawn deposited each year.

Albert Lewis Tester

Tester was to remain at the station for a number of years, but in 1948 he moved to Honolulu to become professor of zoology at the University of Hawaii, a post which he held until his death in 1974, except for a few years with the United States Bureau of Commercial Fisheries, first as director of Pacific Oceanic Fishery Investigations in Honolulu and then as the bureau's chief of biological research in Washington, D.C. In Hawaii Dr Tester studied bait fishes used in tuna fishing, responses of tuna to stimuli, and the senses and behavior of sharks.

The task which faced Tester on his arrival in Nanaimo was to investigate herring. The Pacific herring spawn near low tide level, preferably on seaweed, and preferably near sand or gravel beaches where seepage from land reduces the salinity somewhat. Such spawning areas are distributed at intervals along both sides of Vancouver Island and the Queen Charlotte Islands, as well as the mainland coast. However, most of the Strait of Georgia herring appeared to come in from the open ocean each winter, and, if so, they were presumably mixed with the west coast stocks on their summer offshore feeding grounds.

Vertebral counts
and metal tags

The first question, then, was whether the various spawning stocks were part of one big population in the ocean, or whether they maintained some degree of individual identity from year to year. Tester's first approach was to count the number of vertebrae in samples taken all along the coast, and he in fact found that small but significant average differences existed even between adjacent regions. About 12 stocks were tentatively recognized, and the more northern ones tended to have the more vertebrae.

However, this finding did not exclude the possibility of considerable movement of individual fish from one stock to another. For direct evidence a tag was needed that the fish would carry and, just as important, a means of recovering such tags from the millions of fish that were being caught. A metal internal tag had in fact been developed in California for pilchards, and Hart had used it to tag pilchards off Vancouver Island. Recoveries were made on powerful electromagnets installed in the meal line of reduction plants. As it turned out, California pilchard tags were obtained in British Columbia, and British Columbia tags in California, confirming the migrations that had been postulated for this oceanic fish.

During the late 1920s British Columbia gradually relaxed its ban on reduction of herring, and by 1936 it was permitted everywhere subject to regional quotas for total tonnage taken. Thus recovery of herring tags became feasible and extensive tagging experiments were carried out. These suggested that there was a significant homing of the fish to specific sections of the coast as well as considerable wandering. These results became more precise when electrical induction detectors were developed and installed in the intake escalators of a few plants, so that the tags obtained could be associated with an exact fishing location.

The development of the reduction fishery was accompanied by a sharp decrease in average age of the herring stocks, indicating a heavy rate of exploitation, but

151

recruitment increased to compensate for this and abundance remained high until late in the 1960s. Year-to-year variations in recruitment did occur, but were very much less than among the Atlantic herring stocks.

PACIFIC OCEANOGRAPHY

When Neal Carter began work at Nanaimo there was already considerable information about the Strait of Georgia, but very little was known about the oceanography of the numerous fjords and inlets that indent the British Columbia coast. Yet it is through these channels, some of them 50 to 100 miles long, that millions of young salmon go out to sea and later return as adults. Carter resolved to make the inlets a prime research target, and in fact gathered much information about them before he transferred to Prince Rupert. The inlets typically have shallow 'sills' near the mouth, and some of them contained stagnant bottom water that had probably not been renewed for years. An even more curious discovery was the 'Bute Inlet wax,' apparently derived from pine pollen, that collects on the surface of that inlet during exceptionally cold winter weather and is supposed to make an excellent boot grease.

Neal had not been at Nanaimo a year before the demands on his time snowballed. In Clemens' words (1958): 'About this time a number of pollution problems came up. Further, the Station was asked to carry out a series of analyses of salmon, lingcod and oysters in relation to food values, and to obtain some basic data on the salting of herring. The amount of work was so great that an assistant became essential and Mr J.P. Tully was appointed. Mr Tully came to the Station with a burst of boisterous enthusiasm which seemed a combined product of the prairie environment and Irish ancestry, which added a sparkle to the Station group.'

John Patrick Tully

Jack Tully was born in 1906 in Brandon, Manitoba. Upon receiving his bachelor's degree from the University of Manitoba in 1931 he joined the staff of the Nanaimo station. Later he was to become oceanographer-in-charge of the Pacific Oceanographic Group (1946), to gain a PH D from the University of Washington (1948), and to move to Ottawa to succeed Hachey as oceanographic consultant with the headquarters staff and secretary of the Canadian Committee on Oceanography. Tully worked on various aspects of the oceanography of Pacific marine and estuarine waters, and during the war was enrolled in the navy doing hydroacoustic work. He was elected a Fellow of the Royal Society of Canada, and was awarded the Albert I of Monaco Commemorative Medal for his work in oceanography.

Carter has told the story of Jack's first encounter with the ocean:

I had a lot of correspondence with Tully as one of the applicants for the position of my assistant in the oceanographic work. He lived in Winnipeg at the time and had never seen salt water, but he assured me that he was interested, and in the course of the correspondence it turned out that he had a wooden leg. I felt a little reticent about the idea of employing him because in oceanographic work you have to be on a boat in rough weather when the decks are wet and slippery, and I wondered how his wooden leg would behave. He assured me that it wouldn't be any trouble, so I asked him to come to the station.

He arrived when I happened to be away on the boat on a week's oceanographic cruise. When we came chugging up to the dock at the station on Friday afternoon, here was this individual whom I had never seen before walking down to meet the boat in a resplendent yachtsman's uniform, complete with brass buttons. He figured he had to have a uniform if he was going to work on a boat, not knowing that we didn't go for uniforms. When the boat docked, I had on a dirty old sweater and was carrying some of the bottles of seawater ashore.

152

He asked where Dr Carter was. The skipper of the boat pointed to me and said: 'That's Dr Carter.' Tully's face fell. I never saw the uniform again.

In oceanography, investigation into the Strait of Georgia was resumed, and in addition some synoptic surveys were made on the west coast of Vancouver Island. Probably one of the most interesting achievements took place during a study of the Alberni Canal. Clemens tells the story (1958):

The Alberni Inlet project

A company desired to construct a pulp mill at Port Alberni and asked the Department of Fisheries for instructions regarding the liberation of the effluent into Alberni Canal. The Department asked the Board to investigate and the problem was assigned to Mr Tully. Mr Henry Vollmers, with his large new trolling boat, was engaged for the project. The fish hold was fitted up for a laboratory, and on this boat Tully and Vollmers lived and worked. At a later stage the *A.P. Knight* was used.

After accumulating and analysing a mass of data, Mr Tully decided he needed a model of the upper end of the Canal in order to interpret and confirm his conclusions as to the water movements. With the assistance of three young lads, R.L.I. Fjarlie, H.J. Hollister, and W. Anderson, and with plaster of Paris, buckets, pulleys, hoses, an electric fan, and parts of alarm clocks, there appeared a model approximately 6 by 4 feet complete with tides, river flow and winds, all recorded by well-devised gauges. I was intrigued by all this activity and eventually found myself perched on a stool on a table, looking down on the model and recording on a diagram the flow of dye introduced to represent the pulp mill effluent. Needless to say, I was replaced by a camera. So the model idea was introduced to the Station. Later a larger and more effective model was built on the hill above the Station but the little model in the Chemistry building will always remain as a symbol of vision and ingenuity.

As a result of this study, the dispersion pattern of the pulp mill effluents in the inlet was predicted, and measures to reduce damage to the fisheries were recommended and carried out.

NANAIMO REMINISCENCES

As did F.R. Hayes for St Andrews 10 years earlier, W.E. Ricker has recalled something of the atmosphere of the Nanaimo station in the years just before the residence was closed. Ricker himself worked at Cultus Lake, but he had to come out of the woods occasionally to make chemical analyses and use the station's library.

In those days the station still had a foreshore that went dry at any really low tide, exposing a clam bed with two or three kinds of starfish scattered about, the curious egg collars of the moon-shell, and rocks under which were conglomerations of small crabs, blennies and clingfish. In the water underneath the dock you could see yellow shiners and white seaperch, rock cod, schools of young herring and salmon, pulsating jellyfishes, and sometimes the awkward contractions of a *Melibe*, a sea slug trying to become a medusa. Punctually every afternoon Mr and Mrs Berkeley would come down for a swim, to be joined by some of the visitors. However, it was better fun to dive off the dock at night, when your body would be surrounded by great clouds of phosphorescence caused by *Noctiluca* and comb-jellies.

One year Cliff Carl arrived at the station with a new accordion, on which he played and we sang in the residence living room after dinner. Both Cliff and I left after two or three weeks, and Earle Foerster told me later that Professor F.D. White of Manitoba expressed his profound relief when the noise subsided. Another summer visitor was Professor Leslie

Saunders of Saskatoon, who studied the abundant intertidal midges. During two or three winters Jim Munro of Okanagan Landing, who was federal migratory bird officer for the west, spent time at the station. He was studying stomach contents of fish-eating birds, concerning which he and Dr Clemens produced a couple of papers. He always had his binoculars at his desk to be ready to check on anything unusual that appeared on the water outside. On one occasion, I was told, a group of killer whales swam right by the dock and around the island in front of the station.

During my time Charlie Mottley, Andy Pritchard, Al Tester, and Jack Tully all lived in the residence for a while; then one by one they moved to houses in town, or else replaced Roy Elsey or Neal Carter in one of the units of 'Doctors' Row' at the head of the bay – a group of three dwellings that had been built for Powder Works personnel during the First World War. Among the students and junior employees present at various times were Van Wilby, Edgar Black, Ross ('Rosie') Whittaker, and J.L. ('Laurie') McHugh, today a professor at Stony Brook on Long Island. I knew Laurie well because he had helped us at Cultus Lake in counting migrants at the salmon smolt fence and measuring the samples later.

Three graduates from the University of British Columbia brightened the Nanaimo summers considerably: Mildred ('Scotty') Campbell worked with copepods; Verna Lucas's interest was in ostracods, and eventually she became the wife of Morden Smith at St Andrews; while Josephine ('Babs') Hart studied crab life histories (in fact she still does) and later married Cliff Carl. Evelyn Keighley and Ethel Robinson looked after the library and did all the typing, while Tommy Russell, a graduate of the Nanaimo coal mines, showed great ingenuity in keeping the station's physical plant in smooth operation.

For several years the Nanaimo station and the Friday Harbour station collaborated in an annual field day, its location alternating between the two establishments. I happened to be at Nanaimo for one of them. The Washington contingent arrived in their new research vessel the *Catalyst*. Included in their party were oceanographer 'Tommy' Thompson, Professor Guberlet, and Professor H.B. Ward, a visitor from Illinois. We sailed over to Gabriola Island for the field events, at which Dr Ward, who was over 70 at the time, sparked the visitors' ball team to a resounding victory over the locals. After dinner there were speeches and jokes at the residence, including Charlie Mottley's presentation of a 'Mic-a-Mic Medal' to Professor Thompson. Other copies of this medal, cast in lead in a hand-carved mold, had been given to two or three celebrities previously, always accompanied by Charlie's long-winded account of the accomplishments of the man shown in profile on its face, Professor Doctor Mic-a-Mic of the University of Ungava.

Some of the volunteer people stayed at the station throughout the winter, perhaps getting a little casual employment from time to time. Among these was Helge Fasmer from Copenhagen, who studied brittle stars, and G.H. ('Old Daddy') Wailes, a hefty Englishman who drew beautiful illustrations of protozoans and assisted with projects varying between plankton and fur seals. Another amateur much interested in the ocean and its life was Ira Cornwall. As engineer at the quarantine station at William Head, near Victoria, for 20 years he made daily oceanographic observations for the Nanaimo station. He became a self-taught expert on barnacles, including the specialized whale barnacles which he obtained on a visit to a west-coast whaling station. The last issue of the Board's Pacific Fauna series of publications was his review of the local barnacle species.

PRINCE RUPERT

Orville Denstedt, who was a graduate of the University of Manitoba, like Finn and Brocklesby, came to Prince Rupert as Brocklesby's assistant. He recalled that Finn started operations in a room 6 by 6 feet, with a microscope. Finn, whom he had known since boyhood in southern Manitoba, was a persuasive speaker and a first-class diplomat. When Brocklesby, in collaboration with Denstedt, produced the first bulletin on fish oils, neither Cameron nor Huntsman wanted to publish it in its initial form. They thought it contained too much elementary chemistry. Brocklesby refused to publish it in a shorter version. Meanwhile Finn had been taking his PH D at Cambridge University. When he returned, he read the bulletin and liked it. Then he went to the next Board meeting and persuaded Cameron and Huntsman to withdraw their objections to its length. They agreed to publish 1,000 copies. When the bulletin appeared, there was an immediate demand from the industry for copies, and within six months it was out of print. The demand came at once for a revised edition. By that time Denstedt had left for McGill, and Brocklesby prepared the second edition in collaboration with several other members of the staff.

Denstedt recalled Finn as a first-class administrator who knew how to handle scientists. Once when Denstedt was working on a particularly frustrating problem and seemed to be at an impasse, Finn told him: 'Don't worry. Take a couple of weeks at it, a couple of months, a couple of years.' Relaxed, Denstedt solved the problem the next day.

Neal Carter replaced Finn as director at Prince Rupert when Finn was moved to Halifax to assume charge there in 1933. He said: 'Dr Finn had been in charge as director for about seven years before I took over, and he had done an excellent job in establishing good relations with the fishing industry in Prince Rupert and, to some extent, in Vancouver as well. Consequently, I felt that my greatest challenge was to maintain those good relations with the members of the fishing industry so that they would appreciate we were working for them as well as doing research on certain problems of our own.' Neal Carter succeeds Finn

One of the big problems that had yet to be tackled was the method of handling the catch. The use of ice to keep fish fresh until they were landed was a poor way to maintain quality; the temperature wasn't low enough and the fish at the bottom of the pile received severe abuse. Better and more novel methods of preserving the fish in the boat were needed as well as more modern methods of unloading and handling the fish in the plants.

Fish oils was another study, already well-launched. Which fish had the best yields of vitamins A and D was still not known, nor was enough known about the chemical nature of the oils themselves. So the station specialized in refrigeration and fish oils.

Considerable advances were registered in the greater utilization of fish oils, particularly pilchard oil, which had drying properties similar to the linseed oil used in paints. Paints, varnishes, and stains were developed from pilchard oil. The fish-oil paint had the property of being able to expand and contract along with metal as the air temperature changed, making it ideal for painting bridges and other metal structures. It didn't crack the way ordinary paints would on a metal surface.

There was the incident of the new varnish, described by Carter in 1972:

Dr Bailey was in the process of compiling a section of a bulletin on the use of fish oils and fish oil products, and just about then Dr Brocklesby devised a method of making varnish out of fish oils, and we decided to paint our library shelves with this varnish, because they needed painting in any event. Dr Bailey was looking up a lot of references in *Chemical Abstracts* in connection with the compilation of this bulletin, and he used little pieces of paper for bookmarks as he looked up the different articles. He must have had about 200 pieces of paper sticking out of some 30 volumes of *Chemical Abstracts*. At the end of the day he pushed them back onto the shelves so he could recommence work the next morning, but unfortunately our

W.A. Clemens, A.H. Hutchinson, J. Dybhavn, A.T. Cameron, J.J. Cowie, and N.M. Carter at the Prince Rupert laboratory, 1938

varnish did not have proper drying properties, so next morning when Dr Bailey came to take the *Chemical Abstracts* off the shelf, all the little pieces of paper had stuck to the varnish and were pulled out of the books. So he had to look them up all over again. However, we did get the varnish improved so that it dried a little better, but I don't think much commercial use was made of it because, not many years later, the pilchards disappeared rather suddenly from the British Columbia coast, and pilchard oil became a historical product rather than a commercial one.

John Dybhavn
a stalwart

Neal Carter recalled John Dybhavn, representative of the British Columbia fishing industry on the Board:

He was the owner of the Royal Fish Company in Prince Rupert. He was Norwegian by birth and a most genial gentleman. And when I say gentleman, I mean gentleman in every sense. As a member of the Biological Board ... and as owner of the fishing company in Prince Rupert, he was a real stalwart of strength in helping the station establish good communications with the fishing industry, not only in Prince Rupert but also in Vancouver. He was available for advice whenever some knotty problem turned up. As members of the Board, he and I used to travel back and forth to Ottawa for the annual meeting by train – there was no plane service at that time – and we had many and many an interesting discussion during the four-day trip between Prince Rupert and Ottawa. And not only discussions: there were also interesting episodes. On one occasion Dybhavn and I helped to remove pieces of flesh, skin, and entrails from the locomotive's headlight and valves after a collision with a moose on the way to Jasper. We were all very sorry to hear that Mr Dybhavn had passed on shortly after I left Prince Rupert. I think that a lot of the work and success of the Prince Rupert station was due to him.

Carter also spoke about Hugh Tarr and Otto Young:

Dr Tarr became associated with the station while we were in Prince Rupert, to take charge of all the bacteriological work ... The bacteriological work of the station was very important because it was one of the best checks we had on what progress we were making in our efforts to advance the better quality of fishery products.

Hugh Tarr joins the station

Before Dr Tarr was appointed during my directorship up there, Dr [R.H.] Bedford had been the bacteriologist, and had instituted some very interesting methods of making quick bacteriological examination of fish. When Dr Bedford left, Dr Tarr picked up and continued with Dr Bedford's work, plus some novel ideas of his own. Among these were his noteworthy investigations that led to the commerical exploitation of the then recently developed antibiotics and antioxidants for preserving the quality of fresh and frozen fishery products ...

Then there was Mr Young, who was already on the staff as refrigeration engineer. The work that Otto Young did was of the greatest importance, because he was interested in the refrigeration part whereby we hoped we could overcome some of the difficulties I mentioned earlier about how roughly the fish were treated while on board the fishing vessel. And it was largely due to Otto Young that we got into the possibilities of completely revising the methods of refrigeration on the railway cars that were used to ship fish from the British Columbia coast to the interior and eastern part of Canada.

Young's work of greatest importance

Young recalled that his first assignment when he came to Prince Rupert in 1929 was as supervising engineer of the new building. He also worked with Brocklesby and Denstedt, who 'had little engineering problems that I participated in.' Then he developed a new type of sampling bottle for Dr Bedford, who wanted to make a study of the vertical marine bacteria distribution in the ocean.

When the building was completed, Young said that he was

given two assignments and they were really my own choice. One was to work on some means of cutting down the dehydration in cold storage rooms. And I worked out a curve on the amount of dehydration that would take place in a room under various conditions: that is, using one coil, two coils, three coils, or practically filling the whole thing with coils. The more cooling surface you got in the room, the less dehydration there was ... The amount of evaporation depends on the vapor pressure, which depends on temperature. So temperature is the thing. What I was doing here by increasing the coil surface was reducing the difference in temperature between the surface of the ice and the surface of the coils. The greater the difference, the greater the evaporation ... This led up to the idea of keeping the air that would pick up the heat coming through the walls apart from the product that you were storing and that, of course, led to a jacket. Out of these studies came the jacketed cold storage room.

Huntsman was the originator of it. When I started digging in the oatmeal, so to speak, here was Huntsman's idea. So I wrote to Huntsman and asked him for his work on it. Huntsman had no control mechanism set up. He had just conceived this idea, had put it into practice, but he didn't know really what the efficacy of it was. But he was very interested and he really encouraged me ... So I did the work under controlled conditions, comparatively speaking.

Young credits Huntsman

The other problem was to determine the effect of freezing on fishery products, and that led to some very extensive studies, actually resorting to the microscope to determine what effect the rapidity of freezing had on the physical structure of the muscle. As everyone suspected, the faster you froze fish muscle, the smaller the crystals were ... and [the] less [the] physical damage. I tried to determine where the crystals were forming, whether it was in the cell itself or between the cells, intracellular or extracellular. I was never able to determine that, but I did try to find a critical rate of freezing ...

157

The Pacific Fisheries Experimental Station at Prince Rupert, around 1938

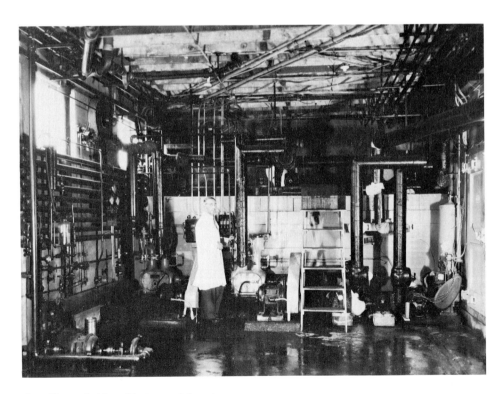

Otto Young in his cold-storage laboratory

A chemical laboratory at the station

I tried to determine the rate of freezing for given temperatures, and I worked out freezing curves for different fish and this led me to the pith of the whole thing, which was: What damage, if any, from the taste point of view, does freezing do to fish muscle? If it doesn't do anything, then why worry about the speed of freezing? So I then carried on a very extensive experiment to try to work out, separately, the effect of the rate of freezing on the edible acceptability of fish muscle and the temperature of storage.

Young then arranged to set up two taste panels; one representing a cross-section of the public, and the other a select group chosen for their ability to detect qualities like salt, bitter, sweet, and other characteristics in very dilute solutions. These panels enabled him to determine the effect of freezing on the two main fish of commercial importance, salmon and halibut.

Taste panels provide answers

He discovered that, although the rate of freezing was important to the eventual edibility of the product after thawing, the temperature of storage was even more important: 'The temperature of storage was so important that if you wanted to keep fish for three months to carry it over from one season into another, well, there was no way you could do this under the conditions that were being used at that time in storing fish. They were storing salmon at about 14 degrees above zero Fahrenheit ... halibut was kept from 16 to 18 degrees.' Actually fish muscle doesn't freeze at the freezing point of water or anywhere near it, as Neal Carter discovered by using a novel application of the expansion of the water in the muscle as it turns to ice. The true freezing point turned out to be about $-5°F$ $(-20°C)$.

Another discovery was that with both salmon and halibut the flavor actually improved through storage and icing: 'When I would give a family a sample of fish right out of the water and fish that I had stored three days on ice ... statistically, the sample stored three days was superior to the other. So when you are drawing up a quality curve for frozen fish, you've got to start with fresh, and that would be something like 85 percent of the acceptability, and it would go up to 100 in about three days and then it would drop off, and the rate that it dropped off depended on storage temperature as well as the freezing rate.'

Young investigated another problem:

What about these fish that come in from the halibut banks that are 14 days old? They are way down on the [acceptability] curve to start with, and I got the processors to admit that any fish that could possibly get to New York or Boston in the unfrozen state would still be acceptable, even though the quality would be way down. And if there was no possibility of getting it through that period, and there was another say 7 or 8 days on top of the 16, they would put it in the freezer. In other words, what the public was getting was fish that went into the freezer way down at 35 or 40 percent acceptability, and then freezing did the rest; so there is no doubt about why frozen fish was not acceptable in 1929.

Young tackles refrigerator cars

Complaints about frozen fish thawing in transit got Young involved in an examination of the standard refrigerator car that was being used in Canada and in most of the United States at that time. These were equipped with bunkers at each end, which were loaded with ice through the roof, together with salt: 'Everybody knew the effect of adding salt to ice, but that had temperature limitations. You can't go below the eutectic temperature, which is about minus 6 Fahrenheit. And with minus 6 in the end bunker, there was no way that you could keep the center of the load of frozen fish beyond Winnipeg from thawing.'

Young learned that these bunker cars were supposed to be dual-purpose. In the summer, ice was used in the bunkers; when carrying vegetables and fruits in winter a heater was put through a trap door in the bunker, and the circulation would be reversed. 'When I queried the officials of both the CN and CP about this hopeless method, they said it was necessary; they had to have a dual-purpose car. When I suggested a top-cooling system ... they said it just wouldn't work.'

In order to show the difference between an end bunker car and the one he was proposing, Young made two small models of each and took temperature curves. When the results were published, a representative of the CPR visited Prince Rupert and offered to build a car along the principles developed by Young: 'The upshot of that was that it subsequently became the standard refrigerator car in Canada; it is today the standard car.'

He goes a step further

However, Young's concern with lowering the temperature in cold storage rooms to improve the quality of the product led him inevitably to the same problem in transportation: 'When we got the cold storages to go to zero and sub-zero temperatures, this was a big step, but ... we couldn't let up even in the transportation phase of it ... As a mechanical engineer, of course, I was all for a mechanical system ... Controversial discussions were going on between us and the railway companies when they pointed out to us that fish were peanuts in the transportation business and that they were not in a position to spend millions of dollars changing cars for two or three percent of the traffic.'

Young then appealed to the National Research Council, which organized a meeting with the railway companies: 'Out of that came a request from the council to the Board that I make a survey in the States of what they were doing about it.' At this point, the railway companies were not happy with Young: 'I was going for

160

something better and they were just converting to the overhead car. They had put in big orders for cars and the Americans immediately swiped them. Most of them went across the line and they didn't come back to Canada. They had reciprocity.'

Young got in touch with the ThermoKing group in Minneapolis, who were making refrigeration units for trucks, and they agreed to make two units to Young's specifications for railway cars. Then the CNR provided a metal shell of a boxcar with the normal side doors but with no bunkers.

The replaceable unit developed

I wanted a unit that, alone, would handle the whole load with about a double safety factor, then I wanted to boost that up to four by putting in another unit in case this failed, so that if we were not running on time, we would have another to fall back on ... These units were so designed that inside of five minutes, you could take one out and put in a reconditioned one ... And I visualized a system that could be carried on continuously with no breakdowns, because if something broke down, a man just pressed a button on the other machine, and the one that failed was taken out and another put in ...

I pounded this into the people at ThermoKing, and they came up with this lovely idea, so that you pushed a unit in just like a drawer. You'd pull out the drawer; you'd put the unit on it, you pushed it in, pulled down a clamp, and that brought the top down on it, so that now it was sealed in. Then you just hooked the battery to the machine; you hooked on the fuel; and you pressed the button ... Now, you would set the temperature you wanted, 5 below say, and that thing would keep it at 5 below even if the boxcar went through 40 below weather. Actually it would heat then, because it would turn on the heat ... It had enough capacity that if it were 90 degrees outside, it would give you 5 below inside. This unitized system had one very great advantage: the connections were reduced to a minimum, and any motion was resisted by the whole unit, not by a part.

Young demonstrated the effectiveness of the system by taking a carload of fish from Vancouver to Halifax, and showed that it could be used in winter to keep fruit and vegetables from freezing. However, the system never came into wide use in the railways because of the shift from railway cars to trucks and piggy-backs for the shipment of fish and other perishable foods: 'The trucks can come to your cold storage doors, take a load of fish, and deposit it at the door of the consignee, and the railways couldn't do that. So, in order to stay in business in the heavy food field involving carload lots, they had to piggy-back. And this is only a step in the continuing development. I'm quite sure that eventually everything will be containerized; a container of this and a container of that, so that you can get mixed loads protected from adverse temperature.'

Young's forecast

These developments in refrigeration occupied Young between 1929 and 1949, in Prince Rupert and, later, in Vancouver.

SHIFT OF DIRECTORS

Meanwhile, Finn had been approached to move from Prince Rupert to Halifax, to replace Leim, who would move to St Andrews to succeed Huntsman. Finn recalled: 'I knew they were having some difficulty down there in the sense that the industry was not particularly satisfied, and ... Professor McMurrich ... asked me whether I would consider going to Halifax to take over, which I did. It was in 1933. Leim was rather unhappy in that sort of job. He was a biologist, and he didn't make the transition which is necessary to handle rough individuals in the fishing industry ... to put up with the guff, etc. Of course I'd been brought up handling that sort of

161

thing ever since I was a kid. I had a knowledge of the industry, so therefore I was more at home.' So Leim moved back to St Andrews where, in the following year, he replaced Huntsman.

Alexander Henry Leim

Leim was born at Fergus, Ontario, in 1897, received his BA from the University of Toronto and, in 1924, his PHD in marine biology. He began his study of the life history of the shad in 1919 and it formed the subject of his doctoral thesis. He followed this with studies of the life history of shrimps and the effect of light, temperature, salinity, and pH on certain copepods and mysids. He investigated the smelt in various lakes and in the estuary of the Magaguadavic River, and cooperated in improving trout and salmon hatchery and fry planting methods.

In 1926 he was appointed assistant director of the Board's Fisheries Experimental Station at Halifax, where he directed investigations into the practical aspects of the fisheries, personally studying such problems as the prevention of drying in cold storage, the preparation of cod liver oil by freezing, the prevention of freezer burn during storage, and the smoking of fillets and haddies. He was closely associated with Huntsman in the problems of rapid freezing of fish products and in the introduction of quick-frozen fillets to the Canadian market.

At Halifax he organized and directed courses for hatchery officers, fishery inspectors, and fishermen, and lectured in fisheries science at Dalhousie University, where he was associate professor from 1929 to 1946. He was also editor of publications for the Biological Board between 1926 and 1929.

In 1930 Leim became director of the Halifax station, and in 1934 he returned to St Andrews as director. In 1941 he became chief biologist at St Andrews, concentrating on research on the fishes of the Canadian Atlantic. During 1944–50 he headed a study of the Atlantic herring and, following the publication of a report, began preparation of a book on the fishes of the Canadian Atlantic. He died in 1960, before the book was completed, but it appeared in 1966, *Fishes of the Atlantic Coast of Canada*, by A.H. Leim and W.B. Scott.

H.B. Hachey said in 1965: 'Those who worked closely under or with Dr Leim appreciated his modesty and quiet dignity, his broad knowledge of the fisheries of the Canadian Atlantic, and his kindly sympathy with the problems of his staff and co-workers. He was ever ready with a helping hand.'

14 Changing goals: the later Cameron period, 1937–47

Effect of the new act

Although no drastic changes took place immediately after the Biological Board became the Fisheries Research Board of Canada in 1937 – the important changes had taken place during the previous decade – the new Board with its new act was now definitely on the road of applied research. Chairman A.T. Cameron hailed the new act with approval in his annual report for 1937; it embodied the recommendations proposed by the Board itself. It was designed to make the Board workmanlike, to prevent life appointments to its membership, and, above all, to enable the election to the Board of scientists representing the major sciences involved in the Board's work: the actual institution they represented was a secondary matter.

So the Board as constituted in 1937 consisted of nine scientists, including four instead of two industry representatives and two representatives of the Department of Fisheries. The nine scientists included four biologists, one anatomist, one bacteriologist, one chemist, and two biochemists, which roughly corresponded to the extent to which each of the various sciences was represented in the Board's work. Significantly, the three scientists making up the executive of the Board along with J.J. Cowie, representing the department, and Handfield Whitman, representing the industry, were Cameron, G.B. Reed, and Monsignor Alexandre Vachon – a biochemist, a bacteriologist, and a chemist respectively: not a biologist among them!

One of the departmental representatives was the honorary treasurer, F.O. Weeks, and Miss Nora Grimes was honorary assistant secretary. Nora Grimes became something of a legend in her time; efficient and helpful, it has often been claimed that she was the headquarters staff of the Board for many years.

Cameron's main task for most of his tenure was one of survival and consolidation. He had to survive the effects of a severely restricted budget and he had to consolidate the position of the Board with the Department of Fisheries and the fishery industry. At the same time he had to maintain an effective scientific instrument capable of meeting the demands of the world war and the expanding postwar era. It would provide the foundation for the rapid growth of the Board in the postwar years as it expanded to meet new challenges and responsibilities.

Attrition

The Board's frozen salary situation, which continued while other institutions were going through a post-depression recovery, made it vulnerable to raiding and a number of excellent scientists left the Board to enter private industry or other organizations where the salary scale was more generous. Thus, in 1937, Dr N.E. Gibbons, assistant bacteriologist at Halifax, moved to the National Research Council; Nanaimo lost Dr C. McC. Mottley to Cornell, Dr C.R. Elsey to British Columbia Packers, and Drs Foerster and Ricker to the Salmon Commission; and at Prince Rupert Dr O.F. Denstedt fled the eternal rain for McGill, and Dr R.H. Bedford moved to the Atlantic Coast Fisheries Corporation.

Cameron thought that this ill wind boded some good: 'While the loss of five of our experienced scientists in one year may seem to be so heavy as to involve handicaps to our work, I feel, personally, that the good outweighs the evil, since when it becomes recognized that excellent service with the Board not only results in reasonable promotion, but also gives good opportunities for securing good positions elsewhere, the attraction to the Board's services of young and aspiring scientists will be all the greater. Stagnation in any service is bad, nor shall we ever be able to promote all our junior men to the higher ranks in ours.'

What Cameron understandably neglected to mention in his 1937 report was the demoralizing effect on staff members of the restricted budget that was reducing the Board's operations in many areas to a holding one, and the frozen classifications that were making the Board scientists among the lowest paid in the country. The Board was to continue to suffer attrition of its staff as it provided a happy hunting ground for recruits to universities and private industry. This was a function of the Board that was not spelled out in its new act.

There were some promotions: in 1937 Needler was promoted from assistant to associate zoologist; R.H. M'Gonigle at St Andrews from assistant to associate pathologist; Otto Young at Prince Rupert from assistant to associate research engineer; and at Nanaimo, both Hart and Elsey, from assistant to associate biologists, although the promotion didn't prevent Elsey from joining British Columbia Packers.

Cameron the man for the job

It was not an easy task that faced Cameron. The situation within the Board, together with the economic depression of the 1930s, might have overcome a softer, kinder man, but he was neither. He was a strong chairman, and he ran a tight ship.

S.A. Beatty, who was chief biochemist at Halifax in 1937, said in 1972: 'He was an excellent chairman. He knew how the labs had to be run. He knew that the ideas came from the bench where the men worked, and he found out what every man was doing. He knew his field; he read a lot and he wrote textbooks, yet he still managed to spend all of his summers and a great part of his winters on Board work. In fact he had to carry on as though he was the Board ... With Huntsman gone, the Board lost its university contacts, though Cameron made it a far more effective instrument in the services of the industry.'

A.W.H. Needler said:

A.T. Cameron was ... much more businesslike than McMurrich, or at least he tried to make the Board into a businesslike corporation. Sort of tidy up the administration and watch the expenditures. He was a rather dominant individual. I did a lot of driving around with him. He used to make a tour of all of the stations once a year when he was chairman, and this involved

164

A.T. Cameron, chairman of the Board
1934–47

him in driving something like 2,500 miles. I have driven A.T. Cameron all over the Atlantic coast. He always liked to be very precise.

I remember with great amusement when we had the opening of the Gaspé station ... There was a meeting of the Atlantic subcommittee at Matapedia, and Dr Cameron had a timetable laid out with the exact minute for what we would do after this, when we would arrive at Paspébiac where a Mr Gyon was going to entertain us for a drink, and how we were going to go on somewhere else for supper, and the time we would arrive at Grande Rivière [for the opening].

Well, it was most amusing because I was in the car that was supposed to lead the procession. Dr Cameron and Weeks – who was the honorary treasurer of the Board – and a couple of others were in the second car, which was driven by a fishery officer who was possibly the fastest driver on the Atlantic coast. The other [third] car was just a bit more normal. But the car I was in had J.R. Dymond and [Professor Georges] Préfontaine and one or two others. We went two or three miles and Préfontaine had some little errand, an acquaintance he wanted to see, so we left the road. The second car [which missed our turn-off] was trying to catch up with us and arrived at Gyon's place about an hour and a half before it was supposed to. The third car arrived at just about the right time, but by that time we had made a lot of little side trips to visit all of Georges Préfontaine's friends, so we arrived about an hour and a half late.

When we left Gyon's place, there were two exits, and Préfontaine was in the car that was supposed to lead the way. It went out one exit, but the other two cars went out the second exit. That put them ahead of Préfontaine's car, so they not only drove very fast to catch up with Préfontaine's car, but they went, I suppose, about ten miles past the restaurant where we were scheduled to have dinner. We were all late, anyhow, but the other two cars arrived an extra hour late.

When we finally got to Grande Rivière, most of the hotel reservations had been taken up by a party of CBC reporters. They had simply said that they came from Ottawa, and occupied the rooms reserved for Dr Cameron and the others. The whole thing was a schmozzle that seemed to annoy him greatly, though it was enjoyed by everybody else. He was very punctual and very precise, a bit didactic ...

Dr Cameron, I think, actually spent a large proportion of his time working for the Board ... It might have been close to half. He certainly spent over a quarter, just by the time he spent on meetings and preparing for meetings and the field trips would account for at least a quarter of his time. I don't think any other chairman had spent that much time, and I don't think that Dymond and Reed [in the succeeding era] felt themselves in the position to either.

D.B. Finn recalled Cameron as a professor of biochemistry at the University of Manitoba:

Very precise in all his work. He was quite a good biochemist, too, and he pioneered the book, *Recent Advances in Endocrinology*, a world standard. But he was a bad lecturer. My goodness gracious, you couldn't hear him. A bunch of us in medical school, where he used to lecture, got together and hired a stenographer and seated her right by his rostrum so she could take down his lectures. Then we had this mimeographed so we could study, and we gave him a copy. This was the start of *Recent Advances in Endocrinology*. But, above all, he was a very precise man, because of his military training. Students were very much in awe of him, even the medical students.

Of Cameron's appointment to the Board, Finn recalled:

In the first place, Clemens was really responsible for his choice. Clemens and I talked about the problems that we faced, and particularly at the experimental stations, because there wasn't a single technician on the Board, no chemists, no anything except biologists. So I said, 'You've got to get people on the Board who are chemists, at least.' And Clemens said, 'I think that is a good idea.'

We weren't on the Board, but we attended Board meetings. I wanted to be understood by the Board. When we talked about engineering and chemistry and things of this sort, we wanted somebody who knew what we were talking about ... Charlie O'Donoghue of the University of Manitoba had gone to Edinburgh and he had to be replaced, and I think Clemens or Hutchinson talked to the people in the university and suggested that it would be good if they had a chemist, and this was the way Cameron was appointed.

R.E. Foerster found him different from Knight:

My first association with a chairman of the Board was with Dr A.P. Knight of Kingston. He was the head of the Department of Biology at Queen's University, a very courtly gentleman, fine, dignified. I always enjoyed conversing with him, and any correspondence that we had was really outstanding, I thought ... That type of person you just don't see nowadays.

Dr Cameron was different. He was loyal to the Board, I guess, but a dogmatic chap ... You never knew when he came to visit the station just what his attitude was going to be. On some occasions he would be sympathetic, very interested in the staff and in the welfare of the station. But I recall one time when he came, he started out by saying that he didn't want to meet any of the junior members of the staff; he was just interested in my impressions of how they were getting along and how the station was doing. And in spite of all my protestations, he absolutely refused to meet any of the senior staff, which puzzled them a great deal, I know, and which caused me a certain amount of embarrassment. But there was nothing one could do about it. He made the pronouncement, and that was that.

I think that was the only time I really resented his visiting, but I do recall one time when he and Major Sutherland were visiting the station, this must have been in the summertime, and he was standing at the back door of the director's residence waiting for Major Sutherland to bring up the car that they were using at that time. And while he was waiting, chatting with my

166

wife and myself at the back door, he saw a peculiar light across the road among the trees and he said: 'Now, look, Foerster, there are fireflies! I never thought there were fireflies in British Columbia!'

I said: 'No, I don't think they are fireflies.' And he said: 'They must be. Look at them flitting here and there.' Well, I just gave up then, and my wife, I could see, was laughing quietly to herself. I had decided to plant a few tomato plants on the bank behind the house and on the other side of the road, and I had put some tin cans at the foot of these to hold water. It was the moonlight or maybe the light from the back door shining on these tin cans that produced the firefly effect in which Dr Cameron was so interested. But I never did tell him. I don't suppose he ever knew that they weren't fireflies.

John Hart, who succeeded Foerster as director of the Nanaimo station, remembered Cameron with mixed feelings:

Mixed feelings about Cameron

A.T. Cameron was a very arbitrary man. From some points of view, I suppose, he was a good executive and administrator. I think he was welcomed by some of the industry people on the Board in that he was effective and perhaps took a good deal of pleasure in putting the directors in their places. He made it quite clear that the policy of the Board was going to be controlled by the Board without any, or at any rate excessive, interference by the directors, and made this as an announcement at the first meeting. He gave very little consideration to the personal lives of his employees, and particularly his senior employees, I would say, because during those days there was no airplane travel and he used to call meetings between Christmas and New Year's, which meant that directors and Board members who lived on the west coast went for many years without ever spending New Year's at home with their families, sometimes Christmas as well.

He provided good administration of a sort, but it is my belief that in some cases his arbitrariness left real deficiencies. I am not sure to what extent he was responsible for the great dissatisfaction among the younger staff in the days immediately following the war, when young men came back after leave of absence and found themselves in intolerable positions financially. I know on one occasion I was summoned to the director's office when Cameron was there, and I was asked to make a choice as to which was the most effective of three of these young men ... I pointed out that each one of them had their merits and perhaps demerits to some extent, and that it was probably impossible to say which one was absolutely the best without specifying what for, but this was not a sufficient answer. I think that I finally came up with the idea that one was the best, but I wouldn't be sure of that. But I regard this as being a rather improper thing for a man in his position to be doing, but this is what he did.

Hart believed that Cameron, in spite of or because of his own military background, was unsympathetic toward staff members who wanted to join the armed services during the Second World War, and was equally unsympathetic toward those who had served in it. Those who had been in the services found themselves, on returning to the Board, without their regular seniority and paid less than younger men who had been hired meanwhile. Hart himself was refused a leave of absence when he wanted to enlist.

If Cameron was not precisely a man for all seasons, there seems to be general if sometimes rather grudging acknowledgment among those who knew him and worked for him that for the circumstances in which the Board found itself prior to and during the Second World War, he was the right man to head it up.

Cameron was born in London, England, in 1882. He received his MA degree from the University of Edinburgh in 1904 and his BSC in 1906. He had two years of study at University College, London, under Sir William Ramsay, and a year at the

Alexander Thomas Cameron

167

Technische Hochschule of Karlsruhe in Baden under Fritz Haber. Originally a physical chemist, during the course of certain experiments he discovered that traces of lithium appeared consistently in a solution where the only metal should have been gold, and he concluded that he had transmuted these two elements. Although checked and rechecked, this sensational discovery was later disproved: the source of the lithium was cigarette ash that he unconsciously let fall into the containers.

Soon afterward Cameron turned his back on inorganic chemistry, and on England, when he was appointed lecturer in physiology at the University of Manitoba in 1909. He became involved in the ill-defined field of physiological chemistry, in which his experience with the methods of physical chemistry enabled him to cope with the problems. There, too, he began the study of what were then called the 'ductless glands' and, starting with an investigation of the distribution of iodine in plant and animal tissues, he laid the foundation of his reputation as an endocrinologist. Except for a summer semester spent in research at the University of Heidelberg and three years during the First World War as a chemist officer for water purification with the British Expeditionary Force in France, his subsequent career was at the University of Manitoba. In 1923 he was appointed to the newly created chair of biochemistry in the Faculty of Medicine. He was the author of over 100 published papers as well as four textbooks in biochemistry. In 1925 he was awarded the degree of D SC by the University of Edinburgh.

Active in a number of scientific societies, Cameron was elected a Fellow of the Royal Society of Canada in 1920, and was president of his section in 1929–30. He was a Fellow of the Royal Institute of Chemistry and a past president of the Canadian Institute of Chemistry. He was awarded the CMC in 1946 in recognition of his services during the Second World War. But tobacco was to deal him another blow, this time fatal, for in 1947 he died of lung cancer.

The late Dr G.B. Reed paid tribute to Cameron:

His genius for organization came to the fore in the Fisheries Research Board ... In the years just preceding his Chairmanship the Board emerged from a rather academic organization into a broad field of fisheries research. During Cameron's regime these advances were consolidated and extended ... The operating budget of the Board in Cameron's first year as Chairman was approximately $175,000, in his last year in office it was approximately $800,000; in his first year in office the Board employed 34 scientists and 54 other ranks, in his last year there were 110 scientists and 150 other ranks ...

Cameron was a man of strong convictions, indomitable will, tireless energy and sterling honesty. He was completely intolerant of half-measures. It was not always easy to work with such a personality but anyone capable of standing the pace was immeasurably stimulated. Canada was enriched by his life, his work, and his organisation. His example continues and some of his personality lives on in his many co-workers and associates. [See Hachey 1965: 433–4]

Board member O.F. MacKenzie of Halifax wrote, following Cameron's death: 'Eminent scientist though he was, Dr Cameron was also possessed of high executive ability, a combination of qualities that unfortunately is extremely rare. He also had an almost uncanny memory and knowledge of details, invaluable in dealing with the manifold activities of the Board's Stations from coast to coast. His devotion to the Board's work, long-range plans for future work, and the qualities referred to above, make his loss to the Board at this time well nigh irreplaceable.' (See Hachey 1965: 434–5)

Although the Board and its stations were able to deal with a number of problems and actually face the war's end in better shape than at the outset, the major and persistent problem throughout the war years was the loss of key personnel. These losses were to the armed services, private industry, the universities, government departments, and to the Department of Fisheries itself. That the Board survived them and emerged after the war in a stronger position than before was a tribute to Cameron's leadership.

Retirements,
resignations, and
appointments

In 1939 he had to report: 'During 1939 we lost the services of Dr L.I. Pugsley, Assistant Biochemist at the Prince Rupert Station, who obtained a position with the Department of Pensions and National Health at Ottawa last July. We have also loaned to the Department of Fisheries the services of one of our Directors, Dr D.B. Finn, as Chairman of the Salt Fish Board, and are glad that his long and excellent record with the Fisheries Research Board has led to his selection for this important post.'

The loan turned out to be something more than a loan. Finn did a brilliant job with the Salt Fish Board in a very short time. Particularly his skill in heading off countervailing American duties against the subsidies to fishermen that the Salt Fish Board instituted must have brought his activities favorably to the attention of the minister of fisheries, for Finn shortly afterwards found himself deputy minister of fisheries. He functioned effectively there, too, during the war years, and then went on to become head of the Fisheries Section of the Food and Agriculture Organization of the United Nations.

Another loss was J.J. Cowie, secretary of the board. Cameron reported:

Mr J.J. Cowie retired from the Department of Fisheries at the end of March, 1940, having been Acting Deputy Minister since December, 1938. At the same time he retired from the Board, relinquishing its Secretaryship which he had held for fifteen years. The Board owes much to him. When he succeeded Dr Prince as Secretary, under the Chairmanship of Dr Knight, the then Biological Board was a relatively small body, with Stations at St. Andrews and ... at Departure Bay ..., while the Board's annual Grant was just under $47,000. In the first year of his Secretaryship it rose to over $105,000, and then increased steadily to just over $386,000 in 1930–31. At the time of his retirement the Board operated five all-year Stations, and the expended Grant was $238,444.12. Before Mr Cowie's tenure of office much, though by no means all, of the Board's work was academic, carried on without much thought of practical application. More and more, during his Secretaryship, its plan of work was devised towards solving the problems of the Department and of the Industry, and much of its progress has been due to his commonsense guidance of its work to matters of immediate importance. We hope that he will be long spared to enjoy well-earned leisure.

J.J. Cowie died in 1943.

The Board also lost Monsignor Alexandre Vachon, who resigned when he was appointed Archbishop of Ottawa early in 1940. He was replaced by Professor J.-L. Tremblay of Laval, and Major D.H. Sutherland, assistant deputy minister, took over as acting secretary. When Finn became deputy minister, on 15 February 1940, Dr S.A. Beatty was confirmed as director of the Halifax station. Then, in May 1940, Dr Arthur Labrie, director of the Gaspé station, was appointed associate deputy minister of fisheries for the Province of Quebec and was replaced by Dr Aristide Nadeau.

That same year the redoubtable Clemens, who had headed up the Nanaimo station, resigned to take the zoology chair at the University of British Columbia. Cameron said: 'During his tenure of this position he cooperated wholeheartedly in the increasing trend of the Board's policy towards researches with immediately practical aims. His Station buildings were increased and substations set up where necessary. He collected a well-trained and loyal staff, organized first-class programs, and saw a great deal of it carried towards completion. The members of the Board have no doubt that he will be equally successful in his new sphere.'

Dr. R.E. Foerster, formerly assistant director at Nanaimo, who for the previous three years had been on the staff of the International Pacific Salmon Fisheries Commission, was appointed to succeed Clemens as director at Nanaimo.

Attrition continues In 1941 Cameron reported the retirement of A. Handfield Whitman as the Atlantic fishing industry's representative on the Board. He had served since 1923 and was replaced by O.F. MacKenzie of Halifax. Cameron also noted: 'In accordance with his wishes to devote himself entirely to research work, Dr A.H. Leim was appointed last September as Chief Biologist at the St. Andrews Station, while Dr A.W.H. Needler was appointed his successor as Director of that Station. In his position in charge of oyster culture work in the Maritime Provinces, acting both for the Board and for the Department of Fisheries, Dr Needler has already gained considerable experience in administrative work.'

Then he gloomily observed: 'A number of members of the staffs of the various Stations have joined the armed forces, and those staffs have thereby become depleted to the stage that further requests to serve with the forces, if granted, may seriously interfere with essential work. Our Directors have been advised, nevertheless, that when such further requests are made they are not to label work as absolutely essential without grave thought and consultation.' This was followed by the now old-fashioned comment: 'Replacement by women may be possible in some degree, but there is much field work for which they could not satisfactorily replace men.'

The following year he again recorded: 'During the year several scientists and technicians of our staff have resigned, or have been given leave of absence to join the armed forces. A few replacements have been possible, mostly by women. As in every branch of scientific activity, suitable men and women for research work are becoming scarcer all the time.'

Brocklesby leaves Above all, Cameron deplored the loss of Brock:

In Dr H.N. Brocklesby, F.R.S.C., we have lost a brilliant scientist. He resigned last September to accept an important research position with a commercial firm in San Francisco at a salary very much greater than that he received from the Board. During his 16 years' service with the Board he was promoted steadily, and had been for several years Chief Chemist at the Prince Rupert Experimental Station, the highest appointment open to him. He has gradually established a reputation as one of the leading oil and fat chemists of this continent, and his researches on marine fats and oils have proved of great value to the industry and greatly contributed to the present high reputation of that station. It is a regrettable fact that Industry will always be able to buy first class scientists from Government service when it desires, until their salaries in Government employment can be made commensurate with those which Industry is prepared to offer; nor is Industry accustomed to overvalue the services it pays for.

Cameron found some solace in the fact that Brocklesby's resignation made possible the promotion of Otto Young to chief research engineer and Dr H.L.A. Tarr to associate bacteriologist. During that same year (1942) the Pacific Experimental Station's move from Prince Rupert to Vancouver caused considerable disruption of its normal activities. In fact, it provided the occasion for Brocklesby's departure from the Board.

The year 1943 was to prove the last year of attrition of Board members and staff. Last year of losses Cameron reported the loss of another stalwart: 'The Board suffered a grievous loss by the death of Mr John Dybhavn on April 3, 1943. He had been a member for almost twenty years, and during all this period had acted as Chairman of the Pacific Sub-Executive Committee. He fathered the Prince Rupert Station and its personnel, and was deeply hurt when by force of the circumstances of war it had to be removed to Vancouver. The members of the Board and its western Directors had learned to rely on his assistance and judgment at all times.'

He welcomed two new members to the Board: Professor Stewart Bates of Halifax and Dr W.A. Clemens of Vancouver. Thus Clemens continued his long association with the Board. Bates would figure importantly in its later history. Cameron also noted the reappointment for five-year terms of Professor J.R. Dymond of Toronto, Dr Georges Préfontaine of Montreal, O.F. MacKenzie of Halifax, and R.E. Walker of Vancouver. Dymond was a key person in the Board's recruitment efforts to maintain its staff strength.

Concerning defections, Cameron observed: 'Dr B.E. Bailey, a biochemist at the Vancouver Station, has left the Board's services to take a position with an industrial firm at nearly twice the salary which the Board was able to pay him. He has done valuable work. There have been several changes among the junior scientific and technician staff. New appointments at present are limited to women and to men in low medical category,' hence not accepted for military service.

Another blow was struck at the Board that year when the station at Grande Rivière was completely burned out. A sum of $45,000 was voted to construct and equip a new Gaspé station. Meanwhile the disrupted work was carried on as well as possible at Halifax and at the University of Montreal.

By 1944 it was clear that Cameron considered that his chief job of holding the Over the hill Board together during wartime was accomplished. The tone of his annual report was more optimistic than in any since 1938, and he didn't even mention any defections. Instead he pointed out that 'the total staff employed by the Board, as of September 30th last, consisted of 83 persons in full-time employment, and four employed seasonally. In addition during the past summer over a dozen scientists from the Universities were temporarily employed in field work.'

It is instructive to examine just how well Cameron had made out with this major problem from the beginning. In 1939 his permanent scientific staff numbered 41. Despite losses it rose to 47 in 1940, and reached a new high of 54 in 1944. In the last year of the war it would reach 65. So, despite all losses and depredations, Cameron had not only held the team together, but had added to its numerical strength.

The reason for Cameron's optimism in 1944 was to be found in his annual report of that year:

Last year, at the suggestion of the Department, we set up a Development Committee to advise as to investigations which ought to be commenced now, so as to aid properly in rehabilitation programmes and in the due expansion of the Fisheries Industry. Arising from the recommendations of this committee, and financed by money voted in the Supplemental Estimates, a number of new projects have been commenced, and others in progress extended. These include investigations in the North West Territories, extension of refrigeration researches along with the training of young engineers in special problems associated with the refrigeration of fish, and a new investigation of Atlantic ground fish.

Arising from the Departmental enquiry held in 1943 into difficulties associated with present export of whitefish due to parasitic infestation, the Central Fisheries Research Station was established last summer in temporary quarters in Winnipeg, its immediate purpose being to assist in the arrangements necessary for the orderly inspection of this fishery in the Prairie Provinces. There are numerous other problems associated with the commercial fisheries of the region between the Rockies and Port Arthur, and there is no doubt that the staff of this Station can be fully occupied with these for many years to come.

In short, the scope of the Board had been greatly increased, and funds were available to tackle new tasks.

New deal for the staff

It was not until 1945 that Cameron was able to do anything for the Board's long-suffering staff, and he reported:

Like all other research institutions, Governmental or otherwise, the Board has been greatly concerned for some time over the relatively meagre salaries which could be paid to its scientists, as compared with those paid by industrial firms. The disparity is too great. We have lost a number of excellent men through such causes, and, if the disparity cannot be lessened, we shall undoubtely lose more. During the year a special committee was set up to suggest a reclassification of positions and of the salaries attached to them, and it is hoped that this can be approved. Its recommendations are based on the Beatty Report of 1930, and, if accepted, will bring the various scientific positions held under the Board, and their salaries, in closer alignment with those of the National Research Council.

The pensions scheme at present used for the Board's scientists, that of the Teachers Insurance and Annuity Association, has not proved entirely satisfactory either to the Board or to its employees, and can scarcely be used at all for its non-scientific staff. It is hoped that before long a more satisfactory scheme can be adopted.

The following year, the last of the Cameron regime, J.R. Dymond served as acting chairman in Cameron's absence due to ill-health. He advised that the reclassification proposed the previous year had been adopted, eliminating a number of inequalities and generally increasing salaries more in keeping with the qualifications and responsibilities of the Board's staff. At the same time the Board requested the Minister to take steps to have the Board's staff placed under the superannuation scheme of the Civil Service or some similar government scheme.

Dymond also noted that as of 31 December 1946 the Board's employees included 78 scientists and 80 non-scientists. Temporary employees engaged in summer field work included 57 scientists and 28 non-scientists. He pointed out that the increase over 1945 was 18 full-time and 19 temporary scientists, 22 full-time and 24 temporary non-scientists. The postwar expansion was under way.

Fate of wartime committees

Dymond reported on a number of committees which had been formed during the wartime period: the National Committee on Fish Culture, which had been originally sponsored by the National Research Council (NRC) and the Fisheries Research Board, was, by mutual agreement, dissolved. Grants in support of research on fish culture formerly made by NRC on the advice of this committee would be continued by the NRC 'in so far as the Council requests it, on the advice of the Fisheries Research Board.' The annual conference of scientists working in the field of fish culture, representing provincial governments, universities, and the Board, formerly held under the auspices of this committee, would be held under the auspices of the Board.

The Canadian Committee on Food Preservation, sponsored by NRC, FRB, and the

Department of Agriculture, continued to serve a useful function in providing facilities for pooling information on the subject of food preservation.

THE PACIFIC FISHERIES EXPERIMENTAL STATION AT VANCOUVER

Shippers of frozen fish, fruit, and vegetables were seeking temperatures as low as 0°F in railway refrigeration cars and the engineering staff of the station was working on new refrigerating devices and constructional designs to be incorporated in an improved refrigerator car that would be tested in 1947. Other contributions to the theory and practice of refrigeration and cold storage were made by the staff, including a proof to the trade that the 'whitening' of reddish-fleshed fish fillets or steaks was a transitory phenomenon actually indicative of good quick-freezing technique.

Experiments on the nutritive value of fish-flesh protein had reached a stage where the crude proteins of halibut flesh were shown to possess a biological value definitely higher than that of beef. The flesh protein of five other fishes tested was shown to be equal or closely approximate to that of beef, though deficient in some of the B vitamins when compared to beef and pork, particularly the latter.

Studies on the separation, identity, and possible uses of non-fat constituents of fish and fish liver oils were continued. Results were circulated to the industry.

The problem of maintaining high quality of fish products during transmittal from the fishing vessels to the consumer was tackled, and many tests of possibly suitable chemicals were performed to determine their effectiveness as germicidal sprays and deodorants for fishing-vessel holds and fish-handling plants. Similar information was obtained in concluding work on chemically treated ices used for retarding bacterial spoilage in whole or dressed fish.

A plastic transparent model was made to demonstrate the action of the smoke in the smokehouse that had been devised by the Board and which was becoming popular on both coasts.

Specialty products such as pastes, baked loaves, and canned 'niblets' continued to be devised from whole fish flesh or the trimmings resulting from other processes.

Leaflets describing the harmless nature of glasslike crystals and 'curd' sometimes found in canned salmon had to be reprinted by the thousands to satisfy the demand by canning companies, wholesalers, and stores. A special article was prepared to explain to canners and consumers that perfectly sound canned fish might develop bulged ends in cities at high altitude. This had caused a lot of grief to fish canners until the headspace vacuum could be better controlled with the advent of the vacuum-closing machine.

An analysis of the potential value of the nitrogenous products in fish materials such as cannery, reduction-plant, and liver-oil-plant wastes was conducted.

More effective ways of inspecting Prairie Province whitefish for the possible presence of flesh parasites were sought. The station continued to collaborate with the British Columbia Research Council on the production and properties of agar and algin from British Columbia seaweeds, and facilities were made available for the council's investigation of methods of preserving fishing gear.

NORTHWEST FISHERIES SURVEYS

Government plans for post-war development of the Northwest Territories and the

Yukon made it desirable to know more about their fishery potential, and early in 1944 the Board was asked to look into this. Field parties were sent out in 1944 and 1945, mostly recruited from the universities.

Professor V.C. Wynne-Edwards of McGill and Dr Ronald Grant travelled the length of the Mackenzie River in 1944, and surveyed the Yukon River and certain other rivers and lakes of Yukon Territory in 1945. Their conclusion was that the fish available in fresh waters could do little more than supply local needs. The only possible exception was the Mackenzie delta region, where there are considerable runs of anadromous whitefish and ciscoes, but the cost of transport would preclude establishing an export fishery.

Much the most promising prospect for a commercial fishery in the Territories was Great Slave Lake, so it was subjected to intensive work by a party headed by Professor D.S. Rawson of the University of Saskatchewan.

Donald Strathearn Rawson

Don Rawson was born near Uxbridge, Ontario, in 1905, and studied at the University of Toronto, receiving the PH D degree in 1929. His subsequent professional career was at the University of Saskatchewan, where he eventually became head of the Department of Biology. Arriving in Saskatoon, he immediately began a survey of the game fish resources and the limnological characteristics of lakes in the newly created Prince Albert National Park, which continued to occupy him and his students for a number of years. Later he did field work in Banff, Jasper, and Riding Mountain national parks, on Lake Athabasca, on the saline lakes of southern Saskatchewan, and on the large lakes in the northeastern part of that province.

Dr Rawson's association with the Fisheries Research Board included the first year of the Prince Albert Park work and two summers in British Columbia studying interior trout lakes prior to his involvement in the survey of Great Slave Lake in 1944–5. Later (in 1959) he was to become a member of the Board, on which he served until his sudden death in 1961. The many students that Don Rawson encouraged in aquatic studies include Dr P.A. Larkin, a former director of the Nanaimo station, now dean of graduate studies at the University of British Columbia and a member of the Board, and Dr J.C. Stevenson, the Board's present editor of publications.

Great Slave and Great Bear

The Great Slave Lake study continued for two years; in addition to Dr Rawson, its personnel included Dr J.G. Oughton of the Royal Ontario Museum, Dr W.H. Johnson of the University of Western Ontario, and several students. Headquarters was a laboratory on a barge that was towed to different parts of the lake, while the sampling was done mostly from skiffs with outboard motors. The shallow western part of the lake proved to be moderately rich, biologically, because it warmed up in summer and received a large volume of water from the sedimentary regions to the south. As a result of the first year's work commercial fishing was started in 1945; it has continued ever since, although present levels of production are less than the original estimates.

Quite different is the situation in Great Bear Lake, which has a rather small drainage area situated entirely in Precambrian shield rocks. Its water is very clear and very cold, and some ice remains on the lake well into August. In 1945 a party that included Dr R.B. Miller of the University of Alberta and Dr W.A. Kennedy of the Board's Winnipeg Station assessed its fishery potential and concluded that it could support only the subsistence fishing of the local people and a limited sport fishery. The latter, however, was very attractive, for there were many trophy-size lake trout in the lake, and today there are several fishing lodges that cater to enthusiastic sportsmen.

174

At the Nanaimo station two major investigations were initiated during the Cameron regime. At Cowichan Lake on Vancouver Island a small hatchery was turned over to the Board for research on game fish production, particularly the reproduction of rainbow and brown trout and of coho, chinook, and Atlantic salmon. The last species was included because a number of local anglers yearned nostalgically for the kind of fishing they used to have in the rivers of Scotland. Concurrently a study was made of the trout and salmon stocks of the river system, their food base, and their reactions to environmental conditions.

Cowichan River study

The work was supervised by a recent recruit to the Nanaimo staff, Ferris Neave. G.C. Carl, who was later to become director of the British Columbia Provincial Museum in Victoria, studied the limnobiology of Cowichan Lake in relation to the salmonid fishes in it. C.P. Idyll, who dealt with the invertebrate foods of trout and young salmon in the river, subsequently worked for the Salmon Commission, the University of Miami, and for FAO in Rome. Neave himself established that rainbow trout from the lake and steelhead trout from the sea constituted separate races of fish, although they both spawned in the river at much the same places and times. His summary of the game fish and fisheries of the system is a model of concise and informative presentation.

In 1937 the International Pacific Salmon Fisheries Commission was getting ready to begin its intensive study of the Fraser River sockeye salmon. The Board felt that similar attention should be given to the next largest sockeye system, the Skeena, and did some preliminary work there in 1938 and 1939. However, the project had to be shelved during the war, and it was not until 1944 that a full-scale Skeena investigation was launched under the leadership of Dr A.L. Pritchard. Dr J.R. Brett was his chief lieutenant in the field, and later the author of a number of scientific papers concerning the work of the investigation. The original survey was to last five years, because Skeena sockeye often live to five years of age, but certain aspects were continued for a longer period. Pink salmon as well as sockeye were studied, but less attention was paid to the other three species.

Skeena River salmon investigation

A group of promising recent graduates and students was recruited for the field work, quite a number of whom continued working for the Board in later years: D.F. Alderdice, K.V. Aro, T.H. Bilton, H.J. Godfrey, J.G. Hunter, H.D. Fisher, D.R. Foskett, J. McDonald, D.J. Milne, D.N. Outram, M.P. Shepard, and F.C. Withler. Others found a career in conservation and development, notably E.W. Burridge, W.R. Hourston, and Dixon MacKinnon. All of the sockeye-producing lakes in the watershed were studied limnologically, and their salmon runs estimated. Special attention was given to Lakelse because it was accessible and of a convenient size for experimentation, and also to Babine Lake, much the largest in the system. Counting fences for salmon were maintained in both of these so as to have accurate information on population trends. A tagging program in the fishery gave information concerning rates of migration and utilization, and smolts were marked to indicate rate of survival at sea.

The survey accomplished as much as a one-shot effort can. It provided tentative recommendations for management of the river system and an excellent basis for continuing study. However, its report proved a disappointment to many in the industry. There had been hopes that something would be discovered in the environment that could be remedied quickly, and that this would permit a quantum jump in production of sockeye and/or pink salmon. But the report's verdict was that 'the commercial fishery must be held mainly responsible for the decline in the sockeye salmon populations,' and it could offer no panacea for a quick upturn. This seemed to contrast with the situation on the Fraser River, where sockeye runs were increasing two-, three-, or four-fold per generation, with no end yet in sight. What was overlooked was that the Skeena was producing sockeye catches equal to more than half of its historical maximum, whereas the Fraser was far below its maximum and hence still had a great potential for increase. Twenty years later the sockeye yields from both systems had reached a plateau relative to current management practices, and the weight of catch taken, per square mile of accessible lake nursery area, was about the same.

Other Pacific studies

The Nanaimo station was involved in other studies. Those on herring had reached the stage where it was possible to predict in some years and in some areas the likely yield to fishermen. Dr Tester was able to advise the industry of the probability of good yields on the west coast of Vancouver Island for 1946, and the prediction was amply justified by recorded catches. A study to learn the relation of the amount of spawning to the number of young fish added to the stock was planned for the coming year.

In the study of problems in salmon conservation in the territory between the Fraser and Skeena rivers emphasis was placed on the pink and chum salmon. A study of chum salmon at Nile Creek on the east of Vancouver Island, by Ferris Neave and W.P. Wickett, revealed that fry production amounted to only 3 percent of the eggs deposited. Much of the loss was attributed to the tendency of spawning fish to mass in the lower reaches of the stream and hence their failure to utilize apparently suitable gravel beds higher up. Means suggested for overcoming the resulting loss included stripping and artificially fertilizing eggs, developing them to the eyed-stage, and then planting them throughout the stream. This method gave good results except when floods shifted the gravel extensively, which happened quite frequently.

John Hart's study of the trawl fishery revealed that about 30 species of fish were caught, of which at least a dozen were commercially important. Since each species had its own peculiar habits, rate of growth, size attained, distribution, spawning time, etc., it was unlikely that any regulations would be equally satisfactory for the conservation of all species. It became a matter of determining what regulations would be the most advantageous for the largest number of the more valuable kinds. Extensive tagging experiments would be necessary to determine movements and other information. A 60-foot diesel-powered vessel, the *Investigator I*, was purchased for this investigation in 1947, and K.S. Ketchen was recruited for the field work and analyses. A study of eulachon and other smelts was made by Hart and J.L. McHugh.

C.R. Elsey's oyster investigation had revealed that successful spawning of oysters occurred in occasional years in Ladysmith harbor. In such years the young oysters, or spat, could be collected there, thus reducing dependence on spat imported from Japan. Attempts were made to encourage oysters to spawn slightly below their normal temperature by releasing into the water milt from oysters held at a higher temperature, but these were not particularly successful. However, some unusually warm years around 1940 put a swarm of oyster larvae out into the Strait of Georgia, where many settled and grew in the intertidal zone, making oysters available to all comers.

176

The Pacific Biological Station fisheries research vessel *Investigator I*, acquired in August 1946 (transferred to the Vancouver laboratory in December 1972)

The *Sackville*, 1942–

The *Harengus*, 1946–

An important part of the work of the Gaspé station included studies aimed at improving the quality of the fishery products of the Gaspé peninsula. Fish plants from Grande Vallée to Gaspé on the north shore were investigated. Important sources of contamination were found on the wharves and during transportation from the landing wharves to the processing plants. None of the eight plants inspected were found to be entirely satisfactory. Suggestions for improvement included provision of an insulated cold room, a filleting room large enough for a straight line production, and a separate room for packing. Investigations carried out on offshore fishing boats showed the possibility of improving the quality of fish aboard with only a few modifications.

In 1947 a thorough study of what happened to the codfish muscle when treated with salt was in progress, and methods of artificially drying and storing capelin had been devised. Investigations of improved methods of utilizing various by-products of the fishing industry were continuing. Control of the quality of frozen fish, in cooperation with the Quebec government, was maintained: 68 per cent of the samples collected were found to be of first quality.

Series of lectures were given by staff members at the School of Fisheries, Ste Anne de la Pocatière, and to the French-speaking people on the Caraquet coast of New Brunswick.

THE ATLANTIC FISHERIES EXPERIMENTAL STATION AT HALIFAX

Most of the studies in progress at the Halifax station in 1947 were designed to effect improvement in the methods of handling fish from the time they were caught at sea until they reached the consumer. Efforts were being made to overcome the causes of deterioration aboard ship by improving the insulation and the refrigeration of ship holds. Means of desliming round fish on a commercial basis were being investigated. Studies were in progress to determine how long frozen fish could be kept at 10°F without loss of quality.

The tunnel smokehouse developed at the Halifax station was now in general use for the smoking of canned herring fillets, and practically all the small herring canneries were so equipped. The air-drying unit for salt fish also designed at the Halifax station was likewise adopted by the trade. It was anticipated that by the end of 1947 drying units would have been installed of sufficient capacity to dry the whole output of the Maritime Provinces.

A number of other miscellaneous studies included one to discover the sources of cholesterol from fish and marine products, and one to determine the drying characteristics of Irish moss. Several other potentially valuable researches could not be undertaken for the lack of qualified personnel.

ST ANDREWS

During the 1940s herring were still the major underutilized resource in our Atlantic waters from Newfoundland south. In the Gulf of St Lawrence they were plentiful alongshore four to six weeks each summer, and then were gone for the rest of the year. Interested persons in the Atlantic Provinces and Newfoundland got together with federal officials to see if means could be found to extend the fishing season and make greater use of stocks available. This led to the establishment, in 1944, of a

The Atlantic Herring Investigation Committee

committee on which New Brunswick, Nova Scotia, Prince Edward Island, Quebec, Canada, and Newfoundland all had representation, each contributing money or personnel to the investigation. Headquarters were at St Andrews, Dr A.H. Leim was in charge, and the investigators included L.R. Day, Louis Lauzier, and S.N. Tibbo. Vessels used included the *Ahic*, the *Gulf Explorer*, and from 1947 the specially built *Harengus*.

The samples taken showed that throughout the area 7 to 12 ages were present in the stocks in significant numbers, showing that utilization was still light. Trials showed that midwater trawling could take considerable quantities of herring, and purse seining was possible on pre-spawning concentrations in a few areas.

Atlantic salmon
At St Andrews investigations of the Atlantic salmon by A.G. Huntsman and P.F. Elson showed that on the Pollett River in New Brunswick, despite very heavy plantings of underyearlings in the lower open water, there was almost complete disappearance of the fish as yearlings before they were ready to descend to the sea; mergansers were believed responsible for this. With continued low water in 1946 the adult Petitcodiac salmon failed to ascend to the upper spawning streams, at least up to the time these became frozen a month or so after the usual spawning time.

Meantime Dr Huntsman made a test of the idea that summer freshets would bring more salmon into fresh water and improve angling. In 1942 a dam was built on the Moser River in Nova Scotia and used to store water for artificial freshets in the latter part of July, which were followed by the entry of numbers of salmon. Later in the season freshets were much less effective. Some years later a similar experiment was conducted for the Nova Scotia government on the LaHave River by F.R. Hayes. On both rivers the effect of the freshets was not necessarily reflected in improved angling, for this became much worse in hot weather regardless of the salmon present. Also, it was found that salmon sometimes ascended the river in numbers when there was no freshet. Coupled with flooding problems created by the dams, the uncertainty over its efficacy has prevented the general adoption of this method of management.

Other Atlantic studies
Lobster studies, which were important because the lobster fishery was the greatest source of income to the inshore fishermen of the Atlantic coast, were continued. Any improvement brought about by regulations based on the station's investigations benefited the Canadian fishermen alone. Among the results obtained in 1947 was a demonstration of the value of a large size limit in certain areas (9 inches total length compared with 7 inches), in increasing the catch of market lobsters.

Studies to discover methods of oyster farming adapted to new areas were carried out at Shippegan and at Shediac in New Brunswick, and at Malagash and Orangedale in Nova Scotia. Clam studies made possible the development of a clam-farming technique suitable to Canadian conditions. Plantings on a small commercial scale were yielding promising results. Studies were continuing on the plankton organisms responsible for paralytic shellfish poisoning.

Two objects in view in the groundfish investigation were the recognition of overfishing, should it occur, and the accumulation of information that might be needed as a basis for regulating the fishery. Since the fisheries for cod, haddock, and flatfish were carried on in offshore waters as well as inshore waters, international action might become necessary for the proper regulation of the fishery. It was necessary for Canada to have exact knowledge concerning these fisheries.

Trout studies revealed great variation in the yield from lake to lake, depending mainly on differences in the fertility of the water. These ranged from less than 1 pound per acre to more than 40 pounds per acre. The possibility of increasing yields through the artificial fertilization of waters of low fertility was under investigation.

THE CENTRAL FISHERIES RESEARCH STATION AT WINNIPEG

In 1947 the station's director, K.H. Doan, reported on a survey of the Nelson and Hayes rivers in northern Manitoba, particularly the anadromous brook trout populations.

Progress was reported toward solving the whitefish infestation problem. Intensive netting of pike was conducted for consecutive seasons in a small lake in northwestern Manitoba, with a consequent reduction of the number of parasites in whitefish by half.

Research by W.M. Sprules revealed that gill nets of the size currently used were taking a disproportionately high number of immature goldeyes. Some modification of the operation of dams controlling the water levels in the marshes of the delta of the Saskatchewan River was suggested to permit the escapement of the young fish from the controlled areas.

The station cooperated with the Manitoba government in determining the optimum size of gill nets for the commercial fishery in Lake Manitoba, and studies continued on the tullibee.

It is clear from these summaries that the Board was making rapid strides in major areas. The biological stations on both coasts and the new station in Winnipeg were deeply engaged in discovering and measuring the resource. Technology was making rapid strides at the experimental stations in both Vancouver and Halifax, while at the Gaspé station the emphasis was on improving the technique of processing. Every assistance was being offered the industry for commercial exploitation of the resource, from the highly informative fish-oil bulletin of Brocklesby to the new smokehouse and air-drying units developed at Halifax.

Progress on all fronts

Resource management would inevitably flow from the discovery and measuring of the resource. This was already well advanced on the west coast, where only one other country, the United States, was involved. But it was clear from the St Andrews report of 1946 that similar resource management must inevitably be in store for the groundfish in international waters, as it already existed in the valuable lobster fishery that involved only Canadian fishermen. The Board clearly saw that international agreements would have to be reached, and it was already hard at work compiling the background information that Canada would need when it came time to attempt to reach such agreements. The horizon was widening.

15 The joint chairmanship, 1947–53

Queen's and
Toronto combine

During the years from 1947 to 1953 G.B. Reed and J.R. Dymond, from Queen's and Toronto universities respectively, guided the Board as chairman and vice-chairman. It was a period that contrasted sharply with the preceding Cameron regime, just as it would contrast with the succeeding Kask era. There were shifts in direction of the short-term goals of the Board, as well as considerable expansion, and authoritarian leadership was replaced by sweet reasonableness. The period saw the establishment of a permanent Ottawa headquarters. It also saw increased effort in assessing commercial fish species on both coasts, stimulated in part by the formation of the International Commission for the Northwest Atlantic Fisheries (ICNAF) in 1949. Attention was again given to unutilized species, fish physiology, and fish parasites, and fish behavior and fishing methods were studied seriously for the first time. Increased study was given to the quality and nutritive value of fish products, improved processing methods, and by-products.

The Reed-Dymond leadership, with the former concentrating on technological research and the latter on biological studies, was of an easygoing type under which the powers of the station directors grew, although there was some apprehension of eventual greater control by the new Ottawa headquarters. Both Reed and Dymond were long-time Board members, having served their apprenticeship as volunteer workers. Both, by all accounts, were fine scientists and fine men.

Guilford Bevil Reed

Reed was born in Nova Scotia in 1887 and received his primary and secondary education at Brunswick, Nova Scotia. He graduated in agriculture from the Nova Scotia Agricultural College at Truro in 1907, spent two years at Acadia University, and then spent two undergraduate and three postgraduate years at Harvard, where he received his B SC. *cum laude* in 1912, his MA in 1913, and his PH D in 1915. His PH D research problem concerned the respiratory enzymes and won him the Bowdoin Prize.

A.G. Huntsman (second from left), curator and director of the Atlantic Biological Station, St Andrews, 1911–33, with J.R. Dymond (left), vice-chairman, and G.B. Reed (second from right), chairman of the Board, and Stewart Bates (right), deputy minister of fisheries, on the occasion of Huntsman's retirement from the Board in 1953

Reed returned to Canada in 1915 as assistant professor of biology at Queen's under W.T. MacClement, but spent most of his time from then until 1919 on army leave as a captain in the Canadian Army Medical Corps, first as hospital bacteriologist at the Queen's University Military Hospital and later as bacteriologist for Military District no. 3. In 1919 he was appointed professor of bacteriology and head of that department at Queen's, a position he held until his death in 1955.

He was consultant in bacteriology to the Kingston General Hospital from 1919 to 1940, and consultant in dairy bacteriology to the Ontario Department of Agriculture from 1919 to 1930. In 1934–5 he went to Cambridge to investigate the problem of variation and inheritance of bacteria and to the Pasteur Institute in Paris to work on the serology of tuberculosis.

During the Second World War he was scientific consultant to the Canadian army and a member of the joint American-Canadian commission on the control of rinderpest, serving as chairman of the commission from 1944 to 1946. In 1947, the year in which he became chairman of the Board, he also became director of the Kingston Laboratory of the Defence Research Board, and from 1953, until his death, he served as its full-time superintendent. He took an active part in many scientific organizations, including membership on associated committees of the National Research Council on tuberculosis, infections, food preservation, and type cultures, and on committees of the Defence Research Board; medical advisory, special weapons, infections, and bacteriological warfare.

Reed began his long association with the Fisheries Research Board in 1921, and spent six successive summers at St Andrews investigating problems involving the canning of lobsters. During this period he carried out what was generally recognized to be the best bacteriological work done in this field prior to the opening of the Halifax station. He stimulated numerous investigators to enter the field of bacteriology and biochemistry of fish spoilage, and in this way founded those aspects of the Board's work.

John Richardson
Dymond

Dymond, the vice-chairman of the Board during Reed's chairmanship, was born in Middlesex County, Ontario, in 1887. He received his BA in biology from Toronto in 1912, together with the Victoria College Gold Medal in Natural Science. In 1920 he received his MA from Toronto and in 1950 an honorary DSC from the University of British Columbia.

After six years as botanist with the Canada Department of Agriculture, Dymond became a lecturer in zoology at the University of Toronto in 1920 and head of that department in 1948. From 1922 to 1949 he was first the secretary and later the director of the Royal Ontario Museum of Zoology.

His association with the Fisheries Research Board began in 1926 when he spent several months at Nanaimo collecting and studying the fishes of British Columbia. He went there again in 1928, and from this came the Board's Bulletin no. 32: *Trout and Other Game Fishes of British Columbia*. From 1938 to 1958 Dr Dymond was a member of the Board, and in the 1947–53 period as vice-chairman he had final responsibility for approving the channels into which increased biological research activity was directed. However, it is generally recognized that Dymond's greatest contribution to the Fisheries Research Board and to the fisheries of Canada consists in the many young people whom he influenced and stimulated while they were students at the University of Toronto. By 1968 some 40 members of the Board's staff had experienced this contact.

Ronald Hayes has said of Dymond and Reed:

Hayes compares
them

Dymond was a charming fellow with a surprising sense of humor; very friendly and unaffected, and much opposed to overorganization. I remember hearing him say once that the Board had been run with the idea of easy administration rather than with the idea of getting good science done, that we supported a fellow if he was easy to administer, but that you often run across the most interesting scientific people who are the most difficult to administer ... Anyway, I had a high respect for Dymond, although his work is not my kind of work. He was essentially a systematist, and as far as I know, he was very good at his own line.

I would regard Reed from my own prejudiced point of view as a more modern man than Dymond. Dymond I would regard as a classical biologist and Reed as a modern biologist. Reed was the opposite of Cameron in that he was an easy man to get along with, and cooperative. He didn't quarrel and he was a good scientist who gradually became snowed under by getting mixed up with the Defence Research Board in chemical warfare during the war. He ran a Defence Research Laboratory and got involved in so many things that his ability to carry out personal research gradually disappeared. But he had done some good work in his youth and had launched a good many students. He was a likable fellow. So was Dymond.

Alfred Needler has added: 'Dr Reed, of course, was an eminent bacteriologist. My impression of Reed was of a man who was very direct and who had absolute honesty and clarity of thinking. He was honest and able; of course, he wasn't full-time like later chairmen were, but he was a very good chairman. Dymond, of course, was intensely interested. I think that Dymond didn't tend to be quite as simplistic as Reed. He tended to take a more complicated, qualified approach to things than Reed would, but they were both good men and they worked well together.'

Dr L.W. Billingsley, former associate editor of Board publications, recalled a celebrated Reed mannerism: 'He would always have a lit cigarette drooping from the corner of his mouth. The ash would drop down onto his lapel, and when your eye would inevitably follow this cascade of ashes, he would absent-mindedly reach up with his hand and brush the *opposite* lapel. I never could figure out whether he did this really absent-mindedly, or just to confuse you.'

Two major factors were at work to cause the vastly increased scope of the Board's activities in the post-war years: demands of the fishing industry, and the need for factual information to support Canada's position in international fishery negotiations. Needler, who was in the middle of both developments, said in 1972:

Impetus to research

I think you might say from the years 1945 to about 1960, and even to 1963 and 1964, research was the magic work in government finance. Research, on the whole, received more assistance than anything else. It was allowed a higher rate of increase ... The activity [at St Andrews] expanded very quickly from a budget of $55,000 or so in 1941 to I suppose over half a million in 1954, maybe more than that. I don't recall exactly. That was one thing.

There was a tendency in the early days of the Board's history for industry to be very skeptical of the value of any of the research being done ... People had the feeling that they were doing research but it wasn't appreciated, or it wasn't being applied, and while they believed in what they were doing, they felt a feeling of frustration. Well, sometime during this period ... the balance swung the other way, so that industry in the early 1950s was wanting more things to be done than the Board was able to do ... This was really quite a definite change.

Meanwhile the Board was faced with some internal problems. One of them, known as 'The Revolt of the Seven,' was a carry-over from the tight money policy of the Cameron regime, which caused seven members of the Nanaimo staff to draw up a petition and forward it to the Chairman and other Board members, protesting their salary levels and the lack of any Board action to remedy them.

Nanaimo crisis

According to J.L. Hart's recollection, Earle Foerster, as director of the station, became an unwitting scapegoat for the revolt: 'Foerster had been at loggerheads with the headquarters people, with both Sutherland and Cameron. Cameron's constitution meant he was tight with money, and Sutherland wasn't always completely in command of what was going on ... By this time Foerster had also alienated part of the B.C. industry, and I don't think he was getting any very effective support from Mr Whitmore in the Vancouver office of the Department of Fisheries. So, with all this in mind, I am sure that the Board was anxious to replace him as a director.'

Foerster, for his part, was by this time glad enough to be relieved of a director's frustrations. He recalled developments during that period as follows:

I had a great deal to do with Sutherland. I guess we had quite a few fights from time to time over this and that. He was a typical Easterner, I would say, and he liked to argue, but he was never offensive at all. If you put up a good case, he always saw that you got justice. And he had a fine assistant, Nora Grimes, a very fine girl, and very loyal to all the directors and also to her boss ...

It was pretty grim during the first four or five years of my directorship. I tried to hang on to what staff I could and also keep the work going as well as possible. It was my fervent desire to devote most money to research and as little as possible to administration in general. And sometimes that made me very unpopular. But I think the members of the staff realized that,

185

after all, our function was research, and as much money as possible should go to that, even if we didn't have the facilities to do some of the field work that we wanted to do. But towards the end of the period funds became available for personnel, and we were able to take on quite good, quite competent scientists who were leaving the armed services. I can't recall now how many new men we took on, but there were quite a number. They all proved very loyal, effective, and worthwhile scientists, and have carried on most nobly, not only during the period of my directorship, but since then.

Foerster gave up the directorship, he said, 'largely [because of] the fact that I had to devote all my time to administration, and I wasn't too keen on that type of work. I don't think I was too effective. And since I preferred to do research work and the Board gave me the opportunity of giving up the directorship and continuing my research, I decided to take this step, and I've never been sorry.'

What Hart did
Hart, who succeeded Foerster as the Nanaimo director in 1950, was asked what he considered to be his main accomplishments in the four years that he was at Nanaimo: 'I don't know, perhaps my single greatest accomplishment was when I ... refused to close off the investigation at Babine Lake, the year they had the Babine slide.' He was under pressure to do this because money was needed for a tagging program in Johnstone Strait. Events proved him right, however, because the Babine work was necessary to assess the damage done by the rock slide and the success of the remedial measures taken, and in later years it demonstrated the opportunity for increased production from artificial spawning facilities.

About other accomplishments, Hart said:

One thing I did do, I hope, was to establish ... a system within the station in which staff members and investigation heads could get a decision. Of course I think this was one of the chief difficulties under Earle's regime. They could not get decisions. I think people got decisions from me fairly quickly, and I worked fairly hard at this. I worked extremely hard on one thing, and that was to get out of the station the backlog of manuscripts that were sort of constipating the whole scientific output. In that, of course, I had one very great advantage in having Bill Ricker available in the office across the hall from me. Bill did almost heroic things in getting second-grade manuscripts up to at least A minus, and with a minimum ... of acknowledgment or credit to himself.

Newfoundland joins the team
In the same year that Hart assumed the Nanaimo directorship, the fisheries activities in Newfoundland became a concern of the Board following Confederation. Work had been in progress there for some time. Under an agreement between the British Empire Marketing Board and the Newfoundland government, the Scottish biologist Harold Thompson made a survey of fishery possibilities in Newfoundland in 1930. He recommended a scheme of biological and technological fisheries research to be carried out from a center in Bay Bulls near St John's. This was accepted, and on 1 April 1931 the Bay Bulls Laboratory began operations in the upper story of a fish plant. Thompson was director, and it operated under the control of the Newfoundland Fishery Research Commission appointed in August 1930. The trawler *Cape Agulhas* was chartered from time to time for work at sea. Research was carried out on Atlantic salmon by Sheila T. Lindsay and Thompson, on cod and haddock by Thompson, on fish eggs and larvae and other plankton by Nancy Frost, on capelin by George E. Sleggs of Memorial University College, and on cod salting and drying and cod liver oil by Norman L. MacPherson.

Thompson resigned in the autumn of 1936 to become director of Fisheries Research in Australia. The Bay Bulls Laboratory was destroyed by fire in April

Pacific Biological Station directors, R.E. Foerster (seated), 1940–9, and J.L. Hart, 1950–4

1937, with the loss of many records and much material. By that time most of the original staff had gone elsewhere; the remainder moved to St John's, where they were soon joined by Arthur A. Blair, working with salmon. In 1940 the laboratory secured space in a newly built Newfoundland government laboratory on Water Street, which it shared with the Public Health and Analytical Chemistry Laboratories. William F. Hampton was acting director of the Newfoundland Fisheries Laboratory in 1937–43, and in 1944 Wilfred Templeman became director. He resumed groundfish research in 1946 when the 82-foot *Investigator II* was launched, and Noel Tibbo was recruited for herring studies.

After Confederation the fisheries portion of the government laboratory became the St John's Biological Station of the Fisheries Research Board. The laboratory building was bought by the Board, and after about five years the analysts were finally provided with new quarters and the station was able to use all of the space that had been purchased. Templeman foresaw a major expansion of trawling, both absolutely and relative to the traditional trap cod fisheries, so a major part of the

Meeting of the North American Council on Fishery Investigations, held at the Atlantic Biological Station, St Andrews, 13–14 September 1933. From the left: A.G. Huntsman, director of the station; J.P. McMurrich, chairman of the Biological Board; Harold Thompson, director of the Newfoundland Fishery Research Commission; H.B. Bigelow, director of the Woods Hole Oceanographic Institution and chairman of NACFI; W.A. Found, deputy minister of fisheries, Ottawa

The Newfoundland Fishery Research Laboratory, Bay Bulls, 1931–6. The laboratory formed part of the building in the centre of the picture

laboratory's work from 1946 onward was directed at mapping the distribution of potentially valuable species like American plaice and redfish, as well as the better-known cod and haddock. This work was stepped up after 1949, partly in cooperation with the Industrial Development Service, and many new fishable concentrations were revealed, especially along the edges of the banks. An annual series of oceanographic observations was instituted across the Grand Bank, which indicated the temperature and extent of the cold arctic water that the fish respond to, and is particularly valuable because it has been maintained consistently.

Templeman was a Newfoundlander by birth, having been born in Bonavista in 1908. He attended the Memorial University College of Newfoundland for two years and then went to Dalhousie University, where he graduated with the BA degree in 1930. He then worked at the Atlantic Biological Station on lobsters, which provided thesis material for a PHD from Toronto in 1933. After a period at McGill University as lecturer in zoology in 1934–6, he returned to Newfoundland as professor and head of the Department of Biology at Memorial. During his tenure there he also did research at the government laboratory at Bay Bulls and later in St John's before he left the university to become director at St John's.

Wilfred Templeman

A prodigious worker, 'Temp' spared neither himself nor his staff in his zeal to discover and evaluate actual and potentially commercial fish stocks all around Newfoundland and Labrador. One result was that he became a well-known figure at ICNAF meetings and on its committees. He was also interested in the unusual, and reported many new or scarce fish taken by trawlers in Newfoundland waters. Later, after retiring from the Board at age 65, he moved back to Memorial as Paton Professor of Marine Biology and Fisheries, where he still continues with his killing schedule – on which, however, he seems to flourish like the green bay tree.

Dr Templeman accounts for his consuming interest in fishery science as follows: 'I believe I was prejudiced towards it from the beginning of my interest in biology. I had grown up in Bonavista, where the whole work of all the people was fisheries, and still is pretty much so. My father and all my relatives for two or three hundred years have been interested in fisheries, and I did some fishing with them in a small open boat. Then, after I got interested in biology at Memorial and graduated from Dalhousie in 1930, I was offered a scholarship by the research board to begin work on lobsters, working out of St Andrews.'

Regarding his laboratory's metamorphosis into an FRB station, Templeman observed: 'I don't think there were any great problems at that time ... I think I attended a Board meeting the previous January in Ottawa, when we were to become part of Canada on the first of April ... There was some attempt to make us a substation of St Andrews, but this got some rather violent opposition from me, and it didn't work out.

Alfred Needler got his first taste of the Ottawa scene during this period. He related the circumstances:

Needler goes to Ottawa

When Finn left the deputy ministership in 1946 and went to FAO as director of the Fisheries Department, he was succeeded by Stewart Bates, who was a strong and brilliant man, really one of the most brilliant that I ever worked with. Bates took the job of deputy minister of fisheries very seriously. He also kept his finger in some of the top-level economic thinking of the government in one way and another. But he took this seriously and he felt a lack of anybody whom he could rely on, who really knew the Atlantic coast. He tried to bring in people from the field who were a little bit stronger ... I think he felt he needed someone who would know the Atlantic picture, and who would also know the ... ICNAF side of it. So he asked me to go up to Ottawa ...

The Fisheries Research Board station at St John's, 1949

Hauling capelin trap at St John's

I started in the job of assistant deputy minister at the beginning of April 1948. I was in it for two years, but I remained officially the director of St Andrews, and spent about a quarter of my time back there. An assistant deputy minister's job has all the disadvantages and none of the advantages of the deputy minister's job. It tends to be sort of an adjutant's job, you know. But I did it because Bates wanted me to. I said I would only do it for two years. A lot of people said, 'Oh, well, once you're there, you will stay.' I did stay six months longer, as a member of the FRB headquarters unit, and then I came back, dropping back from a munificent salary of $8,000 per year to $5,500.

Between 1947 and 1952, the growth in size of the annual reports gave visible evidence of the increased station activities attendant on increased budget and staff. Over that period the report tripled in size. In 1947 the Chairman's report summarized activities at each station. Thus, it touched on the eight major investigations and several minor studies in progress at St Andrews. These covered lobsters, oysters, soft-shelled clams, smelts, groundfish, and brook trout. The work of the Atlantic Herring Investigation Committee was reviewed, and Huntsman's Atlantic salmon investigation was reported. At the Atlantic Fisheries Experimental Station four major studies were in progress: quality control of fresh and frozen fish, salt fish drying, fish smoking, and fish oils. At the Gaspé station there were two main studies: Gaspé-cure cod and seal oil. Similarly, at the Central Station in Winnipeg, there were two main studies: whitefish infestation and the life history of the goldeye. The northwest surveys were proceeding. At the Pacific Biological Station the four major studies were the Skeena River salmon investigation, the general salmon investigation, the herring investigation, and the trawl fishery investigation. At the Pacific Fisheries Experimental Station the major work was on sanitation of vessels and fish plants, refrigeration, antioxidants, smoking, fish scrap recovery, oils and vitamin oils, and processing freshwater fish.

Station activities summarized

Reed also touched briefly on the Joint Committee on Oceanography; the Board's advisory committees, which had been functioning for years on both coasts; the Canadian Committee on Food Preservation; FAO, which had held its third meeting in Geneva, attended by Dr D.G. Wilder of the St Andrews station as a member of the Canadian delegation; and the Board's property, which was going through a process of enlargement to accommodate the growing staff and increased tasks.

In 1948 the formation of a new committee was announced: the Canadian Committee on Freshwater Fisheries Research. Sponsored by the Board, it had the task of carrying on and extending some of the functions of the National Committee on Fish Culture, which had been disbanded in 1946. Some 40 members of the committee, representing universities and fisheries departments from most of the provinces, the federal departments of Fisheries and of Mines and Resources, and a number of research organizations interested in one way or another in freshwater fisheries research, met in Ottawa on 3 January 1948. The organization has met annually ever since. It has served an obvious need by bringing together fisheries biologists from universities, the provinces, the Board, and consulting firms. The 1974 attendance was more than 300 scientists and administrators. Over the years *marine* fisheries research came to be included in its agenda, and its name was finally changed to Canadian Conference For Fisheries Research (CCFFR).

A significant new committee

Another new organization was formed in 1949: the International Commission for the Northwest Atlantic Fisheries (ICNAF). It was to play a most important role in future Board activities, and its first headquarters was at the St Andrews station. Its history and accomplishments are described in a later chapter (chapter 19), but this is a good place to introduce its first executive secretary, W.R. Martin, who was to play an important role in the Board's future administration.

William Robert
Martin
Bob Martin was born in 1916, the son of a chemistry professor at the University of Toronto. He was influenced by his father's broad interest in science to enter an honors science course at that university in 1934. He became particularly interested in biology and in his second year came in contact with the Ontario Fisheries Research Laboratory. He spent four successive summers in Algonquin Park working with the laboratory. In his words: 'I was particularly stimulated by the contact with people such as J.R. Dymond and Bill Harkness, and perhaps especially by Fred Fry. Through the years Fred has been most stimulating in encouraging young fisheries scientists to carry through with their undergraduate and graduate work, and they have scattered throughout North America making their own mark.'

Martin stayed at the University of Toronto an additional year to complete his master's thesis under J.R. Dymond. His subject was the arctic char. He then went to the University of Michigan,

which was really a focal point for fisheries work in North America. Carl Hubbs was operating from the museum in Ann Arbor and they had strong people in the Zoology Department and in the State of Michigan, and the federal government of the United States operated a lab at Ann Arbor. All of these people who were concentrated in Ann Arbor attracted me to go there as a graduate student. So I registered with Carl Hubbs, who already had a large number of graduate students: I think there were about 20 of us, all studying various things under his leadership. I took on the job of studying salmonids, and I worked during the summer months at a hatchery in northern Michigan. I grew young salmonids at different temperatures and kept track of the changes in these animals; this led to a thesis on environmental control of body form in fishes that eventually was published in 1949.

The thesis won him a PH D degree.

Meanwhile the Second World War had broken out, and in the autumn of 1941 Martin joined the clinical investigation unit of the Canadian Air Force. His particular work involved the development of anti-gravity devices for airmen: 'The approach taken was the construction of a human centrifuge at our main base in Toronto, and I worked in that lab for a year as a scientist and as a guinea pig. Being tall – well over six feet – I was a beautiful guinea pig, in that I would black out at about 3 G's, whereas the normal person would black out at 5 G's, so I ended up being the most blacked-out person in North America.'

In 1945, perhaps feeling that he had been sufficiently blacked out, Bob joined the Board's staff at St Andrews to head up the groundfish program. At that time there were still stocks of a number of species that were unused or used to much less than capacity. Much of his energy went into introducing new fishing techniques like mechanized longlining and Danish seining, and he recommended that small trawlers be allowed to catch flounders in inshore waters, and so on. In 1948 he attended a meeting of the International Council for the Exploration of the Sea in Copenhagen, and visited European research stations.

This background made Bob a 'natural' for the job of interim executive secretary of ICNAF. When that was over he went back to research at St Andrews, but in 1963 Kask selected him as his assistant chairman in Ottawa, and he eventually held that post under four successive chairmen.

Ricker once said of Martin: 'He's a classical worrywart; that is his strength. He gets everything down on paper that is to be dealt with, and he sees that it's done. He's one of the best men around in a managerial position.'

Guilford Reed felt that the link with the universities was important. Noting that they were the homes of fundamental science, he wrote in his 1949 report:

From the inception of the Board in 1898 until 1934 there were intimate relations between the Board and the universities. Facilities were available whereby university professors and their students engaged in research at the Board's Stations and there was close contact between many Canadian biologists and the work of the Board. It is believed that this close relationship between academic biology and the Board had much to do with the standing of the Board as a scientific organization. For a variety of reasons the close relationship between universities and the Board has not continued although representatives of nine universities are still members of the Board.

In the highest interests of fisheries research it is necessary that fundamental research in aquatic biology be adequately supported. Some of this should be done in the Board's Biological Stations but most of it will have to be done in universities. It is desirable too that university personnel be encouraged to undertake research in the Board's Stations. Unless fundamental research is more adequately encouraged, the quality of applied research will suffer.

Welcome mat restored for universities

However, there was no move to re-establish the residences for visitors at the biological stations, and indeed university people had become less inclined to work in summer without remuneration, especially when surrounded by staff drawing regular pay-cheques. What did happen was that a number of professors were hired seasonally for special projects, some of them of their own choosing, and many students were employed in field or laboratory work of various types.

The 1950 Annual Report noted: 'At the beginning of this year a new arrangement was made to facilitate administration of the Board's work. Two senior men, an engineer and a biologist from the Board's scientific personnel, were stationed at headquarters in Ottawa. These men act as scientific consultants to the Department of Fisheries, as liaison officers between the Board members and the scientific staff and, together with the Executive Director, as an administrative unit.'

Headquarters unit formed

In 1950 D.H. Sutherland was still secretary and executive director. First Alfred Needler and then C.J. Kerswill was seconded from St Andrews, and Otto Young was seconded from Vancouver, to form the headquarters unit with Sutherland. I.E. Turner was honorary treasurer, and the invaluable Nora Grimes was assistant secretary. Otto Young has described his transfer:

It happened something like this: Professor Dymond and Dr Reed visited our station here in Vancouver, I think it was in late 1949, and intimated that there would have to be a change in the headquarters unit. They intimated that the minister of fisheries was demanding that more knowledge be available in Ottawa so that he wouldn't have to get on the phone and get in touch with a station, either east or west, to get an answer to a question that was brought up in the House the day before on some fishery problem, which could be biological, could be bacteriological, could be engineering, or whatever …

They were toying with the idea of setting up a larger headquarters unit … They'd start off with, say, a senior biologist, who would be head of the unit, and then there would be a technologist who would come a year later. There would be a two-year period of headquarters stint, and when the biologist was replaced by another biologist, the technologist would be head of the unit for a year, and he would be replaced and it would go on that way. They started off by proposing that Dr Needler, who was then assistant deputy minister in Ottawa, would be the senior man on the biological side, and they were looking for the technologist … and they asked me what I thought of it.

Well, I was completely ignorant of what went on in Ottawa. I hadn't even visited the Commons to hear the ... arguments that took place, but I could see the logic of the thing, and they intimated that the Board's future sort of depended on this, that the Board was in a critical state, and there was dissatisfaction creeping in from the departmental side, so they asked me if I would give it some thought. I didn't have to answer right away, but eventually the idea was approved at the annual meeting in the winter of 1949–50, and the upshot of it was that I went to Ottawa, supposedly for two years, starting in July 1950.

The following year Sutherland resigned from the Board. The headquarters unit then consisted of Otto Young as chairman, C.J. Kerswill, senior scientist, and H.A. Wilson, executive assistant, with Nora Grimes as assistant secretary. Also attached to the headquarters staff but stationed at Nanaimo was the new editor of publications, W.E. Ricker.

Full-time chairman
forecast The last annual report of the Reed-Dymond period, in 1952, predicted the appointment of a full-time chairman: 'In the last ten years the work of the Board has greatly enlarged in volume of scientific and industrial work, in the number of employed personnel and in the amount of money involved. It will very soon be necessary to make some changes in organization and to consider the employment of a full-time Chairman.'

Alfred Needler recalled how the idea of a full-time chairman developed: 'I think the idea developed during the time when Reed and Dymond were chairmen ... because they realized that some full-time man was needed. And outside people like the deputy minister and others must have felt this way. The deputy minister of those days was Bates, who was a very keen person and who acted as a member of the Board before he was deputy minister. So the idea was certainly ripe during the latter part of the time when Reed and Dymond were chairmen. As a matter of fact I actually was asked at one stage whether I would be interested in being full-time chairman of the Board. It wasn't a definite offer, but it was an inquiry. That was a year or so before Kask was appointed.'

Needler said that he personally was not opposed to the idea of a full-time chairman, but he felt that there was some opposition on the part of station directors: 'They felt that if you made headquarters too strong, they would lose some of their freedom of action. But in those days, the administration of the Board at the time of Cameron and Reed and Dymond, the actual day-to-day administration, was carried on by Don Sutherland, who was assistant deputy minister and secretary of the Board. And Sutherland was a very able administrator who wouldn't hesitate to make a decision and carry it out. A very easy person for the directors to deal with. Whether he had the authority or not, he presumed he could get it, and would say, "Well, if that's the problem, we'll do this and this," even over the phone, and it would happen.'

The growth of the Board, calling for the services of a full-time chairman, was pointed up by the budget estimate for 1953–4 of nearly $2,000,000 and a staff of 379 paid employees, plus a contingent of 77 students and others employed during the summer season.

MANAGEMENT OF THE FISHERIES

The last report of the Reed-Dymond leadership, for 1952, summarized activities in biology, oceanography, and technology. Concerning biology, it stated:

The work of the Biological Stations is concerned with the living fish – with conditions

194

A modern operating table used for maintaining fish under anesthesia while surgical procedures are carried out. A pump circulates dissolved anesthetic over the gills

A modern respirometer for determining the relation of the swimming speed of fish to the water temperature

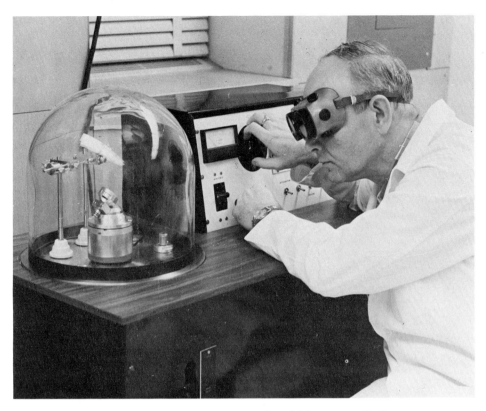

Goldplating specimens for electron microscopy using a high-vacuum shadow-caster

determining the size of their populations, with their movements and distribution, with the extent to which they may be parasitized or rendered unsuitable as human food, with discovering previously unknown stocks and with devising testing and demonstrating more efficient fishing methods ...

Basic to fisheries management is a knowledge of the size and age-class composition of the stocks of all the important commercial species. Getting this information may be compared to the counting of calves, yearlings, two-year-olds and other age classes of a farmer's cattle. Without such information the farmer would not only be ignorant of what he would have for sale in the future, but he would not know whether a failure of a particular age-class resulted from too small breeding stock or from poor survival of young due to bad weather, insufficient food, disease, or other condition.

Fisheries research has shown that in most species there is often an enormous difference in the success of different year-classes. Some years produce up to 60 or more times as many individuals of a species as other years. It is not infrequent for a particular year-class to be a virtual failure. When such failures occur for a number of years in succession there is poor fishing for that species until a successful year-class arrives ...

Population assessments the key

By means of annual population assessments the fishery biologist can not only explain the character of the fishery but is in a position to estimate the probable number and size of fish to be expected in the next fishing season. Their chief value however is in making it possible to take the maximum number of fish without risk of endangering future prospects. An economic loss occurs if fishing is restricted in the belief that decreased abundance is due to overfishing when it really results from natural causes.

The problem of making annual population assessments of fishing stocks is complicated by the fact that there are usually several distinct populations of most species. Thus there are at least seven populations of herring off our Pacific coast, among which there is relatively little mixture. The separateness of these populations has been demonstrated not only by study of their physical characteristics but by tagging. There are several distinct stocks of cod, haddock and herring off our Atlantic coast. Since it is possible to deplete or underfish one without affecting another, in the same way each population must be separately assessed.

In the case of salmon which return to fresh water to spawn, the size of the population is assessed by estimating the number of fish spawning in different river systems. This is usually done by officers of the Department of Fisheries. The Board contributes to the accuracy of the estimates by maintaining a fence-like series of traps across some streams where every fish entering or leaving can be accurately counted.

Since it is impossible at present to increase the productive capacity of the sea, increase in the value of marine fisheries can be accomplished through taking more of the stock so far as this can be done without interfering with future supplies, through locating unused stocks, and through increasing the efficiency of fishing.

Through exploratory fishing new stocks of fish are being discovered. For instance, during the past year the Board, in cooperation with the Department, located large stocks of cod in water 100 to 200 fathoms deep along a stretch of about 250 miles along the northeast coast of Newfoundland. Ninety percent of this stock had not previously been fished. In previous years large quantities of American plaice and cod had been found down the eastern edge of the Grand Bank, and rosefish in Hermitage Bay. In the Gulf of St. Lawrence the existence of populations of large fat herring was confirmed by fishing operations during 1952 ...

One of the ways in which the work of the Biological Stations is adding to the value of the fisheries is through increasing the efficiency of fishing. Experiments on new methods of fishing often go hand in hand with exploration.

The introduction of long-lining with powered haulers is one of the improvements successfully initiated through the work of the Board in cooperation with the Department. The size and type of vessels for use in fishing the newly discovered stocks of fish in Newfoundland waters is another finding of experimental fishing. Danish seining, as pointed out in the report of the St. Andrews Station, was found to have limited applicability because it needs smooth bottoms. It has however proved its commercial worth in limited areas.

The method of harvesting scallops on a bed discovered in the Gulf of St. Lawrence was that devised as a result of trials with several types of gear.

Most of the Board's work has as its object the making of contributions to the management of the fisheries in the more or less immediate future, but some work of a more fundamental character constitutes part of its program.

The control of some aspects of nature requires so much research that if advances are to be made in the foreseeable future, researches must be undertaken now and consistently prosecuted until success is achieved.

The report of the Pacific Biological Station records the initiation of such researches in the behaviour of salmon. Already striking results of a preliminary nature have been obtained which suggest that it may be possible to devise means for obviating some of the danger to the future of these fish inherent in the construction of power dams on salmon rivers.

Many of the studies at the Central Station on the parasite which interferes with the export of whitefish to the United States are of a basic nature. Other researches on parasites at the Biological Stations are bound to be of a long-term nature, although it is hoped that some results which may emerge as the studies proceed may be applied to the improvement of the fisheries.

Some of the investigations on the Atlantic and Pacific salmon as well as on other species are designed to throw light on the factors which limit production.

Exploration for
unused stocks

Improvement of
fishing methods

Basic and long-term
research

197

The *Calanus* at Frobisher Bay

EASTERN ARCTIC INVESTIGATIONS

The First World War had made it impossible to continue the Canadian Fisheries Expedition into arctic waters, so the seas and their inhabitants adjacent to northern Quebec and Baffin Island remained largely unknown. What was needed was someone knowledgeable and enthusiastic about these northern waters, and after the Second World War such an individual turned up in the person of Professor M.J. Dunbar of McGill University.

Max Dunbar came from Edinburgh via Oxford, and received the PHD degree at McGill in 1941. Always interested in the Arctic, for five years he was Canadian consul in Greenland, returning to Montreal in 1946. There he immediately conceived a plan for biological and oceanographic studies in the north, and persuaded

The *Mallotus*, 1951–

the Board to support it. A small diesel ketch, the *Calanus*, was built with something of the strength of Nansen's *Fram* in order to cope with ice conditions. This ship remained in the Arctic from 1948 until 1954, when it was brought to Montreal for a refit. Usually it was hauled out on shore during winter, but on one occasion it was frozen in the ice at Igloolik with a wintering party of two men on board, to make year-round oceanographic observations.

Dunbar trained a small group of students in arctic studies, notably E.H. Grainger, A.W. Mansfield, and I.A. McLaren. Their operations covered the seas around Baffin Island from Cumberland Sound to Foxe Basin, also Ungava Bay and the northern half of Hudson Bay. The ecology and local utilization of ringed seal and walrus were studied in detail, as well as fishes and invertebrates. In 1954 the Eastern Arctic Investigations became the Arctic Unit of the Board, with H.D. Fisher as full-time director, but Dunbar continued to work with it for a number of years afterward.

TROUT STUDIES

The inland fisheries of the Maritime Provinces remained under federal control even after their management had been delegated to the provincial administrations elsewhere. Consequently the Board maintained a trout research program at the St Andrews station. During the 1940s and early 1950s this was focused on small nutrient-poor lakes in southern New Brunswick, under the care of M.W. Smith and

Fertilization and predator control

199

J.W. Saunders. Fertilization of such lakes greatly increased trout growth rates, but the resulting larger stocks attracted additional loons, mergansers, herons, kingfishers, mink, and otters, which harvested the increased supply of trout very efficiently. In order to divert any considerable proportion of the increased production into anglers' creels it was necessary to have someone continually on hand to destroy or scare off these fish-eaters. Eels too ate the smaller trout, and could be controlled, over a period of years, by trapping adults and by preventing elvers from entering the lake.

Later the trout work was extended to the much more fertile waters of Prince Edward Island. In addition to the production in fresh waters, many of the Island trout move out to sea in summer where they grow rapidly and on their return add greatly to the stock available for angling.

SALMON STUDIES

Spawning channels

One aspect of the pink and chum salmon investigations of the late 1940s and early 1950s was to evaluate the success of spawning of these species at several sites, notably Nile Creek on Vancouver Island and Hook Nose Creek on King Island near Bella Bella. Begun by A.L. Pritchard, the work was carried forward by Ferris Neave, W.P. Wickett, J.G. Hunter, and others. In all cases there was a large loss of potential production between the time eggs were deposited and the fry began their seaward migration. Natural survival from eggs in females to emerging fry was never more than 30 percent, and averaged about 10 percent. The process of egg deposition and fertilization was quite efficient, but the eggs subsequently were exposed to floods, droughts, frost, suffocation, fungus attack, and predation by insects, to varying degrees at different sites and in different years. After emerging from the gravel the fry moved quickly downstream, but nevertheless up to two-thirds of them were captured by trout and sculpins, whose stomachs were found to be crammed full of young salmon during the migration period.

To reduce losses in the gravel a two-pronged attack was launched. The hydraulics of water movement through gravel was studied by Wickett and L.D.B. Terhune in troughs and in nature, using dyes introduced at different depths, and best gravel sizes for efficient aeration were determined. To avoid disastrous flooding, a controlled-flow spawning channel was constructed at Nile Creek, and it showed greatly improved egg survival over natural spawning beds. Subsequently this technique progressed gradually from the experimental to the proven stage in successive channels constructed by the Resource Development branch at Jones Creek, Great Central Lake, Big Qualicum River, and two tributaries of Babine Lake. The same plan was adopted by the Salmon Commission at Seton Creek in the Fraser system in order to compensate partially for the destruction of pink salmon spawning beds by a hydroelectric development, and subsequently it built channels alongside four sockeye salmon spawning streams.

There are still problems of silting and algal growth in some channels, and the best means for ensuring adequate movement of water through the gravel are still being studied. Sometimes, too, it is difficult to get the fish to distribute their nests more or less evenly throughout the spawning area. However, fry productions of 50 to 80 percent of the eggs in females are regularly obtained, a two- to five-fold improvement over good natural sites.

The Babine River slide

Sometime early in 1951 the side of a hill about 300 feet high crashed into the Babine River at a point 40 miles below Babine Lake. No one saw it happen, and its existence was not suspected until salmon were observed to be late in arriving at the

Babine fence that year, many in poor condition. According to A.L. Pritchard, the slide 'occurred at a time and place where it could cause maximum damage. A saboteur with unlimited facilities could scarcely have done better.' Nothing could be done to help the fish in 1951, but the Conservation and Development Service immediately began to build a road in to the site, a distance of 60 miles. In 1952 it constructed temporary works that helped some of the fish to pass the obstruction, and planned the permanent removal of rock, which was accomplished the following winter. At the same time a tagging program and other studies were carried out, in cooperation with the Board, at the slide and at Babine Lake, which provided concrete estimates of the damage done. The Board's Babine fence played a key role in this assessment.

The report on the biological aspects of the project, by H. Godfrey, W.R. Hourston, J.W. Stokes, and F.C. Withler, showed that in both of the slide years about two-thirds of the sockeye escapement failed to reach Babine Lake. Some of those that did get there either died before spawning or spawned incompletely. Pink salmon, though never very numerous in the Babine River, were even more seriously affected. The effects of the slide were not permanent, but very damaging at the time. After the obstruction was removed, with adequate protection from fishing, the progeny of the year-classes affected regained a normal level of abundance within two generations.

THE OCEANOGRAPHIC GROUPS

Reed's 1952 report commented on oceanography as follows: 'At the end of the last war a Joint Committee on Oceanography was organized by the Royal Canadian Navy, the National Research Council and the Fisheries Research Board. Up to the present time the Navy has supplied ships and the Fisheries Research Board the scientific personnel. As indicated in the accompanying reports, a large body of oceanographic data has been accumulated. The personnel of the Joint Committee on Oceanography consists of: Dr G.S. Field, Chairman, Defence Research Board, representing the Navy. Dr G.B. Reed, Fisheries Research Board. (Mr O.C. Young, Fisheries Research Board, Alternate member). Dr T.D. Northwood, National Research Council. Mr F.C.G. Smith, Hydrographic Survey.'

This brief statement conceals a major development in the Board's operations. When Board oceanographers were again free to pursue peacetime goals, those at the two major biological stations were organized into two 'groups.' The Pacific Oceanographic Group (POG) was headed by Jack Tully. He recruited a band of recent graduates that at one time or another included F.G. Barber, A.J. Dodimead, L.A.E. Doe, R.L. Fjarlie, N.P. Fofonoff, G.R. Harris, B.S. MacKay, R.A. Pollard, J.A. Shand, Susumu Tabata, Michael Waldichuk, and R.J. Waldie. Oceanographers at the University of British Columbia were supported seasonally, notably W.M. Cameron and G.L. Pickard. The Atlantic Oceanographic Group (AOG) was headed by Harry Hachey, who was also chief oceanographer for the joint committee. Its staff at various times included W.B. Bailey, F.D. Forgeron, H.J. McLellan, and, seasonally, D.G. MacGregor, N.J. Campbell, A.E.H. Collin, and R.W. Trites moved from one group to the other, and St Andrews had another oceanographer, L.M. Lauzier, who had been seconded by Quebec to the Herring Committee and later joined the station's staff.

The work of the two groups, the St John's station, and cooperating agencies greatly improved our understanding of the seasonal and year-to-year changes in the ocean climate along our coasts. 'Ocean climates' is a better expression, for there

Launching of framework and flotation chamber for a large, spherical, plastic container used for experiments on the basis of production of fish food organisms. The sphere allowed plankton to develop under almost natural conditions, but isolated from the confounding effects of changes due to water movements. Pacific Biological Station, circa 1960

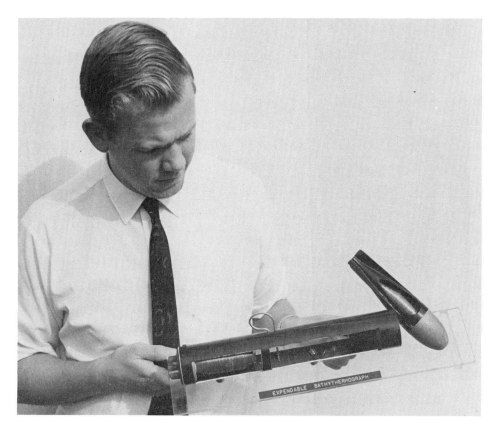

An expendable bathythermograph, which is launched from ships travelling at full speed. As it sinks through the water it sends temperature information back through the wire to a recorder aboard ship

The two rocket-like instruments are recording current meters. In use, they are attached to anchored buoys and measure tidal and other currents in the sea

Measurements at depth

A 'technician' (Dr Tarr) places a special continuous extracting head on a refrigerated centrifuge used for collection of bacteria, phytoplankters, etc. where large volumes of liquid are involved

are tremendous differences between, for example, the Bay of Fundy and the Gulf of St Lawrence, not to mention the peculiar characteristics of smaller bays and fjords. Special attention was given to Canso Strait, which separates Cape Breton from mainland Nova Scotia, to predict changes that would result from construction of a causeway, and also to Vancouver and Nanaimo harbors with reference to sewage outfalls. POG put effort into extending the network of monitoring stations for seawater temperature and salinity, usually at lighthouses, and the new weathership at 50°N, 145°W, provided an excellent platform for continuous observations of offshore waters.

THE TASKS OF TECHNOLOGY

After the war the experimental stations were able to increase their staffs as rapidly as did the biological. Under S.A. Beatty's aggressive direction the Halifax station had 32 scientists on strength by 1952, the leaders among whom included bacteriologist C.H. Castell; biochemists W.J. Dyer and E.G. Bligh; chemists P.L. Hoogland and A. Guttmann; engineers M.A. Foley, W.A. MacCallum, and A.L. Wood; organic chemists F.A. Vandenheuvel and R.G. Ackman; physical chemists J.R. Dingle and D.G. Ellis. The Gaspé station, headed by Aristide Nadeau, included bacteriologists Henri Fougère, H.-P. Dussault, and R.-A. Lachance; chemists L.-C. Dugal and A. Cardin; engineer Rosaire Legendre; and two others. Neal Carter's Vancouver station had 14 scientists, including bacteriologists H.L.A. Tarr and Burnette Southcott; biochemist R.A. MacLeod; chemists L.A. Swain, N.E. Cooke, and P.J. Schmidt; and engineers A.W. Lantz, J.S.M. Harrison, and S.W. Roach.

Chairman Reed's report for 1952 defined the tasks of this constellation of talent as follows:

The purpose of the work carried out at the three Experimental Stations is to improve the quality of fishery food products, to improve methods of processing and preserving fish, to study the nutritive value of fish and fish products, and to increase the range and value of fishery by-products.

Ordinarily two types of approaches are made. In some instances it is only necessary to find means of applying existing knowledge to solve an industrial problem. In the more involved problems it is generally necessary to fill in important gaps in fundamental knowledge before any industrial application can be made.

In the case of fresh and frozen fish it has long been known that the flesh of living fish is sterile and remains free of bacteria for a short period. The problem of quality is primarily one of reducing the penetration and multiplication of bacteria in the flesh. The reports indicate that much has been done and is being done in these respects. It is shown that some current procedures have little beneficial effect and that other proposed procedures have a marked effect. With the best procedures so far devised, the number of bacteria in fillets at the end of the processing operation may be reduced some 90% and the storage life of the fillet doubled as compared with less efficient methods. These results have been obtained by pooling the work of bacteriologists, biochemists and engineers.

How to improve quality

Similar problems arise in the preparation of frozen fish, but in addition the freezing produces changes in the muscle protein. Work is now in progress on the physical chemistry of these proteins with the expectation that a wider knowledge of the nature of the reactions will provide a better basis for commercial procedures.

The experimental results provide new evidence of the necessity for high quality at the time of freezing, with some evidence of advantages of freezing at sea as one solution of the quality problem. Further evidence is also brought forward of the advantage of sub-zero temperatures for transportation and storage, especially if storage is for more than a few weeks.

Much has been done in recent years to improve the quality and economy of salting and drying fish. Previous reports have indicated the wide application which has been made of these newer procedures. Particular attention has been given during the last two years to the problem of light salted fish, as this is a highly important product, particularly on the Gaspé coast and in Newfoundland. The light salting introduces problems not met with in normal heavy salting ... difficulties are gradually being resolved. Mechanical drying of light salted fish also introduces serious problems. The results indicate that this has been done successfully on an experimental-commercial scale. Difficulties remain, but the principal problem has been solved.

Vitamin studies   Considerable attention has been given to the nutritive value of fish. During the last two years particular attention has been given to the content of vitamin B-12 and other growth factors in fish meals. The results ... of detailed feeding experiments with chicks indicate that fish meals, solubles and related products are rich in this group of growth factors. It is, however, apparent that the method of preparation of the meals influences the vitamin content. The report also indicates that a large amount of work is in progress on the origin and nature of the B-12 group of substances.

Until the last few years, extensive work in the Board's laboratories on marine oils was primarily concerned with medicinal oils. With the development of synthetic oil soluble vitamins, emphasis has been transferred to methods of preparing edible and industrial oils. While valuable *ad hoc* results have been obtained, it was apparent three to four years ago that fundamental studies in oil chemistry were essential to progress in these newer directions ... Good progress is being made in these basic studies which should lead to important applications.

16   The Kask decade, 1953–63

The Kask decade was a period of unprecedented growth and international recognition under the leadership of a dynamic personality, Dr J.L. (Jack) Kask, who set out to make the Board an integrated national organization instead of a group of stations run by directors who were working more or less independently in geographic isolation. He built up an efficient, though relatively small, headquarters staff, established that Board members should serve for no more than two 5-year terms, and developed regional advisory committees which differed from their previous counterparts in that they assessed program priorities rather than handling administrative details.

New tasks and new opportunities

This period saw many major changes in Board policy. The technological stations broadened their perspectives of fish products and processing by moving into the field of physiological, biochemical, and bacteriological studies of live animals, and this trend continued under later chairmen with impressive results. Studies on marine mammals were centralized in the new Arctic Biological Station at Ste Anne de Bellevue, Quebec, and freshwater research was moved from Winnipeg to London, Ontario, because of the dominant position of the Great Lakes in Canadian freshwater production and the threat of the sea lamprey. Physical oceanography, which had been the Board's prerogative for half a century, was largely handed over to another government department.

There was a significant increase in the activities of established international fisheries commissions of direct interest to Canada, such as those concerned with northwest Atlantic fisheries, whales, and fur seals. In addition, Canada ratified the International North Pacific Fisheries Convention. The Board was the principal Canadian source of knowledge concerning the fish stocks involved, but a major effort was required at Nanaimo to provide the additional information needed by the Canadian commissioners. Similarly the Great Lakes Fishery Convention required

J.L. Kask, chairman of the Board 1953–63

new information from Board biologists and those of the Ontario Department of Lands and Forests.

Although the Board's staff more than doubled during the Kask decade, Ricker has pointed out:

for the most part this did *not* represent the workings of Parkinson's Law of inevitable bureaucratic growth. Kask's headquarters staff was kept ridiculously small by Ottawa standards, and in the field every new investigation was a response to current or imminent demands for new information, either from industry, from the department, or from some international commission. The only unnecessary proliferation resulted from the cumbersomeness and inefficiency of new procedures imposed on government agencies generally – things like centralized purchasing and personnel services. Such schemes are always introduced with the claim that they will make the government more businesslike. Long-suffering administrators can be pardoned for reflecting that if this is what business is like, no wonder prices keep on rising.

John Laurence Kask

Kask was born in Red Deer, Alberta, in 1906, and received his early education in the Alberta grade schools and in the high school at Prince Rupert, British Columbia. His father died when he was quite young and at 15 Kask found summer employment as a cook on a seine boat. Then he graduated to full deck fisherman, fished for crabs, dug clams, and worked in a cannery to make enough money to see him through university. He graduated in biology from the University of British Columbia in 1928.

His first job on leaving university was as scientific assistant at the Pacific Biological Station at Nanaimo, assisting Clemens in a dogfish survey, and later assisting McLean Fraser in a razor clam survey. Then he went to work for the Halibut Commission, spending two years at Prince Rupert and eight years in Seattle. In

208

1936 he obtained his PHD from the University of Washington and in 1938 he began working for the Salmon Commission, later becoming its assistant director. He was with this commission until 1943, when he left 'because of a small disagreement with the director,' Dr W.F. Thompson. Turned down on health grounds by the Canadian army, he finally joined the California Academy of Sciences as curator of aquatic biology. He held this position until 1947, interrupted by military leave when he served with the United States Army as assistant chief of the section and chief of the Fisheries Division, Resources and Development Branch, in Tokyo.

In 1947 he joined FAO as chief biologist of the Fisheries Division, but when FAO moved to Rome three years later he shifted to the United States Fish and Wildlife Service as assistant director of the Pacific Oceanic Fisheries Investigations in Honolulu. Later he became chief of the Office of Foreign Activities in Washington, DC, and then assistant director of fisheries.

It was about this time, in 1953, that Kask was approached by deputy minister Stewart Bates, who told him that the minister of fisheries, James Sinclair, had decided that a full-time chairman of the Fisheries Research Board was an absolute necessity and that he 'wanted a person who was more administratively inclined than most good scientists seemed to be; a man who could bring these little independent units, the Board's Stations, together, and could develop a central program as well as regional research programs for the Board. They were looking for a person who had had administrative experience, a sort of trouble-shooter rather than a very distinguished scientist.'

Kask changed his mind

At first Kask declined the offer, as he had just taken over the job of senior fisheries officer in the United States, and he felt that he had not been there long enough 'to have earned the right to take another job.' However, this coincided, in 1953, with a change in government in the United States and consequent dismissal of a number of senior civil servants to make room for candidates of the new regime. Kask had a rapid change of heart. He called Ottawa, learned that the post of chairman was still open, offered his services, and was accepted. He has said:

I had followed the growth of the Board very closely ... My first professional job after graduating from UBC was with the Board. When I worked with the International Halibut Commission in Prince Rupert, my office was in the Board's Technological Station for some two years, so that I was associated with the Board and the Board workers during that whole period. And I knew the station director who later became deputy minister, Don Finn, very well. And through him I followed the development of the Board and the Department of Fisheries. I always admired the fact that the Board was separated out as an independent scientific organization, directly responsible to the minister. To me, this was something so fantastically new and so imaginative, that such an organization could be made ... into something that would serve as a pattern for the conduct of research everywhere.

When Kask arrived in Ottawa, he called on the minister and asked him: 'Now that you've hired me, what do you want me to do?' Sinclair replied: 'Well, I do have some instructions for you. I want you to make that godamn thing work. I want you to make the Fisheries Research Board the best fisheries research organization of its kind in the world.'

Independent fiefdoms

As the first continuing head of the Board, Kask set out to find out what the rest of the body was like:

I found a number of independent little fiefdoms stuck in the most peculiar places. Nanaimo, where I held my first job, had very little contact with Ottawa, except that this was the place

Former directors of the St Andrews Biological Station. From the left, A.W.H. Needler, A.H. Leim, A.G. Huntsman, J.L. Hart

where the cheques originated. There was another station at St Andrews that was so hard to get to that you couldn't reach it by any usual public means of transportation. And then there was a technological station in Halifax under Dr Beatty, and there was another technological station in Vancouver in an old Coca Cola bottling works building under Dr Carter. And still another very small technological organization in Grande Rivière, and a small unit up on Great Slave Lake which represented most of our freshwater operations. And when Newfoundland entered Confederation, we had inherited another biological station, complete with staff, in St John's.

Kask's objectives Kask felt that whatever opposition there might be to the idea of a full-time chairman probably was among the station directors who 'were all a little reluctant to have someone placed over them who would start examining what they were doing and why they were doing it. Especially since the new chairman represented constituted authority to make changes if required. So, to this extent, I would say, and it was entirely natural, there was a little resentment at first.'

Kask felt that, as chairman, he had two aims. One was to give Board programs specific objectives: 'The purpose of the Fisheries Research Board of Canada, in my

210

view, is to enhance the value of the Canadian fisheries through scientific research ... All investigations were in charge of very capable people and were slowly expanding general knowledge of these fields, but none of them, it seemed to me, were aimed at a specific purpose. So one of my first jobs was to see that the objectives of each investigation were very clearly defined.'

The second aim was administrative: 'The problem was to unify our effort to expedite our scientific – and I had hoped, our development – mission. Directors and their staff should not be required to spend a large fraction of their time and energies fighting administrative battles. They should be given the assurance that their interests with respect to salaries and research support ... would be fairly taken care of at headquarters ... A corollary to this was to keep the administrative staff at headquarters of very high quality, but numerically small.'

He set out to create a national organization of the Board. Previously the Board members and the directors met together at the annual meeting early in January, and there were one or two subexecutive meetings annually on each coast, usually at one of the stations. Kask felt that the Board was too much involved with administrative matters to give sufficient attention to the research programs at these annual meetings. But the Board's great value, he believed, was precisely program appraisal and review, since the Board was drawn 'from every scientific discipline and from businessmen knowledgeable in fisheries matters.' He believed that the previous system allowed the more aggressive directors virtually to dominate and monopolize the meetings.

The steps that Kask initiated to relieve the situation were (a) to take over all routine administrative measures that fell above the directors' authority, leaving the executive committee and the full Board only matters of high import and policy to consider, and (b) to organize regional advisory committees to appraise and review programs and to suggest new ones. Members of the advisory committees were balanced among industry members and scientific specialists, and any member of the Board could be co-opted for special reviews by any committee. They also helped in setting program priorities. Kask's measures

He said later: 'All of these developments, namely the program reviews by special committees, the cross-fertilization of ideas by bringing Board members from one coast to the other and to the middle of the country to review regional programs, the transfer of staff members and the centralizing of most administration, all helped, I believe, to foster a true national organization. Also, a great help in achieving a national perspective was the fact that the agenda for all meetings was prepared at headquarters and sent to members with background papers, well in advance of the next meeting.'

Kask had strong views on the role of administration. He saw it as a service and not an end in itself. Research had to be done in the field, and could not be dictated or controlled from Ottawa: 'My basic sentiment is, keep administration small but efficient, and delegate all except final authority to the field. I recommend this to governments in both Canada and the United States, and particularly to the Communist governments, who are collecting all authority into the central area. Sooner or later, they're going to fall from their own weight.'

There were those in the organization who felt that the change from subexecutive to advisory committees was a matter more of form than of substance. At any rate most directors and scientists adapted to the new regime without noticeable trauma. Regardless of who did the reviewing and advising, a good idea still had a good chance of being tried out, whether it came from a station, a Board member, or an outside source.

The present laboratory at Halifax

One of Kask's more controversial decisions was to switch the directors of the Nanaimo and St Andrews stations, Hart and Needler. He did this, he said, because of complaints of non-cooperation from the industry on both coasts, and lack of ability or willingness to convey the results of research to industry: 'So I thought, if matters had gotten a little sour on both coasts, that it might be a good idea, in order to retain and re-inspire these two exceptional and slightly bored men, to give them new incentives and new challenges by changing them from one station to the other. It wasn't easy to do because, first of all, you had to convince two very independent men of the value of such a move.'

Both men were opposed to the move originally, but both later agreed that it had worked out well for them. Kask said:

During the first two years that each of these directors spent at their new stations with the new challenges, the new problems, and with the new priorities that the advisory committees imposed, they were very busy. And, in my view, they did outstanding jobs on both coasts. Another thing: there were so many new programs thrust upon us by the international commissions that were being formed. ICNAF, with its groundfish on the east coast, the North Pacific Commission, and when pink salmon was added to the Fraser River Sockeye Commission, and so on, all these moves resulted in new problems for the Board without our having been given in many instances any new resources for their execution. We had to absorb them into the existing organization. And this, again, placed a new challenge on these two very capable directors. Both of them had to really scratch to accommodate their programs to these new demands, and both of them came through very well.

Kask had a more serious problem with Halifax director Stan Beatty:

Beatty was one of the directors who challenged the Chairman's authority from the beginning ... When I first went to visit his station, he told me that he would be very happy to show me around and explain what they were doing, but he hoped he would be left alone and given his

212

A supply of salt water is pumped to the live holding facilities from 60 feet down in Halifax Harbour and approximately 500 feet from the laboratory by one of two duplicate pumps rated at 180 gallons per minute. Water is drawn in via 2–3 inch plastic suction lines, which are weighted every 12 feet by 30-pound concrete circles (see bottom front of picture)

$200,000, in which case he would guarantee to produce the required technological research in his area. That, of course, would make my job very easy, because I wouldn't have to worry, just wait for the results each year. But it wouldn't quite meet the requirements that I had set for myself to make the Board a national organization.

In biological work, there are discrete regional problems and, to a certain extent, there are in technological work as well. But most subjects, like the study of fish preservation, or the development of fish protein concentrate, are universal problems; they are just as important on the Pacific coast as they are on the Atlantic or anywhere else. I found that the Halifax staff, by and large, were rather parochial, and they were more concerned with plant sanitation than with fundamental research. But Beatty was a popular man in the area. He was well-liked. He belonged to the social clubs and went hunting with industry leaders.

Kask said that when he endeavored to develop a national Board program in technology by bringing together the scientists from Halifax, Vancouver, and Grande Rivière, Beatty resisted the effort, so when a foreign assignment with FAO became available, he suggested that Beatty take it for a few years, on a leave of absence. Beatty didn't return.

Beatty had proposed that Henri Fougère, director at Grande Rivière, replace him as acting director at Halifax. Kask agreed to this, and was pleased that a francophone Canadian should become director of a major station. However, it turned out that Fougère, although a charming man with a great sense of humor, could not maintain discipline in the larger station, so Kask did not discourage him too much from also taking a post with FAO.

Then, upon the strong recommendation of Hugh Tarr, who was director at Vancouver, Dave Idler was appointed director at Halifax. It was a happy choice:

Idler moves east

213

The concrete bracket, standing 10 feet high
in an upright position, is being lifted by
crane to its Halifax Harbour location behind
the laboratory, where it will be buried in
the Harbour bottom

'His performance was outstanding. He took a station that had been doing very little
except cleaning up fish plants – incidentally, that was one reason the station staff
was popular with industry: they worked hard in the plants for no pay – and
developed sound and fundamental programs of research within a year. He worked
very hard.'

Kask felt that Idler raised the station to a high international reputation: 'No
question about that. He was getting money from the National Institutes of Health in
the United States, extra monies with which he could bring in outside investigators.
People from China and Japan and other places were sending investigators to Halifax
just to work with him.'

The legendary
Charlie Castell Kask described the assistant director at Halifax, Charles Castell, as a good
bread-and-butter bacteriologist who had an excellent rapport with the fishing industry:

The way he did it was by going directly into the fish plants and being one of the boys. He had
very earthy expressions. He was the original plant sanitation and quality control man. This
'in plant' work was something the Halifax boys did very well. They would go right in the
middle of the plant operations and demonstrate sanitation and quality control techniques,
and Charlie was one of the best. I think that if we had had a bigger organization, and if we had
had the development section within the Board, that is, if we could have had research and
development together, Charlie would have been one of the best bridges between research

and application we could have had. He would have really come into his own in a situation like that.

Legends about Charlie Castell in the fishing industry of the Maritimes are almost as numerous as those about Huntsman. A missionary on personal cleanliness in fish handling, he would interrupt a lecture to ask whether any of his listeners needed to go to the toilet. Those who left to answer the call of nature would be asked by Castell on their return: 'Did you wash your hands?' Usually they would say, 'Yes.' Then Castell would take a smear from their fingers and run it under the microscope to reveal the presence of *Escherichia coli* bacilli. It turned a number of his audience into obsessive hand-washers.

Lecturing skeptical fishermen on the deterioration of fish at the bottom of the catch in shipholds, caused by slime, fish feces, and the sheer weight of the fish above them, he made his point effectively by asking: 'Now, if four of you fellows were standing four-high on each others' shoulders, and you had to have a piss, which one would you rather be?'

Castell ended a very successful summer seminar with a group of university students by offering to 'show them the town.' When, delightedly, they accepted his invitation, he conducted them to the top of Halifax's Citadel Hill and said cheerfully: 'There it is.'

Castell, when he retired in 1971, received the Earl McPhee Award from the Atlantic Fisheries Technologists and a Senior Public Service Merit Award, which carried with it $2,500. Minister of Fisheries Jack Davis was on hand to present the cheque at a meeting of all department managers. In his acceptance speech Castell kept insisting that he knew at least 20 scientists who deserved the award more than he did. Finally an exasperated Davis had to say: 'Charlie, shut up and sit down!'

During the course of the Kask decade, the center of the Board's freshwater operations was moved from Winnipeg to London, Ontario. The reason for this, Kask explained, was the initiation of the lamprey control work on the Great Lakes: 'More than half of Canada's freshwater fisheries were in the Great Lakes, and when lamprey control was thrust upon us, there was nothing else to do but to move down where the problem was ... The Great Lakes [Fishery] Commission was beginning to come into operation and, as I could not get a Board employee on the commission and [W.A.] Kennedy [director of the Biological Station at London] could not get along too well with anybody on the commission, I sort of slipped George [Smith] in as a Canadian scientific adviser to the commission, since he was able to handle the early delicate bio-political problems much better than Kennedy.'

Neal Carter, who headed the Vancouver Technological Station, was transferred to Ottawa. The problem there, as Kask saw it, was that Carter, an excellent scientist, a painstaking, meticulous worker with a profound knowledge of his field, was not an effective director. There were complaints from the fishing industry. They liked Carter, but they didn't think he was the right man for that particular job: 'However, he was far too valuable a man to lose. His meticulousness and his profound knowledge in the technological field, and his interest in editing appealed to me.'

It took some persuasion by Kask but the reluctant Carter made the move and performed a brilliant job there working with Ricker on the Board's growing number of publications (see chapter 22). But Carter made it clear from the beginning that he would stay in Ottawa only until he reached the age of 60. Then he retired and returned to British Columbia and its beloved mountains, though he has done some editorial work by contract through 1976.

Freshwater shift

Carter and Tarr

215

Kask then appointed Dr H.L.A. Tarr to replace Carter, and Tarr turned out to be that rather rare animal, a brilliant scientist who was also an excellent leader.

Hugh Lewis Aubrey Tarr

Hugh Tarr was born in Clevedon, near Bristol, England, in November 1905 and moved with his parents to British Columbia in 1911. He graduated from the University of British Columbia in agriculture and took his master's degree in microbiology in 1928. He then went to McGill, where he studied under Professor F.O. Harrison, who stimulated Tarr's interest in marine microbiology. Tarr spent a summer in Halifax working for Huntsman inspecting 'ice fillets.' Then he returned to McGill and in 1931 received his PH D in microbiology with biochemistry as a major. Two external research grants enabled him to spend three years at Cambridge, where he earned another PH D in biochemistry.

It was at this point, in 1938, that R.H. Bedford left the Prince Rupert station and his position as bacteriologist became open. Carter offered the job to Tarr, who wanted to return to British Columbia, and he accepted. The scientific staff then consisted of Carter, Brocklesby, Young, and Tarr. Tarr said of Carter: 'One thing I shall always appreciate was that Neal Carter never told me what I should do. He never said I should do this or that. He let me work on my own steam, at my own pace, on what I wanted to work on. Of course he knew what I was working on, but he never interfered ... He left Brocklesby alone, Brocklesby developed his work, and Otto Young developed his work, and that was fine. This is the sort of thing I did when I took over Vancouver as director.'

The scientist as an administrator

Tarr accepted the appointment on the understanding that if the chairman was not satisfied with his performance on the job, he could always return to research. He made it clear that he did not intend to give up research. Both Kask and Carter told him that he wouldn't have any time for it, but Tarr quickly found that by turning over routine administrative details to his office personnel, he could still devote 80 percent of his time to research. This was in sharp contrast to the previous period when '90 percent of the efforts of the early directors involved administrative matters, even expense accounts ... I think that the director who has a good administrative staff and uses it, and doesn't try to do all this himself, I think he has got it made.'

Tarr disagreed strongly with the concept of moving good scientists into administrative positions at the expense of their research: 'He shouldn't have to let his own work drop, and he'll know what is going on in the establishment. And what is more important – a lot of deputy ministers and assistant deputy ministers don't realize this – if the staff know that this man is not entirely isolated from them and still is a bench man to some extent, is still embedded in research, then they are going to appreciate him a lot more.' He cited Dave Idler as an example of a working scientist who was also a successful administrator.

Mission-oriented research

Tarr found nothing new in the current government emphasis on mission-oriented research. When he first joined the Board briefly back in 1929, the work at Halifax on fast-freezing and refrigerated sea-water was all mission-oriented, and when he came to Prince Rupert 10 years later, most of the work was mission-oriented. It was not until the late 1940s that a little more fundamental research was done. The work on refrigerated sea-water as a means for keeping the catch in good condition was estimated by a prominent fishing company executive, Dick Nelson, as having saved the industry at least $600,000 a year when it first came into use; the figure is even larger today. Yet 'we had to fight to get this through.'

Developments blocked

One of Tarr's greatest disappointments was the fact that after Canada, through the Fisheries Research Board, had developed 'about the best method of preserving fish chemically that had ever been discovered' through the use of tetracycline antibiotics, the process was discouraged because the Americans would not permit its use. He thought that one reason why the process was not pursued more vigorously was that 'the industry who made the antibiotics were disappointed in the fact that whereas in animal feeds they were using, say, 20 or 25 parts per million and making a lot of money, that they wouldn't do the same in foods like fish,' where only one part per million was needed to secure effective preservation.

And the Americans were interested in radiation pasteurization, which was a complete and utter flop, which has never in any part of the world amounted to anything. They've spent, I suppose, millions of dollars, yens, marks, but nobody uses radiation pasteurization of the flesh foods ... and yet they still [1973] flog away at this. So, with this I feel a bit disappointed ... when I think of all the work that went into this, not only by our own lads, which is probably only a fraction of the whole, but the potential that antibiotics could have, and still could have, in the fishing industry to improve the quality of smelly fish! I go into a store now and I buy some cod fillets or something at an astronomical price of about a dollar and a half a pound, and they smell of trimethylamine, and I still think that if that fish were iced in aureo ice, I could take it home and fry it, all the aureo gone out, and I'd still have a decent piece of fish.

Hugh Tarr was to continue with the Fisheries Research Board until his retirement in 1970, though even in 1974 he was still regularly going to his laboratory at the Pacific Environment Institute to do research on the nucleic acid metabolsim of the starfish, a subject which he chose because it would not interfere with the work of other scientists in the region.

Kask proceeded with his plans for change, but met with failure in one major attempt: the move of the St Andrews operation to Halifax. By that time the biggest single fishery on the east coast was the groundfish fishery, and the majority of both fishing boats and processing plants were in or near Halifax, so the move seemed logical. When questioned about his failure, he said in 1972: 'Well, one of the reasons was that it became a very hot political issue, and I must say that some of our own staff members were heating the political pot. Most of the staff, it seemed, were against the move. They all had houses in St Andrews: "What do we do with our houses?" I tried to work out some way that the Board would buy the houses and recompense them for the move, but then the mayor of the town got in on the act, and at that time we had a minister who didn't relish political heat, so we dropped the idea.'

Popular legend around St Andrews credits the intervention of C.D. Howe, who summered every year at St Andrews, and who took up the cause of the alarmed townfolk – the St Andrews station is the major payroll in the town – and persuaded fisheries minister Angus MacLean to reverse his support for the move. If so, both the Liberal Howe and the Conservative MacLean rose above partisan politics.

Kask also wanted to see the groundfish investigation which was taking place in St John's, Newfoundland, moved to Halifax, but 'it was very hard to do anything there to change them from their 300-year-old history.' Similarly, he could not see the sense of maintaining a technological station at Grande Rivière, 'in a small community where there was no real interest in science and where the scientists would be hard-pressed to find anybody to talk to about their scientific problems.' He had hoped to bring the whole group under one command in Halifax, but failed, though some coordination developed when Idler went to Halifax.

One of Kask's most debated decisions was to limit the Board's role in oceanography:

Kask fails to close St Andrews

A crucial decision on oceanography

The interest in oceanography began to grow about the same time that Sputnik 1 was put into orbit. All over the world an interest in oceanography and science in general was triggered off by this Soviet break-through. The Department of Mines and Technical Surveys at this time had two ambitious builders. One of them was [W.E.] Van Steenburgh and the other was one of our own former staff members, Bill Cameron. They were very anxious to expand and broaden their hydrographic and related geological and meteorological work into oceanography.

At this time it was the Board's feeling – and we discussed this very thoroughly – that we did not want to go into oceanography in a very substantial and big way. However, we did want to remain first in fisheries oceanography. We decided, very deliberately, to limit our interests in the general field. It was my feeling, and this was supported by the Board, that it would not be very long, if we took up the whole field of oceanography, before the oceanographic tail would be wagging the fisheries dog. Our principal responsibility was in fisheries. Oceanography, in all its ramifications, we were sure, would grow enormously and fast. And, as the Department of Mines and Technical Surveys wanted it and had the resources to fund it, we thought, with others in the Canadian Committee [on Oceanography], that a broad oceanographic program would fit better into Mines and Technical Surveys than into the Board. It was done deliberately; we were not surrendering anything.

Lamprey control a waste of money?

Discussing the Great Lakes Fishery Commission, Kask pointed out, in 1972, that though it was ostensibly set up to combat the lamprey menace in the Great Lakes, which was decimating the lake trout, from the American viewpoint the commission had another purpose, 'to get the six or seven states that border on the Great Lakes to work together. The fisheries responsibility in the United States rests with the states. Very often the fishery regulations or sanitary regulations of one state were in conflict with another. The United States was trying to get a forum where these conflicting issues could be more or less reconciled by the United States federal government. Under an international treaty this could be brought about.'

Kask's chief objection to the commission was scientific. He contended that an organism like the lamprey cannot be eradicated, so why undertake it? 'Later, this term, lamprey eradication, was changed to lamprey control and lamprey reduction. It looked to me like even lamprey control would be an operation that would go on forever. But so long as they were going to make it experimental and in Lake Superior alone, I didn't fight it too much.'

But as soon as they got the electric barriers and lampricides working in Lake Superior, the United States proposed extending the operation to Lake Huron and Lake Michigan. Since the latter lake lies entirely within the United States and the Canadian government was committed to one-third of the total cost of the program, Kask opposed the proposal vigorously. He came in conflict with Andy Pritchard, the Fisheries Department representative on the commission.

Kask continued: 'Actually I was opposed to the whole program of lamprey killing, even when I was in the United States, And now, 20 years later and $40 million later, I don't see any reason to change my mind ... Lamprey control, it seems, is going on forever. Even if we find that we do finally get the lampreys under a degree of very expensive control and then decide we don't really want lake trout that much anyway, or we find that lake trout are being destroyed by pollution, it will be sad indeed.'

The role of international commissions

Kask felt that the idea of international commissions was good, but that the commissions had to be flexible enough to change with changing conditions. The most efficient and effective commissions were those which had their own independent research staffs, such as the Pacific Halibut Commission, the Fraser River

218

Sockeye Commission, and the Inter-American Tropical Tuna Commission. In organizations like ICNAF, 'researches are carried out by the national research organizations of the members, and each national section does the research that best serves its own interests. And the underdeveloped members nations do no research at all. I am not at all sure that this is the best way to do it.'

In those commissions with their own research staff, the work is done with very specific research objectives, and the underdeveloped nations share the scientific staff and the results of their investigations on an equal basis, thereby preventing the technically capable countries from dominating the research and interpreting it to their own advantage.

He defined the International North Pacific Fisheries Commission as 'a political commission entirely. When we drew a line down the middle of the Pacific to keep the Japanese on the other – Asiatic – side, and when most of the researches were designed to prove that this line should be continued, then it became a political commission. A lot of very good research has resulted, but always for the wrong reasons. The wrong reasons, because researches are carried out for economic and political reasons and not for conservation reasons. They are designed to protect the catch of salmon in the eastern Pacific for Canadian and American fishermen. This may be a worthy objective, but it is not a scientific objective.'

He thought there was a need to consolidate commissions rather than proliferate and expand them. Commissions should conform to the needs of the species. He cited the proliferating tuna commissions: the Eastern Pacific Tuna Commission, the Atlantic Tuna Commission, the Indian Ocean Tuna Commission, as well as a number of smaller national units. What was needed was a World Tuna Commission, for the tuna roamed all over the world. Research and management could be conducted much more economically and effectively and efficiently if it were under one command.

Kask assessed his successes and failures as chairman of the Fisheries Research Board: 'I didn't do what I set out to do. I do think I did quite well what the minister instructed me to do, which was to make the Fisheries Research Board work as a national research organization, and to make it one of the finest research organizations of its kind in the world. Both of these things, I think, were accomplished. But it was not my doing, although I did serve as headquarters catalyst. It was largely the work of very capable people, with whom, under the very generous recruitment policies of the Board, I was able to help the directors staff their stations.'

Kask's greatest disappointment was the lack of tangible results: 'If our purpose indeed was to increase the value of Canadian fisheries through scientific research, then, after 10 years of good research and after the expenditure of many millions of dollars, we should have a much more valuable fishery than when we started. Actually, this hasn't turned out to be the case. We have improved quality a little bit and perhaps have been instrumental in introducing some conservation measures, but the catch has remained the same or even decreased, and it takes much more effort to make it increase.'

He believed that part of the reason for this failure was the lack of a development arm in the Fisheries Research Board: 'It has been my strong conviction that if you want research to respond to a need, then the development and research have to be under one head ... so that if the desired objectives are not met, the boss, or the minister, can go to one person and say: "Why the hell isn't it working?" '

In this way, Kask believed, a firm fisheries policy could have been developed for the government, and objectives could have been set. He had proposed this development arm to the minister and to the deputy minister when he joined the Board,

Accomplishments and disappointments

Lack of development arm

219

and he thought there was a tentative agreement. Instead, political considerations took precedence: 'Newfoundland had just recently entered into Confederation. She had to be given a rightful place in the Ottawa sun. In fisheries they had a number of people that they wished to second to the department, and they had no intention of seconding them to the lower jobs. They had ambitions right up to the deputy minister, assistant deputy minister, and division level heads. There were a number of candidates that were being proposed for this and Steward Bates, with all his intelligence and capabilities, was not able, like most humans, to withstand the political pressures that were being leveled at him. He was asked to find a job at a high level for a Newfoundlander.'

So the Industrial Development Service was created in the department, and a Newfoundlander, Lou Bradbury, was placed in charge: 'He was a very generous, a very intelligent and loving man, but I was disappointed, because we now became competitors rather than natural collaborators. Also it relieved the research people from responsibility for development, a place where I thought a good part of the responsibility should rest.'

Kask found that the result of this new development was to isolate the research men even more than before. Harry Wilson had commented: 'Lou Bradbury and his Industrial Development Service was guarded by the department because the Board was producing most of the hard news, and without IDS, the department was nothing more than a policeman: protection and inspection.'

Kask contrasted the Canadian situation in fisheries with

A contrast with Russia and Japan the fact that Russia and Japan, both with built-in R&D systems, were dominating postwar high seas fisheries. Russia especially was a case in point. Traditionally a land power, no one seriously believed that in 15 years she would become a leading high seas fishing country. By marrying R&D – and education – and setting realistic targets, she not only increased her fleet and catch, but built new fishing ports where none existed before. She is now number two in world fishing, and soon may be number one. I had hopes that we could do something the same for Canada ...

I think it was the second disappointment that really helped to precipitate my move from the Board. It was my view that Canada didn't really have a fisheries policy, so none of us really knew exactly what we were working towards.

Some quick appraisals Kask made some quick appraisals of people he had worked closely with at the Board:

You can't think of the Board without thinking of Bill Ricker. Bill was truly a man for all seasons. He was not only smart; he was completely honest and dispassionate. For instance, Bill was the only man that the Board could agree on to serve as acting chairman when I left. They couldn't or wouldn't agree on anyone else, including members of the Board ...

One of the very useful men was Alf Needler. He gave me dispassionate advice, no matter how unhappy he was with the situation. Of this I am sure, and he gave it at any time I requested it, and on almost any subject. Another man I leaned on very heavily was John Rogers.

Rogers was successor to Harry Wilson as administrative assistant to the Chairman.

Hayes appraises Kask Kask, and the 10-year period that he spent as chairman of the Board, have been appraised by others, including the next permanent chairman, Dr Ronald Hayes. Hayes found Kask a professional chairman. I think Kask brought his professionalism into the central government, so to speak, into Ottawa, and took over control. This aroused considerable objections from the directors, who saw their positions of dominance losing out to central government, and who were scarcely placated by the Kask bluntness of manner. It was a provincial-federal kind of situation, and here was a professional as a full-time man establishing a full-time headquarters in Ottawa.

What he did in professionalizing was what I would have had to do, or anybody else would have had to do, and I was very grateful that it had been done by Kask. There was no doubt when I came there, except in the mind of Peter Larkin, that Ottawa was running the show. The other directors knew that there was a headquarters, and this was a very solid accomplishment.

Hayes felt that Kask's weakness as an administrator was that he never understood the Board system of administration: 'He was never really prepared to open the meeting up and let the Board make a decision.' Instead Kask would rehearse the proceedings of the Board with his vice-chairman and with his secretary and arrive at the conclusions he anticipated the Board would reach at the meeting.

So under Kask the Board became rather impotent, and this was something that I couldn't understand. The practice was very different from my experience in the NRC or in a university faculty. So I came in with another point of view in dealing with the members of the Board, and the first thing I had to do was to indoctrinate Martin and Claire Duffy and Rogers that we were not going to draw the conclusions of the Board meeting before the Board met.

The second thing that was wrong, which long predated the Kask regime, was that the Board meetings were public meetings of all the directors and of all the fellows who came up to Ottawa to give papers, and the Biological Committee would be sitting in, so that there was virtually no real discussion of policy. So I attempted to shift it from a discussion of detail to a closed meeting of discussion of policy, and I don't think the directors liked this. In fact, they complained very much about this restraint on them.

Bill Ricker, who took over as acting chairman when Kask resigned, said of him: He did a fine job

In general he did a fine job. He had some excellent ideas, as well as some not so excellent (to me), but you don't expect to agree with anyone 100 percent. He was determined to maintain the independence of the Board, and would trot out the act whenever he felt it was threatened.

Unfortunately Jack had a capacity for making people annoyed, sometimes even when they agreed with him; I'm not sure why. He never affected me that way. The coolness between Jack and [Deputy Minister] George Clark was unfortunate, although neither of them let it stand in the way of getting things done. Andy Pritchard too: Andy had regarded himself as the top candidate for the chairman's post instead of Kask, so there was a built-in source of antipathy, but again it didn't prevent their cooperating. In fact the real test of a good administrator, or a good politician, is how well he can work with people he doesn't like.

Kask might be criticized for having turned down the expanded oceanography program for the Board. Whether his decision was the best one remains to be seen; certainly the present effort in oceanography is disappointing from the fisheries point of view.

Needler commented on the Clark-Kask controversy: 'Well, I don't know what Kask and Clark
you should call it. I had the distinct feeling towards the end of Kask's tenure of chairmanship of the Board that he really had ambitions to be deputy minister.' When Kask was asked whether he had any ambitions to succeed Clark, he replied:

221

'No, not really. I had already been [the equivalent to] deputy minister in the United States. The director of the Bureau of Commercial Fisheries has about the same authority as the deputy minister ... Also, as far as I am concerned, I had always thought of the deputy minister's job in the Department of Fisheries almost as a policeman's job. That is, inspection service, protection service, fishermen's loans, and so on. And I'm not particularly interested in being a policeman. I was, however, very interested in who might succeed Clark, for very selfish Board reasons. I was pulling for an alert businessman.' In fact, Clark was succeeded by Needler.

Other views of Kask John Hart questioned Kask's handling of difficulties with his directors, then said:

I do have a lot of respect for his policy of keeping the headquarters staff of the Board small. I think that this was really an experiment in good government. I don't think that the experiment was a success, because other people weren't prepared to go along, and ultimately he lost out by comparison with other competing departments. I would say that, essentially, during Dr Kask's administration, he lost oceanography for the Board. I can't think that this was really a good thing. I don't think that it was done accidentally, and I don't think that it was ineptness: I think it was a matter of policy, and I think that the policy may have been accepted or proposed by fisheries members of the Board who did not want to see their interests submerged.

Harry Wilson said: 'He fought for the independence of the Board and he stepped on a lot of toes. But it was incorrect of him to say that he ... created an organization. The organization was in existence when he got there.'

Finally, Claire Duffy, secretary to four successive chairmen, said of Kask: 'He was tops with me. He did what he thought was right. He was resented by some because he was imposed on them from outside ... He treated the staff graciously and thoughtfully. He was a hard worker; in every morning by 7.30 and the last to leave at night. He remembered every anniversary, every illness. His wife was a kind, kind person. She called people at home if they were ill, or if anyone in the family was ill. The staff were very loyal to him.'

RESEARCH

The research accomplishments of the Kask decade were substantial. Many projects were carried out in response to the needs of the various international commissions, and are described in a later chapter (see chapter 19). At Nanaimo the Babine work showed that the larger part of the lake was underutilized as a sockeye salmon nursery, which set the scene for construction of two spawning channels by Resource Development. The trawl fishery of the Strait of Georgia was analyzed by K.S. Ketchen and C.R. Forrester, and improved management measures were suggested. J.R. Brett concluded several years of experiments in guiding both yearling and adult salmon, while D.J. Milne reported on coho tagging experiments and the chinook and coho salmon fisheries generally. Starting in 1955 the Skeena salmon investigation was integrated with the department's management in that area under a Skeena Salmon Management Committee, consisting of the chief supervisor of fisheries, Pacific area, and the director of the Nanaimo station. The first two directors of investigations for the committee were Board employees, F.C. Withler and Jack McDonald. A small test fishing operation was maintained above the regular fishing zone to give a daily estimate of the number of salmon that escaped the fishery. T.R. Butler tagged crabs to learn their movements and rate of utilization, and also prospected widely for shrimps and prawns. Board studies on Pacific

The Atlantic Biological Station, as seen from the air in 1972. The town of St Andrews and Passamaquoddy Bay lie in the background; the building in the middle ground is the residence

Important in management of fish resources are fish counting fences, which are used to estimate population sizes. This fence and trap has been used for counting migrant salmon on the northwest Miramichi River since 1950

The Vancouver laboratory on the University of British Columbia campus

The Arctic Biological Station at Ste Anne de Bellevue

clams and oysters were resumed when D.B. Quayle transferred his work from provincial auspices to the Nanaimo station.

Two new faces at POG, John Strickland and T.R. Parsons, made a complete check of methods of chemical analysis of sea water, found many inconsistencies, developed new procedures and improved old ones, and eventually produced a *Manual of Sea Water Analysis* that quickly became the world standard. They then went on to experimental and field studies of phytoplankton production. Jack Tully turned his technicians and theoreticians to the task of constructing a new and more ambitious model: of Hecate Strait this time. It was completed successfully, but the scaling problems proved formidable. On the east coast Louis Lauzier's work with 'sea-bed drifters' helped to fill out the oceanographic picture, and Harry Hachey produced a summary for the whole region. Northern oceanographic work was extended to the central Arctic and the Chukchi Sea, with a major cooperative effort in 1962 when five field parties and two icebreakers took part.

The Arctic Unit hired permanent staff members under H.D. Fisher's directorship. The Board decided to base all eastern marine mammal work there, so that D.E. Sergeant's harp seal work was transferred from Newfoundland. The unit's area of marine operations was extended into Coronation Gulf and the Mackenzie delta region. Work on fresh waters north of 60°N was conducted by Lionel Johnson and J.G. Hunter, the latter's party being the subject of a three-week search in 1956 when their plane flew off course and ran out of gasoline. Bearded seals were studied by A.W. Mansfield, and I.A. McLaren completed several years' investigation of the ringed seal, identifying areas of potential overexploitation. E.H. Grainger and A.S. Bursa handled macroscopic and microscopic invertebrates, respectively.

At St Andrews, J.C. Medcof compiled the station's work on oyster farming and summarized it in a comprehensive bulletin. D.G. Wilder and D.W. McLeese brought together their work on storage and shipment of live lobsters, R.A. McKenzie reported his six-year study of the Miramichi smelt fishery, and Neil Bourne summarized the work on scallops. At the St John's station W. Templeman and A.M. Fleming reported on the longlining experiments that had introduced a new type of fishery to the Newfoundland economy, and Templeman found time to review the distribution of redfish and sharks all along the coast. H.J. Squires surveyed shrimp resources, and D.E. Sergeant concluded his study of the pilot whale, which supported a small local industry.

The St Andrews station undertook a major salmon investigation on the Miramichi River under C.J. Kerswill, using fences on three tributaries to obtain exact escapement figures. Later the St John's station did a complementary study on the Little Codroy River in Newfoundland, under A.R. Murray. These studies finally settled long-standing arguments by demonstrating that Atlantic salmon return quite faithfully to their home tributary, that their main foraging area is somewhere north of the Strait of Belle Isle, that grilse contribute very little to the subsequent run of larger salmon, that the season of return to the river is largely hereditarily determined, and that commercial fisheries were taking 75 percent or more of the large salmon and anglers up to half of the survivors.

The Vancouver work with refrigerated sea water and antibiotics has already been mentioned. At Halifax, Stan Beatty and Henri Fougère published a new salt-fish bulletin before their departure, and W.A. MacCallum summarized the work on gutting and holding fish on board trawlers, while improved methods of handling and storing fish continued to be tested. Fundamental work on fish oils and muscle proteins provided a basis for further advances. The technological units at St John's

The Pacific Biological Station fisheries research vessel *A.P. Knight*, acquired in March 1958

The Arctic Biological Station research vessel *Salvelinus* at Starvation Cove near Cambridge Bay, Northwest Territories

The St John's Biological Station research vessel *A.T. Cameron* (1958–) leaving harbor on a cruise of offshore waters

The St John's Biological Station research vessel *Parr* (1957–9)

and London worked with local problems; Bill Lantz and Louis Dugal, at the latter, produced a wide variety of new types of products, or old-style products from new kinds of fish.

NEW BUILDINGS AND NEW EQUIPMENT

During the Kask era a strong effort was made to increase physical facilities at the stations to match the increase in work load and in personnel. An extra story was added to the Halifax station and numerous improvements were made throughout the structure. At St Andrews a new wing was built that more than doubled the space available. The Vancouver laboratory moved to a new building on the campus of the University of British Columbia. Similarly the Arctic Unit moved out of downtown Montreal to a new building on a site near Macdonald College in Ste Anne de Bellevue, and became the Arctic Biological Station. At Nanaimo aggressive action by Earle Foerster had provided a new building as early as 1949, built on the site of the original 1908 station. Subsequently several small structures were erected, and preparations were made for a new wing to be built on the waterfront.

Inside the new buildings new equipment was installed that made possible quicker analyses and also types of work that were quite unknown a decade earlier, for example mass spectrometry. Computers speeded up data processing and made it possible to test for and evaluate the effects of many potential causes of the observed changes in fish stocks.

In 1953 the principal ships available were the *Harengus* at St Andrews, and two small Pacific-type double-gallows trawlers, the *Investigator I* and *II*, based at Nanaimo and St John's respectively. A somewhat larger trawler was purchased for

227

The Pacific Biological Station fisheries research vessel *G.B Reed*, built in Victoria in November 1962

The *Cyprina* (1955–) with hydraulic shellfish escalator

Nanaimo in 1958; it was given the same name, *A.P. Knight*, as the station's work-boat of the 1930s, which had caught fire and burned to the water-line during the war. A new vessel, the *Salvelinus*, was built for work in the western Arctic, and was freighted by barge down the Mackenzie River to Tuktoyaktuk.

The big event, however, in 1958 was when the *A.T. Cameron*, a 177-foot specially built research ship modeled after a commercial side-trawler, was commissioned for work in the Atlantic, based at St John's. In 1963 a sister-ship, the *G.B. Reed*, began work in the Pacific. Both vessels have the capability of going to any part of the ocean where there are fisheries of Canadian interest. The *Cameron* has prospected north to Davis Strait and Greenland, and the *Reed* has worked in the Bering Sea and made a series of trans-Pacific oceanographic observations along the great circle route to Tokyo.

17   The Ricker interim, 1963-4

When Kask left the Fisheries Research Board in August 1963 W.E. (Bill) Ricker, then a biological consultant at Nanaimo, was appointed acting chairman. He was, as Kask later pointed out, the only person upon whom all could agree for the post. At the same time Ricker made it clear that he would serve as acting chairman only until a permanent appointment could be made. His was essentially a caretaker regime, and there were no major changes in policy during his year's tenure.

A man for all seasons

The Fisheries Research Board, over its 75-year history, has provided the opportunity for numerous brilliant and talented scientists to achieve world stature in their field, from as far back as Stafford and Knight to Brocklesby and Young, Finn and Needler, to name a few. But possibly only Huntsman matched Bill Ricker in sheer versatility, and by an odd quirk the two men discovered in 1953 that they had a common ancestor – one of the Hessian soldiers that George III sent to America in his effort to subdue the rebellious colonists. To this quiet, self-effacing individual, so aptly termed by Kask 'a man for all seasons,' fell the task of keeping the Board's program on course during the search for Kask's successor.

Ricker was born at Waterdown, Ontario, in 1908, and took his higher education at the University of Toronto, ending with a PHD in 1936. He spent several summers with the Ontario Fisheries Research Laboratory, working on brook trout ecology under Professors B.A. Bensley, J.R. Dymond, and W.J.K. Harkness. He was also much influenced by Professor E.M. Walker, at whose suggestion he began to take a special interest in stoneflies. He joined the Board as scientific assistant late in 1931, helping Foerster with the Cultus Lake investigations. When the Salmon Commission began operations in 1938 both he and Foerster joined it in order to come to grips with the still unsettled problem of increasing the population of Fraser sockeye. Early in 1939, however, he went to Indiana University as professor of zoology and head of the Indiana Department of Conservation's Lake and Stream Survey. At

William Edwin Ricker

229

Board senior staff at the annual meeting, 1964. Seated (l. to r.): H.B. Hachey, W.R. Martin, W.A. Kennedy, W. Templeman, D.R. Idler. Standing: P.A. Larkin, J.G. Hunter, J.C. Stevenson, J.L. Hart, W.A. MacCallum, J.P. Tully, R.W. Trites, R. Legendre, J.A. Rogers, H.L.A. Tarr, E.G. Bligh, L.C. Dugal

Dymond's invitation he returned to the Board 11 years later as editor of publications, and held this post from 1950 to 1962, while stationed at Nanaimo.

Following his interlude in Ottawa, Ricker returned to Nanaimo, and in 1966 he was named chief scientist of the Board, a title which he held until his retirement in 1973.

A Fellow of the Royal Society of Canada, Ricker was awarded its Flavelle Medal in 1970. In 1955 and again in 1959 he received the Wildlife Society's citation for the best research paper in fisheries, a feat that has not been duplicated before or since. In 1966 the Professional Institute of the Public Service of Canada presented him with its Gold Medal 'for outstanding achievement in the field of pure and applied science.' The University of Manitoba conferred an honorary Doctorate of Science on him in 1970, and in 1974 Dalhousie University made him a Doctor of Laws.

Ricker is best known for his 'green books' – guides to the calculation and interpretation of the vital statistics of fish populations, which appeared in successively updated versions in 1948, 1958, and 1975. While in Indiana he learned Russian in order to read a pioneering treatise by F.I. Baranov; of this he has said: 'It is impossible to overestimate Professor Baranov's influence on my thinking, and on the development of fish population dynamics in general.' One of Ricker's major contributions while editor was to put into English more than 100 Russian research papers and to make these and other translations generally available in an FRB Translation Series.

In 1969 the American Fisheries Society selected this Canadian to receive its first

230

W.E. Ricker, acting chairman of the Board
1963–4

Award of Excellence, which included a medal and a prize of $1,000. At the presentation ceremony, the society's president, Elwood A. Seamon, said, in part: 'Dr Ricker has been called "the foremost fishery scientist in Canada" by his Canadian peers. In the United States he is recognized for his superb and original contribution to: the theory of lake circulation; the methodology of statistically sound sampling in fishing waters; measuring and interpreting the vital statistics of fish populations; new concepts about growth and mortality and predator influences on salmon survival; and relations between parent fish stocks and numbers of surviving progeny.'

Within the Fisheries Research Board there are many conflicting opinions about the achievements and personalities of senior figures in the Board's history. Regarding Ricker, the attitude is unanimously favorable and warm. John Hart said: 'He is an incredible man. He is, first of all, an excellent fishery biologist. But he doesn't stop there. He has written on applied statistics, innovative work. He has produced a Russian dictionary of fishery terms. He's an amateur authority on the native Indians, and on early alphabets; in fact, he's a walking directory. He also plays the bull fiddle. But, above all, he is a man of generous spirit. I suspect that his influence in the Board will loom larger than that of bigger names.'

J.C. Stevenson wrote of Ricker in 1973, in part:

Ricker, who retires in the year of the Board's 75th anniversary, is a remarkable scientist who has earned the respect of colleagues around the world ... His major thrusts have been in the dynamics of fish populations and Pacific salmon biology, but he has made significant contributions to aquatic entomology, lake circulation theory, fish transplants, and the development of primary scientific publication. Countless scientists have benefitted by his stimulation, and his advice has been sought around the world. Probably more than any other individual, he has been responsible for improving communication in aquatic science between the USSR and western countries, through his efforts in promoting knowledge of the Russian language, development of publication exchange agreements, and involvement with several international commissions and agencies.

Stevenson's
appraisal

The Pacific Biological Station barge *Velella*, built in New Westminster in 1963. This floating laboratory can be towed to any point on the British Columbia coast, and can accommodate an operating crew of 14 men for extended periods of time when moored

An excerpt from a recorded conversation between Stevenson and Ricker demonstrates the unassuming modesty of Ricker. After discussing Ricker's work in initiating the transplanting of pink salmon from the Pacific coast to Newfoundland, Stevenson said: 'Well, I think you really stimulated the whole idea of transplants.'

Ricker: 'I don't think I had anything to do with the coho transplants in Michigan. That was the idea of that chap, what's his name?'

Stevenson: 'Wayne Tody. Well, I happened to be in Seattle when he was there, and he made arrangements for the first transplant to Lake Michigan. This would be about eight years ago, or something like that, and he strongly inferred that his idea came from thoughts that Bill Ricker had.'

Ricker: 'Really? I'll be damned!'

Concerning transplants Ricker had a later comment:

A joke on the biologists

The first man after the war to actually do anything about salmon transplants was Bill Harkness. [W.J.K. Harkness had succeeded Clemens as director of the Ontario Fisheries Research Laboratory, and later became head of the Fisheries Branch of the Ontario Department of Fish and Game.] He was very interested to see if Pacific salmon could be introduced into the Ontario rivers flowing north to Hudson Bay, and arranged a tour of inspection for a group of interested people. Originally Andy Pritchard and Bill himself were going to come, but they both had to withdraw. This left Jack Kask, three Ontario biologists, and myself. I was fascinated by my first visit to Ontario's far north; we flew from Sioux Lookout to Attawapiskat Lake, Winisk, Sutton Lake, and the lower Attawapiskat River, checking streams as possible spawning sites and rivers as homes for parr. Actually the freshwater indications were excellent, but there was the old bugaboo of the big bay that young salmon would have to traverse to find water warm enough for them to survive a winter in the ocean. Although our report indicated that success was unlikely, it is worth playing long odds when the prize is large enough, so Bill decided to give it a try. He planted both pink and chum salmon eggs in two or three streams of the north slope. None are known to have survived to adult.

But there was a most unexpected side result. The eggs had been held in transit in the Port Arthur Hatchery, whose superintendent decided to hatch a few thousand pinks to see what they looked like. Eventually he released them into Lake Superior without telling anyone. In the fall of the following year a couple of mature pink salmon were taken in Minnesota, and they have been seen somewhere in every second year since. Today there are spawning runs of pink salmon in several streams that enter the north shore of the lake. Now, of all salmon, pinks are the species that seemed most wedded to salt water. There are no native populations living in lakes, and in a hatchery it is not easy to get them to feed well enough to survive the period when they would normally go to sea, though it can be done. So if anyone had polled salmon scientists about establishing pink salmon in Lake Superior or any other lake, they would probably have returned a unanimous verdict of 'impossible.' But the joke was on us biologists: mother nature had fooled us again.

THE RICKER YEAR

Ricker's summary of activities for the year 1963–4 did not differ greatly from any of the later reports of the Kask era, but it did reveal: (a) how far the nature and direction of Board research had changed over the previous Kask decade from that of the Reed-Dymond era; and (b) how much the character of the problems facing the Canadian fisheries was changing in a new era of increasingly intensive international exploitation of the resource, and the consequent activities of international commissions endeavoring to cope with the consequences of that exploitation.

It is instructive, therefore, to observe some highlights of Ricker's report in these regards. The lines of future developments in many areas may be clearly traced. The year had been an active and productive one for the Board in all three of its areas of scientific interest: fishery biology, fishery technology, and oceanography. Sixteen vessels, ranging from small inshore and lake craft to large sea-going ships were operated for its biological studies, and a new research barge (the *Velella*) was added that year to the Board's Pacific fleet, to serve as a floating laboratory with living accommodation, initially as a base for the study of young salmon in the sea.

Cooperative oceanographic programs were continuing with the Royal Canadian Navy, the Department of Mines and Technical Surveys, and the Department of Transport, and close liaison was maintained with the Institute of Oceanography at the University of British Columbia, the Great Lakes Institute of the University of Toronto, and the Institute of Oceanography at Dalhousie University in Halifax. Cooperative programs were also carried out with the provincial governments and with the International Pacific Salmon Fisheries Commission at New Westminster.

A large proportion of the Board's biological and oceanographic effort in research was devoted to work carried out for the international fishery and sea mammal commissions to which Canada belonged, but which had no scientific staffs of their own.

Ricker examined the state of the resource in both the Atlantic and the Pacific, as well as the inland fisheries. It was here that the shape of things to come was clearly visible.

Looking first at the Atlantic, he observed that in the Newfoundland area total

State of the resource: Newfoundland

233

Special underwater instruments such as these attached to the headline have been in use since 1965 for engineering studies of trawl nets at sea

landings of the principal commercial groundfish species were about four percent higher than in 1962. Haddock landings were 60 percent lower. He attributed the overall increase to increased fishing effort and the sharp decline in haddock landings to decreased abundance. In the annual shore survey of baby cod, the relatively small numbers of the cod of the year caught indicated only moderate survival and settlement of the 1963 year-class.

In the offshore cod surveys it was reported that great concentrations of cod were being fished by a large fleet of European otter-trawlers off the southeast edge of Hamilton Inlet Bank. These fish were generally small, with a high proportion of four- and six-year-olds in the catch. Extensive surveys by research ships failed to find any concentrations of haddock. The basic cause was the poor survival of the young; the last successful year-classes were those of 1955–6, and a new abundant year-class was not yet in sight. Heavy exploitation had contributed to the shortage.

From a planting of 2.5 million pink salmon eggs by A.A. Blair in the North Harbour River, St Mary's Bay, hatching was highly successful and about 2.2 million fry passed downriver between 7 May and 16 June. The survivors were expected to return to the river in 1964. In fact many did return, but the run has decreased subsequently, and there is still no self-sustaining stock of pinks in our Atlantic waters. This contrasts with the success achieved by the Russians in the Barents Sea: they persisted after initial discouragement, and today have a new commercial fishery for this species, not only in Atlantic waters but east to the Yenisei River.

In the Maritimes area, population studies on lobsters continued at key points so as to follow natural changes and the effects of fisheries regulations ... A comprehensive bulletin on the care and handling of lobsters was prepared to assist the industry in applying results from 10 years of research.

Efforts to rehabilitate mainland oyster populations wiped out by the Malpeque disease during the 1950s were going satisfactorily. Advances took place in cooperative work with other branches of government. An oyster hatchery was built for cooperative use by the Fish Culture and Development Branch of the department, and studies of bottom sedimentation by the Department of Mines and Technical Surveys confirmed suspicions of heavy silting on oyster beds and the consequent deterioration of productive grounds.

The Board cooperated with the Economics Service of the department and the Dominion Bureau of Statistics to compile catch and effort data for national use and to meet international commitments through ICNAF.

Surveys of important fishing grounds indicated no major changes in cod abundance on the Nova Scotian Banks but poor recruitment was anticipated for haddock during the next several years. The proportion of fish discarded after being caught, determined to meet obligations under ICNAF, showed conservation measures to be generally effective.

The behavior of cod and other groundfish was being studied to provide a basis for designing more effective fishing gear. Studies were made of levels of fatigue which could be sustained by fish without dying, and of daily movements in relation to light intensity.

Catches of various pelagic species were sampled routinely and the conditions of the fisheries noted. Herring were sampled in the Bay of Fundy partly to meet international obligations.

As a preliminary step in having fishing gear designed specifically to recognize the behavior of fish, gear engineering studies were begun to examine the forces on trawl nets while they were being towed.

Maritimes area of the Atlantic

235

The Babine salmon-counting fence, 1966

Artificial spawning channel for sockeye salmon at Pinkut Creek, Babine Lake, 1972

In the report on salmon studies, a relatively innocent sentence had explosive significance: seven tagged Miramichi salmon had been captured in Greenland, and it seemed that many fish of Canadian origin might be caught there.

Great emphasis was given to oceanography in its relation to fisheries. Water circulation was studied by drift bottles, electrical analogue models, induction effects, and dynamic heights. Temperature conditions in the sea were being monitored on routine cruises and at shore stations. Other properties of sea water such as optical properties and chemical contents likely to have a bearing on fish life were being investigated. To provide a more detailed exploration, a geochemical description of the Gulf of St Lawrence was undertaken. Studies of invertebrates on the bottom of the Gulf of St Lawrence were showing relationships with fish concentrations and bottom conditions.

The Nanaimo station had conducted many investigations largely related to the immediate national interest, providing scientific background for international problems and service, investigation, and advice for management of fisheries in coastal waters. The marine commercial fisheries investigations continued to provide statistical analyses, predictions, and other management information on groundfish, herring, shrimps, and crabs. In addition to conducting population studies on local stocks, new studies were initiated in the Gulf of Alaska, particularly on ocean perch. Herring investigations continued in a monitoring of populations, cooperation in studies of the incidence of salmon in herring catches, and trials of new midwater trawl gear. Studies of the scattering layer, which was an important area of concentration of juvenile fishes and food organisms, were continued.

In the Pacific

Service to management problems was a major feature of the work on Pacific salmon at the station. An important contribution was the part played by the station in the production of a report on the pink salmon in the Strait of Georgia, under the auspices of the Pink Salmon Co-ordinating Committee appointed by the governments of Canada and the United States.

Salmon management

A particularly large management-type of activity was embodied in the Skeena River salmon investigations, which supported the participation in every aspect of Skeena salmon research management. Studies of sockeye at Babine Lake continued to provide information on growth and abundance of young from spawnings of known size and distribution. Pink salmon fry production was estimated in Lakelse River.

The investigations on salmon propagation and lake sockeye, though less concerned with matters of day-to-day management, were concerned with questions applying to current work of the Fish Culture and Development Branch of the department, and liaison with the department was maintained at a good working level. The salmon propagation study at Lakelse was attempting an assessment of an artificial sockeye hatchery in conjunction with studies on critical aspects of early life-history biology. The sockeye studies on Babine Lake had led to the setting up of a 'Babine Development Program' aimed at strategic location of artificial spawning channels to ensure better utilization of the lake's rearing potential. The work of Drs W. Johnson and C. Groot on the migration of young sockeye was awarded a prize, as paper-of-the-year in fish ecology and management, by the Wildlife Society.

Much of the salmon group's work was devoted to the international aspects of the salmon. So the work on early sea life of salmon in Burke Channel by R.R. Parker and colleagues produced valuable data on growth and mortality of young pink salmon, as well as estimates of total marine growth and survival. The latter findings were particularly significant to international problems, supporting the view that

The international salmon

237

Arctic ringed seal caught at Brown's Harbour, Cape Parry, Northwest Territories. This animal was one of many branded and then released for studies on migration

Dissecting a large male gray seal at Basque Island, Nova Scotia

238

high-seas fishing for pink salmon was less productive of total yield than shore-based fisheries taking mature fish.

During 1963 chinook and coho salmon studies were re-established to provide information on new problems arising from sport fishing and the international aspects of the fisheries for these species. In addition, this investigation would undertake the Canadian commitment in the program for evaluating the contribution of Columbia River chinook salmon hatcheries.

The Nanaimo station was heavily committed to conducting research related to the problems of the International North Pacific Fisheries Commission (INPFC). In 1963, extensive tagging in the Gulf of Alaska confirmed the general picture of distribution of salmon stocks in the eastern part of the North Pacific, allaying concern that Japanese high-seas fishing was taking large numbers of salmon of Canadian origin. Scientists engaged in salmon stock assessment had been called upon to prepare arguments documenting the Canadian viewpoint that salmon should not be harvested on the high seas, and were being rationally exploited to their biological capacity. Similarly, staff working on marine commercial fisheries participated in the documentation of the case for abstention for halibut and herring.

The Pacific Oceanographic Group had also been involved in INPFC activities, contributing to the collection, synthesis, and presentation of the oceanographic material as background to understanding North Pacific fisheries problems. The general program of the group covered a wide range of oceanographic research, reflecting the fact that they were currently undertaking the Canadian requirements in oceanography in the Pacific. _{New objectives in oceanography}

An important new development for the group was the reorientation to new objectives in fisheries oceanography and marine ecology in the first part of 1964. A program was established to cooperate with existing fisheries investigations in assessing the existing state of knowledge of the environment and fisheries, to derive correlations where possible, to define what information was lacking and how it might be obtained, and to provide systems of monitoring the environment that would aid assessment of the fisheries. Some transfers of staff were effected in the station as a whole, and further developments in this area would be reported the following year.

In work on the inland fisheries evidence continued to accumulate that a program to control the parasite *Triaenophorus crassus* in Heming Lake, which terminated in 1960, succeeded in eliminating the pest completely; apparently it had not been re-established. In Lake Superior, the Canadian catch of lake trout began to increase again, as did the average size of individual fish and catch per net. These were encouraging signs of improvement in the condition of the lake trout population, an improvement attributed to a recent sharp decline in sea lamprey abundance. _{The inland fisheries}

The Great Lakes Fishery Commission's 1963 program required chemical treatment of 15 Lake Superior streams and surveys of 630 streams in the Great Lakes watershed. In 1963, as in 1962, counts of spawning-run sea lampreys were much lower than counts for several years prior to 1962. Apparently there had been a dramatic decrease in lamprey abundance as a result of control efforts.

Ricker reported, for the Arctic, that an aquatic biological survey of Great Bear Lake, devoted principally to limnology in its initial year, found primary production to be generally very low, with a maximum occurring in the relatively shallow southern arm. Fish abundance paralleled the primary production findings. _{In the Arctic}

In salt water, the feasibility of maintaining a fishery on the west coast of James Bay was investigated, but was found to be unprofitable. In Coronation Gulf Greenland cod were found abundantly, and could support a limited fishery.

Tagging a harp seal pup on the ice off the Labrador coast

Sea otters in a tank during re-establishment of colonies on the British Columbia coast

Netting narwhals at Koluktoo Bay near Pond Inlet, northern Baffin Island

240

Life-history studies of the narwhal were started in the Pond Inlet area of Baffin Island, and sampling and a census of the white whale population of the Churchill area of Hudson Bay were completed.

An aerial census of molting and breeding harp seals and breeding gray seals was carried out, and young harp seals and gray seals were tagged and/or branded in order to determine exploitation rates and migration patterns.

Planktonic studies to determine arctic water masses and currents were carried out, and the Arctic Biological Station actively participated in an international program designed to study the drift of fish larvae and eggs in relation to plankton distribution and hydrographic conditions on west Greenland waters.

On the east coast, programs at Halifax, St John's, and Grande Rivière were now integrated, and the technological work at London was coordinated with that of the laboratory at Vancouver. In the Atlantic technological research program, a commonly employed diet for lobsters held in pounds was seen to be little better than starvation in maintaining lobster blood cell counts at the level found in a natural environment. Liver supplement corrected the situation. The tolerance of lobsters to various woods and fluoridated water was assessed; western cedar was found to be extremely toxic.

Malpeque-disease-resistant oyster stock was shown by serological tests to possess an antigen absent in susceptible Cape Breton oysters; whether this was associated with resistance still had to be determined. Dimethyl sulfide was identified as the substance responsible for the off-odor of fillets from Labrador cod feeding on the pteropod *Limacina helicina* ('blackberry'). The sample thickness was found to be an important variable in determining texture characteristics for freeze-dried cod. A mascerated muscle residue from commercial filleting operations in Newfoundland produced a freeze-dried product with much better texture than that of fillets.

Fish protein concentrate (flour) was prepared from filleting scraps and whole fish, and the process was modified for application to oily species such as herring. Nutritional and esthetic characteristics of all products were very good. Important strides were made towards defining the influence of physiological and environmental factors on quality of fishery products.

Ricker reported that in the Pacific significant progress had been made in devising novel equipment for unloading comparatively large fish, like salmon, by either the air-lift principle or by employing an ingenious device using alternate suction and pressure cycles. Also new engineering developments had been applied to the refrigeration systems on vessels engaged in tuna or halibut freezing at sea, in tuna canning (the vacuum principle), and in bulk handling of herring meal.

Other continuing investigations included: attempts to prevent 'belly-burn' in non-eviscerated salmon by holding them in sea water at about $25°F$ ($-4°C$); studies on development of oxidative and hydrolytic rancidity and its significance and control in frozen fish; the value of tetracycline antibiotics in shrimp preservation; control of drip formation in thawed fish; and the effect of sugar removal on browning of heated fish flesh.

Biochemical problems associated with live fish absorbed much research effort. Significant progress had been made in elucidating the nature of food and home stream odor attractants for salmonid fish, in the assay and purification of salmon pituitary gland hormones, and in investigating the post-spawning survival of Pacific salmon. A number of modified techniques had been applied to separation of certain of the muscle, blood, and serum proteins of fish, and it was concluded that there was now no doubt that this method would prove an invaluable addition to existing

241

An entangled narwhal is extricated from a net at Koluktoo Bay

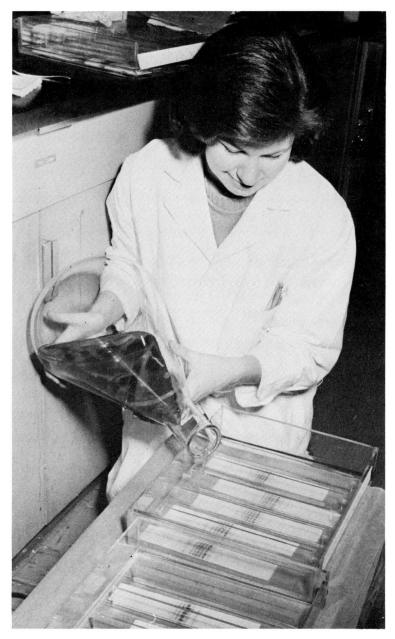

The electropherograms in which proteins are separated in speciation studies are developed chemically, and appear as discrete bands characteristic of the biological species

methods of fish classification. Several fish enzymes had been studied for their ability to form rare chemical compounds, and basic chemical studies had been made of novel marine invertebrate and fish nucleic acids and nucleotides.

At London new programs were concerned with improvement in quality of lake fish and the possible use of rapid spray chilling for this purpose. Research on speciation of freshwater fish by use of protein patterns had been initiated, in cooperation with the Vancouver laboratory.

A project that continued through the terms of several chairmen was that of improving hatchery procedures for fry release. Production of Atlantic, chinook, coho, and sockeye salmon is limited mainly by the extent of natural freshwater nursery areas or of artificially provided rearing ponds. However, our pink and chum salmon have no such requirement: they go to sea directly after emerging from the stream bed. Consequently it seemed likely that there could be a major role for hatcheries in increasing stocks of these species, which do not require expensive and disease-prone rearing facilities. What was needed was a fool-proof and economical hatchery technique that would produce fully viable fry.

One trouble with the traditional hatchery was that it required too large a labor force; all winter long it was necessary to pick dead eggs out of the baskets in order to avoid the spread of fungus that would quickly suffocate the live ones. This problem was solved during the 1940s when satisfactory fungicides were discovered in the United States; added periodically to the water supply, they kill fungus without damaging the eggs. This made it possible to hold eggs on trays stacked vertically, greatly reducing the amount of floor space required in a hatchery.

There remained the question of the viability of hatchery fry as compared with natural fry. Work at Nanaimo during the early 1950s by Dr W.S. Hoar of the University of British Columbia suggested a reason why hatchery fry were not up to standard: being held under ordinary illumination while the yolk sac was being absorbed, they lost the negative reaction to light that makes wild fry emerge at night and move downstream under cover of darkness. Facilities for testing this view were constructed by the Board at Kleanza Creek, a Skeena River tributary. M.P. Shepard kept pink salmon fry there in complete darkness until they set off downstream after the yolk sac was absorbed, but the returns of adults were still unsatisfactory.

The study was resumed at Nanaimo by R.A. Bams, who observed that fry massed in troughs or tanks stimulated each other and so were almost continuously in motion, both by day and by night, whereas fry in gravel are separated and move only intermittently. As a result, trough fry had lost a good deal of body weight by the time the yolk sac was absorbed, so they were smaller than natural fry.

Both the weight loss and the maladaptation to light could be corrected by hatching the eggs in gravel – an idea that had been strongly promoted as early as 1919 by Alexander Robertson of the Fish Culture Branch, although unfortunately he failed to convince his superiors that it should be generally adopted. Field trials were made at a tributary of the Tsolum River on Vancouver Island. These showed that pink salmon fry hatched in gravel boxes are the equal of naturally spawned fry in producing adult fish. Substitutes for gravel that are easier to handle have since been developed in Alaska, and the system is ready for widespread application.

18   The Hayes regime, 1964-9

Several major changes and developments took place during the five-year tenure of F. Ronald Hayes, the Board's second full-time chairman. Two concerned policy. The first of these involved a sharp increase in university participation in Board activities. This was brought about by the establishment of a grants program to develop centers of excellence in aquatic science in Canadian universities, by encouraging universities to use Board facilities, and by promoting graduate student and postdoctoral fellowships at Board stations. Hayes also endeavoured to increase the number of senior Board scientists holding honorary appointments on the staffs of universities. One of the tangible fruits of the new spirit of cooperation was the establishment of the Huntsman Marine Laboratory adjacent to the St Andrews station in 1970.

The other major policy change concerned Board meetings. In contrast to the practice of previous regimes as far back as Reed and Dymond, Hayes sharply curtailed the participation of staff members, with a consequent increase in the influence of Board members. Hayes held most meetings in camera, and staff participated only when requested.

Some of the changes in the direction of Board activities resulted from the growing awareness of water pollution and lake eutrophication and the emerging interest in resource enhancement and aquaculture. These new urgencies led to an unprecedented diversity in the Board's research, and greater emphasis on experimental investigation and the ecosystem approach to studying aquatic life. The research thrust was directed towards productivity studies, effects of heavy metals and organochlorine pesticides on aquatic organisms, fish diseases, and the role of hormones in fish physiology. Highly sophisticated water facilities were developed at several of the stations.

Another change was the disappearance of the traditional functional distinction

A new approach and new problems

245

The Board at the annual meeting, 1965. Seated (l. to r.): I. McT. Cowan, W.R. Martin, A.W.H. Needler, H.J. Robichaud (minister of fisheries), F.R. Hayes, D.B. DeLury, C.W. Argue. Standing: W.M. Sprules, G. LeBlanc, G.L. Pickard, M.K. Eriksen, S. Sinclair, O.F. Denstedt, J.M.R. Beveridge, G. Filteau, R.L. Payne Jr., R.G. Smith, F.E.J. Fry, H. Favre, M. McLean, H.A. Russell

Board senior staff, 1966. Seated (l. to r.): C.J. Kerswill, L.M. Dickie, J.L. Hart, W.E. Ricker, K.S. Ketchen, F.R. Hayes, W.R. Martin, G.I. Pritchard, J.C. Stevenson, J.S. Willmer, W.E. Johnson. Standing: J. Rogers, W. Templeman, D.R. Idler, H.L.A. Tarr, N. Tomlinson

246

The Marine Ecology Laboratory occupies part of the Bedford Institute's main building at Dartmouth. The MEL Fish Laboratory and trailer complex are at left middle ground. From the left, the ships are the *Carino*, *Dawson*, and *Acadia*

between the biological and technological stations, as the latter tended more and more to deal with living animals. A close liaison between biology and oceanography was required to deal with pollution problems. This was evident in the planning of the Marine Ecology Laboratory at the Bedford Institute of Oceanography in Dartmouth, Nova Scotia, and the Pacific Environment Institute at West Vancouver, British Columbia, which incorporated elements of the former Atlantic and Pacific oceanographic groups, respectively. Similarly, when the London station was relocated and became the Freshwater Institute, strong emphasis was placed on the relation of physical limnology to biological processes both at Winnipeg and at a satellite station at the Canada Centre for Inland Waters at Burlington, Ontario.

An ominous development took place when government policy dictated the loss of the Board's separate status as an employer, a feature which Board members believed to be an important factor in contributing to the high caliber of Board scientists. Despite strong objections by Hayes and other Board members, as well as the directors, the move was carried out and, at the same time, government budgetary restrictions together with a shift in emphasis in government science policy from natural science research to sociological and economic fields brought a virtual halt to Board expansion.

A series of broad government administrative changes began with the merger of the departments of Fisheries and Forestry, accompanied by a policy shift towards 'mission-oriented' programs. Clearly, more changes were in store and these were disquieting times for the Board. But Hayes, with the able assistance of Deputy Chairman W.R. Martin, who served in this capacity during four regimes, left the chairmanship with the quality of Board research undiminished.

Frederick Ronald
Hayes
Hayes was born in Parrsboro, Nova Scotia, in 1904, but moved with his parents to Halifax five years later, where his father practiced medicine. Hayes had intended to become a doctor himself, but as a result of his experiences at St Andrews in 1924–7, as already related, switched to biology. He received his master's degree in histology and embryology at Dalhousie and then, in 1927, went to the University of Liverpool, England, to take his PH D. After two years there and a year at the University of Kiel in Germany, he returned to Canada not, as he had originally expected, to join the Board, but instead to replace his old mentor, J.N. Gowanloch, as professor of biology at Dalhousie University.

He remained at Dalhousie, working in the summer, at first with Huntsman and later with M.W. Smith, on salmon studies and on lake productivity. He also worked with the Nova Scotia Research Foundation, studying the chemistry of the mud-water interface. He felt that the productivity of lakes lay in the exchanges between the bottom and the water: 'I regarded the surface sediments as an integration of what had been going on in the lake during recent years, and so began to study the mud-water interface from a bacteriological and chemical point of view.'

At Dalhousie, he set up the PH D program in the Division of Biological Sciences and became chairman of that division, and then set up the PH D program in oceanography and was made chairman of that division as well and vice-president of the university. It was at this point that he was appointed chairman of the Fisheries Research Board.

Following his term with the Board, Hayes returned to the university in 1969 as Killam Professor of Environmental Science and wrote the book *The Chaining of Prometheus*, a study of the evolution of a power structure for Canadian science, in which he traced the changing relations between government and the scientific community in Canada. He was able to write from first-hand experience. A man of sharp and nimble wit, Hayes dedicated his book 'To my colleagues in the public service with the sincere hope that they will find this book too long to have it Xeroxed by their secretaries and distributed free.' He once described a colleague who tried to stretch over too wide a field of science and administration as 'the performing flea of Canadian science.'

Rapprochement
with the universities
Probably the outstanding accomplishment of the Hayes regime was the renewal of close ties between the Board and the universities. Hayes said of this development (in 1972):

Well, I had grown up, of course, in the golden days of St Andrews when it was at its best, and when there was very little distinction between Board people and university people. In fact, they were the same individuals who, as members of the Board, came down and worked in the stations for the summer. That was deliberately removed by A.T. Cameron, who thought it was a waste of money. I was interested in trying to restore some of the working relationships between the communities of the government and the universities, and I still believe this is a very important thing to do. So we got going in the Board a system of grants. I had also had experience as a member of the National Research Council in seeing the methods of [E.W.R.] Steacie [president of NRC, 1952–62], a man whom I greatly admired and who represented perhaps the most successful person in developing university and government relations that we've had in Canada. I copied some of Steacie's ideas in the Board system of doing things, such as saying, for example, that if a student is doing a thesis in a Board laboratory, the Board will pick up the tab for whatever the university would normally subsidize him, and this type of thing. These were Steacie's ideas.

I would have preferred to do more. I would have preferred to have something of the order of 10 percent of the Board's assets going into university support, and also I would have preferred to see the free use of government equipment, such as ships, made available on a basis of university use. That ship business does exist in Halifax with Energy, Mines and Resources and now, of course, that is Environment. That has developed very well.

The Board has never done quite so well elsewhere. Nanaimo gave some ship time to the universities on the west coast. Templeman was rather stingy with a ship: I don't think Memorial University ever got much ship time out of him. So it varied considerably. We have had great cooperation, I think, in Winnipeg. Nanaimo was difficult because of its geographical situation; you cannot have a close day-to-day relationship between the University of British Columbia and Nanaimo. The Technological Lab on the campus of UBC was also rather a loss, perhaps because Tarr thought that the university people were too abstract in their approach.

Hayes felt that there was a limit to the effectiveness of the case-history approach at stations run by classical biologists:

Templeman was doing the systematics of the fishes of the North Atlantic, and general population studies without a quantitative basis. This costs a great deal of money. You have to keep ships at sea. And I felt from my experience living next door to a medical school all my life that the case history method of repetitive observation is of very little value in science beyond a limited number of years. I suppose a classical example of this would be the weather. The meteorologists have been telling us how hot and cold it is every day for the last hundred years, and they can't deduce anything from it. I feel the same way about charting the Pacific salmon runs, for example, in every little stream. If you know there is going to be great variability and you get the level of variability, then that is as far as you can go, except for a specific purpose.

Now the idea of repetitive observations had a great hold on the Board. Even good experimenters like Fred Fry would say: 'You keep on surveying things for 20 years.' The trout catch in Great Slave Lake or somewhere should go on and on. I have felt by analogy with medicine and everything I know about elsewhere that you don't make progress by repetitive observations, except, of course, for immediate use. So I was interested in experimental work and getting it going and I was pushing for the kind of thing that Dickie was doing.

Hayes had made two senior appointments in his first year of office. Dr C.J. Kerswill was made director of the Arctic Biological Station and Dr. L.M. Dickie was made director of the Marine Ecology Laboratory, setting it up as a separate research establishment that reported directly to Ottawa headquarters. 'I had been rather amused. Templeman told me that if the Board ran into measures of economy that the best place to economize would be to wash out Dickie's station altogether. Dickie told me exactly the same thing about Templeman's kind of business. Well, I happen to be a Dickie man on that particular question.'

Meantime, during the Hayes regime a major confrontation took place between the Chairman and the director of the Nanaimo station, Peter Larkin. Hayes discussed this with typical candor:

Well, I may say that I was a great friend and admirer of Larkin ... I used to see him at scientific

Classical versus experimental science

Confrontation with Larkin

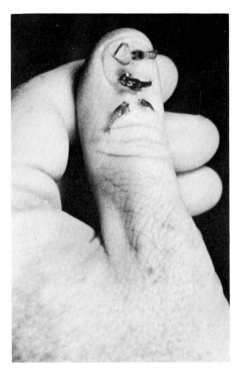

Deep-sea stocks of lobsters have recently been discovered in the Atlantic 50 to 100 miles offshore of Nova Scotia. This 20-pound specimen shows its formidable claws

Lobster larval stages I to IV on a scientist's thumbnail

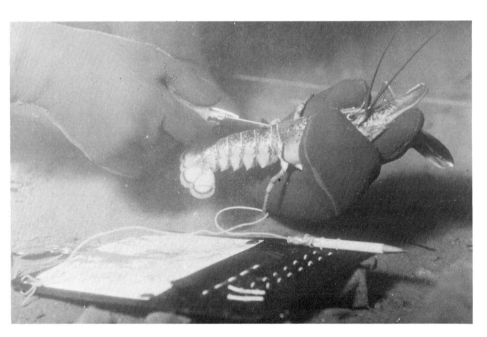

Diver tagging a lobster to obtain information on movements, growth, and survival. Underwater tagging minimizes the effect on the lobster

Scientists lower a fine-meshed net to study changes in abundance, distribution, growth, survival, and behavior of lobster larvae

Fatty Basin Field Camp (lobster project), Pacific Biological Station, mid-1960s

meetings, and he was a very lively fellow. So I was very pleased to know that he had taken the job when I arrived, because I thought he was a grand guy. I think that Larkin must have taken the job without ever sitting down and considering that to run a three million dollar enterprise is a full-time job, and involves some sacrifice of one's own research career. Also he had just left the academic and research autonomy of a university department.

Now Jack Kask, in the face of considerable opposition from some directors, had established the policy of full-time Board professionalism at the national level; there was a headquarters, and policy was made in Ottawa. In Kask's time Stan Beatty had resigned as Halifax director over local autonomy. Larkin was the only director who in my time was prepared to challenge the central policy.

So, Larkin directly challenged me as to whether the policy of the Board was made in Ottawa or in Nanaimo. It had to do with a particular program of high seas fisheries that was very costly and which we had gone into carefully at the Board level, and Larkin decided – as Beatty and the rest of them would have decided in the old part-time chairman era – that it was none of my godamn business. I think too that Larkin got some bad advice from his own staff. Anyway he went ahead, and we came to a showdown ...

Eventually Larkin resigned, and I think that a few sincerely expected that the whole Board was going to come to pieces. Probably the Cabinet would fall or something. And I think they were extremely surprised later. Within 24 hours of the time that this blew up we had appointed Bill Ricker as acting director, we went out and found a director, and things quieted down.

Hayes has spoken about a number of his associates, including Needler:

An appraisal of Needler

Alfred Needler is a big man physically, handsome, impressive, and his face in repose appears to be casting a smile, a tremendous asset to a diplomat. He's also able to entertain people by singing and piano playing, and he likes people, and he has no nerves. So he is able to go to a meeting with the Russians or Japanese or whoever, and sit it out for days on difficult international fisheries treaties and the like, something which I'm sure I couldn't do. He's a natural selection as a chairman of international groups because he never loses his calm, grasps easily the essence of the problem, and has the patience to go on for years without much visible progress.

As deputy minister I think he started out with the idea of conducting considerable reforms and he was stopped by Newfoundland politics. He really intended to make over the department when he came there ... I think Needler sort of lost his reforming zeal when he got licked on several occasions by political forces beyond his control, a very natural and understandable thing ... Latterly the inroads of international travel claimed more and more of his time.

Other appraisals

Hayes commented on assistant chairman Bob Martin: 'I think Martin is one of the most completely honest men that I have ever met. He's absolutely straight, and you can't say that about very many people. I would trust him. There are only about six men I know that I would trust as much as Martin, and so we never had any real problems. I think Martin's difficulty was that he couldn't delegate authority. He was so conscientious that he had to tell a fellow to go and do something, and check up on him to see whether he was doing it.'

On Dave Idler, the aggressive director at Halifax, with whom Hayes had some differences:

I think the best thumbnail remark that I know about Idler was made by Ronald Smith ... who was the head of National Sea Products. After we had some trouble with Idler's general aggressiveness, involving simply trying to put his station ahead and get more money than anybody else, Smith said: 'You know, he reminds me of our highliner skippers; the fellows who catch the most fish are also the biggest SOBs.' I've often thought of that about Idler; he was working very hard for his own station.

He is a compulsive worker, day and night. He used to run his administrative affairs during business hours and then go downstairs to his private lab and work until 11 o'clock. He rationed one night a week to spend at home with his family. Work was his life and I expect he'll go on this way until he kills himself.

John Hart, who was director of the St Andrews station during the Hayes regime, said of him:

Hart looks at Hayes

Well, as an individual, I think he was a very charming and interesting man and amazingly adroit in punch and tongue, which may have run away with him sometimes. But as long as he wasn't talking about me, it was always interesting. As chairman of the Board I think he was a bit inclined to avoid what I would regard as major responsibilities ... But he proceeded to direct all the Board's activities towards fostering and helping universities and in following academic-type subjects. As a result, by the time he retired, I think he left the Board in a position where there wasn't much left for the Board that could not be done equally well by either the National Research Council in the field of pure science, or in the field of application by the Department of Fisheries.

There might be one exception to this, and this would be the work the Board was doing for the international fisheries commissions and, of course, this was something that he would have a tendency to deplore. So I think that maybe Dr Hayes had a concept of the Board's function which was not really intended in the statement of the Fisheries Research Board of Canada Act.

Alfred Needler was diplomatic: 'Well, Dr Hayes is a very able, clever man. His concept of the Board certainly leaned towards the academic side rather than application. This has been Dr Hayes' whole background and career, and he is a very good research man. I think, on the whole, Dr Hayes could be expected to see that there was a high quality, a high caliber of research carried out, and not to put enough effort to contribute too much to the improvement of the application. I think this is actually what happened.'

Needler the diplomat

Ricker said of Hayes:

He had a strong interest in the university association and was instrumental in establishing the system of university grants that still operates. He felt that the Board should make an effort to get close to the universities. John Anderson put this into practice at St Andrews with the Huntsman Marine Laboratory. Surprisingly, Hayes didn't seem too enthusiastic about this at first; if so, he changed his mind later.

Ricker's views

However, Ron Hayes and I differed in respect to the emphasis that should be given to systematic collection and interpretation of data. Fish populations are not static entities that can be studied thoroughly for a year or two and then you know how to manage them for all time. Each and every one of them changes continuously from year to year, in response to a variable environment and in response to the fishery. The more important a population is commercially, the more exposed it is to changes: in abundance, rate of growth, age composition, and so on. As I see it, one of the Board's main responsibilities is to analyze the state of our fish stocks on a continuing basis, and particularly to sound a clear warning if overexploitation is threatened. Ron seemed to think that this was someone else's job, if it needed doing at all, and that the Board should restrict itself to short-term experimental work. Fortunately the needs of the international commissions prevented any major erosion of Board involvement in this area.

Ricker thought that Hayes had done his best to save the employment clauses in the Fisheries Research Board Act, but had to yield when collective bargaining was introduced:

The first year of collective bargaining for scientists' salaries was a tremendous waste of time for all concerned: no one on the Board's staff was experienced in such negotiations, and the other side was in much the same fix. It makes no sense to go through this rigmarole every year for so small a unit as FRB. A lot of us felt that we should have got tied in with the National Research Council, which up to that time had been the norm for FRB salary schedules. But Board scientists were never offered that option. A majority of them had joined the Professional Institute, never dreaming that it would some day become a bargaining unit. Yet the new law stipulated that any employee association having 50 percent or larger membership automatically had that role. No vote was taken to ask the scientists: 'Do you want to enter into collective bargaining?' And so the Board lost an important aspect of its independence, which the National Research Council and Defence Research Board still retain.

The honeymoon over   But there were deeper and more serious problems arising for the Fisheries Research Board and its chairman than the personality of Hayes or even the loss of independence in the Board's employment policy. The honeymoon that science had enjoyed with government in Canada, which Needler has described as beginning in the postwar period and ending in the early 1960s, was coming to a close. This was reflected in a slowing down of the rate of expansion for the Board, restrictions on budget, which had begun in the Ricker interim, and restrictions on the hiring of staff.

Although spending on research and development in Canada still lagged behind that of other 'developed' nations, the government decided to take a hard look at its scientific creations. The Royal Commission on Government Organization, the Glassco Commission, had been set up by the Diefenbaker government in 1960 and in 1963 it reported to the Pearson government. Like all public review bodies, it had to recommend extensive changes in order to justify its own existence, without being in a position to assess their ramifications and eventual costs. Its Report no. 4 recommended, in effect, that the Fisheries Research Board be absorbed within the Department of Fisheries.

Kask, commenting in 1972 about this recommendation, noted that it was in direct contrast to the commission's recommendation concerning the Defence Research Board, for which it proposed wider powers and complete control of development as well as research activities. He said:

Bringing the research assignments under one head, of course, has merit; but the developmental and policy recommending aspects introduced in the new DRB were completely lacking. This, in my view, was a grave error. And by recommending that the Board be placed in what the commission itself labels 'a department whose functions are predominantly operational and regulatory' gave the genius of the founding fathers, who provided for a flexible, politically buffered, independent research organization, a sound trouncing, whereas the same commission strengthened and honored the concept of the independent research board in the case of the Defence Research Board. In my view, this recommendation was a retrogressive step. One of the reasons this recommendation occurred at all, I think, is because people who deal with financial resources tend to gravitate towards formulas that work for them and with which they are familiar.

Hayes, in his first annual report, for 1964, showed his awareness of the ways the government winds were blowing towards science, and in particular, towards the biological sciences:

Hayes senses
the winds

Government planning for science in Canada is critical for the biological sciences at this juncture. Lack of an appreciation of the influences involved may permit establishment of a pattern that would impair, for many years, the development of areas of marine and fresh-water study with which the Board is concerned. The Second World War brought a marked distortion in the balance of research fields. The dramatic outcome was the atomic bomb. In Canada, the National Research Council program became weighted towards physics, especially nuclear physics, and engineering. Later, separate research groups were formed for nuclear physics in its peacetime applications: Atomic Energy of Canada Limited, and for defence – the Defence Research Board. World emphasis on medical science was reflected in Canada by the establishment of the Medical Research Council, an outgrowth of the National Research Council.

In April 1964, a report by Dr C.J. Mackenzie on the recommendations of the Glassco Commission – Royal Commission on Government Organization – concerning science was tabled in Parliament. This report endorsed the Glassco recommendations that a secretariat or 'Central Scientific Bureau' and a 'National Scientific Advisory Council' be established, the latter to represent federal, provincial, industrial, and academic science. The permanent federal seats on this proposed committee show a preference for physics and engineering rather than research areas associated with food and biological sciences generally.

The last annual report produced by Hayes was in 1968, and it reflected his concern about developments outside the Board which might well determine its future. About 'Policy Planning' he wrote:

The Board under
scrutiny

The year 1968 has been a time of reassessment and review of purpose for the Fisheries Research Board of Canada. Early in April the Government-University Committee of the Science Council visited the Fisheries Research Board. In the same month the Marine Science Committee of the National Research Council, on which the FRB was well represented, examined the possibilities for university use of our laboratories. Teams from the Organization for Economic Cooperation and Development visited the FRB in March and in November; the Committee's report on Canadian science policy is expected in 1969. Following the announced fusion of the federal Departments of Fisheries and Forestry, a government task force looked into the FRB and its functions. In September, the Science Council visited the Winnipeg Laboratory. In December, Board representatives appeared before the Senate Committee on Science Policy and a 139-page brief was presented. The Science Council is developing plans for reports on Canadian marine science and on fisheries and wildlife. All these reviews are in addition to those regularly carried out by the Board and its committees.

Hayes met implied criticism head-on when he stated that management proposals contained in a recent Science Council of Canada report would not require changes in 'present FRB practice.' He warned: 'The recent economy drive has brought to decision a direction of government planning that has been increasingly noticeable for several years. Science is now in rugged competition for funds with other activities. The rationalization that is taking place is to define major national goals, which are not in themselves science-based but which, in today's world, are likely to have a component of science and to require large multidisciplinary, mission-oriented research projects.'

The Pacific Biological Station, Nanaimo, after foreshore development in 1966. The Taylor wing, standing partly in front of the main building dating from 1949, is shown in this aerial view of the station complex. The research vessel *G.B. Reed* is in the center foreground. The *A.P. Knight*, the barge *Velella*, and the *Caligus* are also alongside

The courtyard of the Taylor wing, circa 1960

Then, noting that the Science Council's report had enumerated some national goals, Hayes listed a few of his own for the Board that placed the Board squarely in the national picture. These were:

National sovereignty over coastal waters: The resources of the continental shelf; their description, conservation and use.

Recreation and use of leisure: Aside from national parks there has been little federal acknowledgement of responsibility in this area, but some concern is likely to develop for water quality in relation to sport fishing, boating, and bathing.

Balance of payments: Fisheries products are largely exported to the United States. Continued success in foreign markets depends on improved quality and guaranteed supply.

Education: Major facilities essential for training scientists, such as seaside establishments and ships, may best be managed for national use by a government agency.

Fitness of the environment: Fish are the best indicators of water quality; except in navigable waters, the only effective federal laws are those that protect fisheries.

Foreign aid: FRB contributions so far have been based on informal individual arrangements.

The Just Society: Technical advances will help to equalize economic opportunity in all regions.

National defence (marine): Defence and fisheries interests overlap in environmental oceanography programs.

Multiple use of resources: The land-water complex involves river basins for fish, power, and forests.

Having made his case for the role of the Board in future government plans, Hayes set out to show how pertinent were its present activities, and how flexible:

Many of the goals are tied closely to those of the Department of Fisheries, which has responsibility for fisheries management, and continuous vigil is kept to assure that FRB is responsive to the research needs of the Ministry. To keep abreast, FRB participates in departmental and interdepartmental task forces aimed at defining the needs of tomorrow, examining such diverse topics as water pollution, Atlantic salmon, herring and information retrieval. There is growing concern that some stocks fished on the high seas are approaching full exploitation and that the catch will be in need of regulation. Such needs must be assessed by international studies, and FRB has continued to share in these responsibilities. The demand for pollution control has created an urgent need for research to understand the biological mechanisms in eutrophication, and to define indices and criteria of water quality. Many Canadian fisheries undergoing rapid change, such as the Atlantic herring, queen crab, seaweeds, and lake fisheries, are receiving special attention.

As to recent program changes, the Dartmouth Laboratory is now well launched on new studies of environmental oceanography and the production chain. The Arctic Laboratory has strengthened its marine mammal studies and reduced its concern with northern fresh waters. The Freshwater Institute, now established in Winnipeg, has revitalized its program on food technology and on freshwater fish populations. It is also well advanced with its investigation of eutrophication, and has a branch laboratory at Burlington on Lake Ontario.

A new laboratory site was acquired at West Vancouver for product research under field conditions, and to serve as the western base for antipollution research and development. At Nanaimo, a major extension is scheduled for completion in the summer of 1969 and planning is well advanced for new laboratory complexes in Winnipeg and St. John's. A vessel, the *E.E. Prince*, was built in 1966 for pelagic studies, and is now operating out of St. Andrews, reflecting the increased interest in herring in our Atlantic laboratories. The FRB fleet is used

257

The submersible *Pisces II* and diver in a rubber raft, August 1970

A PLC 4B shelf diver on board ship, after completing a dive, August 1969

to capacity, and recruitment of a vessel manager has improved coordination of the Atlantic ships.

FRB has long encouraged its scientists to publish their studies, and its scientific publications such as the *Journal of the Fisheries Research Board of Canada* have international renown. The number of contributions to the *Journal* has again shown a large increase, particularly from Canadian Universities. With the publication in 1968 of a Bulletin comprising a cumulative index and list of titles to all FRB publications since 1901, information on its studies in the past can be easily retrieved and made available to a large variety of users ...

During 1968 the scope of university support provided by the Fisheries Research Board was increased by the conclusion of a working arrangement with the Department of Zoology, University of Toronto. The year also saw an increase in the program for visiting investigators at the St. Andrews Biological Station. The third year of the University Grants Program was marked by its emergence into a period of more clearly defined functions. Policy development led to an increase in the proportion awarded as Development and Block Grants, and in support of collaborative FRB University projects.

The Board continued to operate under the cloud of financial restrictions. In 1968 Hayes wrote: 'The Fisheries Research Board has tended to respond to the current economy drive as a short-term emergency. If the fabric of the FRB continues to erode 5 per cent to 10 per cent per year, more radical treatment will be indicated, possibly reducing both the number and extent of projects.' *Budget blight persists*

Concerning staff, he observed that although the Board had 853 full-time positions authorized for the year, 'a freeze applied to the entire Public Service during the past year reduced the number of positions to 834 and limited the casual employment to 88 man-years.' This actually represented a decline of three persons from the previous year.

RESEARCH

Dealing with research programs for the year, the report revealed the increasing sophistication and widened range of subjects under investigation. Concerning products, fish protein concentrate (FPC) was receiving increased attention with plans announced for a commercial plant in Nova Scotia; ultrasonic instruments developed in the Winnipeg laboratory promised to be valuable tools in fish inspection; vacuum-sealed air-tight membranes improved both quality and taste of frozen groundfish blocks and fillets; pigmentation of the flesh of juvenile pink salmon, cutthroat trout, and rainbow trout produced a canned and processed product that had the appearance of coho salmon; a color-sorting instrument had been developed for the sockeye and coho canning lines which successfully predicted the color of canned salmon by testing the raw fish; scallop meats frozen pre-rigor were found to be markedly superior in both taste and texture to those frozen in-rigor or post-rigor; ice-making machines and prefabricated insulated sheds were engineered to be readily portable by plane for fishing remote lakes; the first commercial model of the patented 'airlift' pump was installed at a Vancouver cannery for unloading salmon. *Product research*

The Board participated in the work of seven international commissions. During the year new ways were sought to study the depths of the ocean. Scallops and their behavior relative to trapping techniques were studied by scuba divers. New instruments were developed to record the action of otter trawls and determine how to improve their effectiveness. The activity of Atlantic lobsters transplanted to the Pacific was tracked by remote control instruments. The submersible *Pisces* was used on the Pacific coast to observe the performance of and recover instruments *Harvesting and management*

Scientists of the St Andrews staff attach a TV camera to a scallop dredge to monitor the effects of the dredge on a scallop bed, 1971

Scallops on sea bottom

A towed underwater automatic camera
sled is hauled up during a cruise in 1970

A towed underwater research plane, 1970,
designed and built at the St Andrews
station

that had been moored to the bottom for a year to record sound, to survey bottom features, to collect samples, and to identify scattering layer organisms.

In the Atlantic, a two-man submarine was used to study invertebrates and their behavior in a natural environment. Acoustic echo-sounding equipment was adapted to count bottom-living fishes and provide information on their sizes. First attempts to study fishes associated with sound-scattering layers showed that catches were not strongly correlated with echo-strengths. Continuing studies on the reaction of fish to noises showed that the ambient noise of the ocean can mask gear noises to some degree, but fish are able to detect gear noise at least as well as man.

Areas of shrimp abundance on the Scotian Shelf could be predicted from the type of sediment on the ocean floor. Prawn tests revealed the superiority of oblong traps with sides of plywood and plastic sheeting to other materials, and exploratory longlining in the Gulf Stream off Nova Scotia from March to June revealed substantial resources of yellowfin tuna that were not being exploited.

Infrared aerial photography was used to delineate seaweed beds; on floating vegetation or vegetation exposed by the tide, different species showed varying infrared reflectivity.

Atlantic salmon were collected northeast of Labrador and from both the inshore and offshore Greenland fishery to identify fish originating in Canadian waters.

The investigation of protein characteristics supplemented conventional techniques in identifying the 55 Pacific rockfish species; the technique was found to be most reliable.

Research was under way on both coasts to find new industrial uses for marine plants, which were rapidly becoming important commercial commodities.

Hatching of petrale sole larvae was attained in the laboratory for the first time.

It was found that the pituitary gland of salmon was intimately involved in sexual development, in the start of migration, in the cessation of body growth once the fish

261

Small-egg incubator, used with marine species such as flatfish, cod, herring, and pollock

Rainbow trout fish farming in a prairie pothole lake

enters fresh water, and in the degenerative changes associated with spawning.

The raft culture method of growing Pacific oysters was demonstrated on a commercial scale throughout the British Columbia coastal region, including the Queen Charlotte Islands. Increasing the resource

Culture methods for hatching and rearing hybrid Pacific salmon were developed to the point where young hybrids survived as well as the pure stocks. Time from fertilization to hatching is inherited from the male. Hybrids with one sockeye parent were found to be more tolerant of direct exposure to sea water at the time of complete yolk absorption than were the pure sockeye.

Rainbow trout stocked in June in a Manitoba lake that had undergone 'winter kill' grew from 1.5-inch fry to harvestable size by October. Mortalities were high, probably due to predation by salamanders. Considerable success was also achieved from stocking young whitefish.

Frozen sperm of Atlantic salmon were used successfully to fertilize eggs, but the conditions necessary for freezing with predictable results and long-term preservation still remained to be established. By contrast, the sperm of cod and herring were both frozen successfully and used for fertilization of eggs under conditions where success could be predicted.

Survival rates and chemical composition of sockeye salmon fry from spawning channels constructed by the Department of Fisheries were the same during lake residence and at the time of seaward migration as for fry spawned in natural streams.

THE DISCOVERY OF THE ENVIRONMENT

Three social phenomena had a world-wide impact during the 1960s: the student revolts, the return of beards and long hair, and the public's sudden concern with its environment. What connections there may be among these will exercise historians for a long time to come, but there is little doubt that the proximate cause of the environmental movement was that the smog level in major cities had finally become intolerable. Once aware of this, it was not hard to discover numerous other disturbing effects of increasing populations and accelerating gross national products: for example, more pollution and new kinds of pollution in rivers and lakes, and even in the ocean; receding forest lands and encroaching deserts; wildlife everywhere in retreat; great salmon runs exterminated; and the mighty cataract of Niagara, once a haunt of honeymooners, reduced to a second-class waterfall that stinks of sewage. On the horizon there loomed ominously irreversible shortages of oil and certain minerals. All in all, the environment appeared to be worsening rapidly.

Government agencies were quick to respond to the new climate and the Fisheries Research Board was no exception, although it was able to point out that it had already been concerned with the aquatic environment for more than half a century. The Board's most conspicuous response was a quantum increase in its work in fresh waters, where the effects of pollution were most serious. A key figure in this change was the colorful Wally Johnson, a man of driving energy and tremendous enthusiasm.

Johnson was born in Wisconsin, but had a strong Canadian attachment through his mother, who came from an Alberta homestead. He studied at the University of Wisconsin, where he was greatly influenced by the eminent limnologist Dr Arthur D. Hasler. While studying the biological productivity of a series of bog lakes that had been treated with lime Wally encountered what he described as 'one of the Waldo Eugene Johnson landmark publications in this field, the first measurement of real production of fish per unit area or volume of water, done by Ricker and Foerster ... It was the most influential paper in my graduate thesis work ... The whole model of the calculation of the production was based on this fundamental paper.' Later on Johnson met Ricker when the latter came as a guest lecturer to the University of Wisconsin: 'I'll never forget my first meeting with Ricker. I had studied this man's papers, which had been published over a period of 20 years, and I visualized him almost as a young theology student would picture Jesus Christ. And here Bill Ricker appears on the scene, the most unassuming, young looking, kindly man I had ever met in my life, and it was completely overwhelming. I just couldn't believe it.'

Ricker's impression must really have been Messianic, for Johnson promptly applied for and obtained a postdoctoral fellowship from the United States National Science Foundation that enabled him to spend a year with Ricker and Foerster at Nanaimo. That was in 1954, and the following year Johnson was taken on the scientific staff of the Fisheries Research Board. Concentrating on sockeye salmon research, he developed a net that proved to be the first effective method for catching lake-dwelling salmon fingerlings, and then made the startling discovery that Babine Lake, which accounted for 80 percent of the sockeye run in the Skeena River, was being very incompletely utilized by the growing salmon, to less than 25 percent of the lake's plankton-rich capacity.

Johnson's continuing research confirmed his original estimate of the lake's plankton potential and traced the migration patterns of the salmon through the lake. His published report earned a Wildlife Society award for the year's best paper concerning fish management and fish ecology; but, more importantly, he said in 1974, 'One of the greatest satisfactions and perhaps the greatest satisfaction of my research career has been the fact that, as a result of my work on Babine Lake, in the mid-60s, they began an enhancement program to create, at the cost of 10 million dollars, some artificial spawning channels which would produce a much bigger population of young fry entering these upper regions ... of Babine Lake ... This has been the culminating satisfaction of my life's work, really, to see that paying off, and the Indian and other fishermen on the Skeena River are going to be three or four times more wealthy in the next two years, and from then on, than they ever were before I did my work.'

Ottawa maneuvering

The call from Hayes to Johnson came after he had spent a year with Jack Tully at the Pacific Oceanographic Group; he was wanted in Ottawa. Later he spoke about the move:

I departed for Ottawa early in 1965, to spend a year there; the task being one of looking at what sort of planning we should be making for the future work of the Fisheries Research Board in freshwater fisheries, and for work generally in fresh waters ... I was already informed when I came there that the Fisheries Research Board had discussed the future of the then freshwater establishment at London, Ontario, for some time, and it had made the decision that this laboratory ought to be moved to Winnipeg, which was the headquarters of the Fisheries Service, now Fisheries Operations. They said they were investigating possibilities of setting it up on some university campus, hopefully in Winnipeg ... Discussions with a number of prairie universities had shown that the University of Manitoba would be very pleased to have on campus a government institution with which it would establish close relations by appointing our senior staff associate professors or honorary professors, much as they had done for 30 or 40 years with the Department of Agriculture ...

The year 1965 was really the beginning of public and political recognition of environmental pollution problems. In that year the problem of pollution in Lake Erie and Lake Ontario was referred to the International Joint Commission (IJC). I was asked to represent the Board and be adviser to Bill Sprules, who was a member of the IJC board ... Well, I was simply astounded at this meeting. Here was a group of Canadian government department representatives talking about the pollution problem on the lower Great Lakes, and I had to spend almost one hour explaining what eutrophication was, and why it was likely the most important pollution problem in the lower Great Lakes. This was the beginning of a strong involvement of the Fisheries Research Board in problems in the area of eutrophication. As this went along, I was continually called on by this IJC group to advise them on what kind of program Canada should have for coping with the pollution problem in the lower Great Lakes. I spent ... the next six months in drawing up a program that the Fisheries Research Board could do on a continuing basis to help Canada cope with this growing problem ... Eventually a proposal was submitted to Cabinet that the Fisheries Research Board be a prime participant in Canada's program to deal with the pollution problem in the lower Great Lakes, which called for us establishing a staff of 100 people at a cost of two million dollars a year. This passed through Treasury Board in a matter of a few months ... This greatly increased the resources which would be available to a new freshwater laboratory which was being contemplated for Winnipeg.

Johnson was not bashful about his interest in becoming director of the new laboratory: 'I made it known to Bob Martin that I was keenly interested in being a candidate. I felt I could do a better job in setting up such an institute than any other candidate in Canada or any place else. I don't know how much water this cut, but I was offered the job and I accepted it.' The name chosen for the new station was the Freshwater Institute.

Johnson still had a couple of months to spend in Ottawa, but he immediately composed a letter announcing the creation of the new institute.

I was looking for staff, and primarily for three section heads. One, a section on eutrophication and limnological work to cope with this problem of overfertilization and eutrophication of lakes; another section to be concerned with freshwater fisheries problems; another to be concerned with technological problems of freshwater fishes. There was already a technological group at the London station and there was a determination by the Board to make this laboratory one that included all aspects of fisheries work in the inland regions. I've always felt very strongly that if you want to do a good job you get the best people you can, and you should pay any money you have to get them. I sent off something like 250 copies of this letter to every freshwater laboratory and every university in the world that I knew of that trained freshwater people and to many individuals personally. I had a tremendous response ... I wanted people who were outstandingly qualified, and who had a view, as I did, that we ought to apply the best kind of science to solve some of the relevant real world freshwater fisheries and freshwater pollution problems. That was the criterion.

The Freshwater Institute was set up at a time when the Fisheries Research Board was still an independent employer. In 1974 Johnson said: 'The advantages were obvious now when we look back. Once you're in the Public Service Commission, you're subject to a lot of bureaucracy in hiring of staff. When I became director of the Freshwater Institute, I did my own advertising for staff. Once I was satisfied I had a man I wanted, I just got him on the telephone and said: "I offer you so many thousand dollars a year; when can you start work?"' He contrasted this with the situation in 1974, when he was director of the Nanaimo station: 'First of all, we're part of this government system, the Public Service Commission and everything that applies in the government. Secondly, and even more important, is the fact that I no

The Freshwater
Institute at
Winnipeg

265

The Freshwater Institute, Winnipeg, opened in 1973

The main entrance

The stairwell above the main entrance

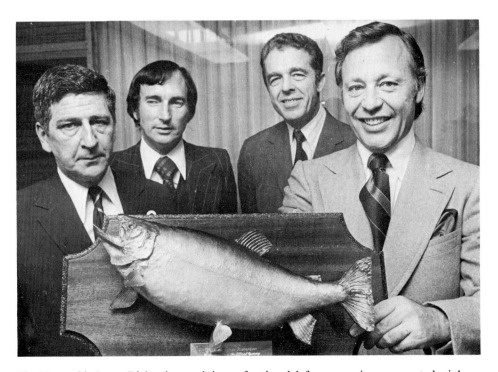

The Honorable James Richardson, minister of national defence, receives a mounted rainbow trout at the official opening of the Freshwater Institute. With Mr Richardson are, from the left: G.H. Lawler, director of research at the institute; G.R. Douglas, director of operations; and W.E. Johnson, director of the Pacific Biological Station at Nanaimo and former director of the institute

longer have any personal staff responsible to me. They are part of a centralized personnel service for the Department of the Environment. They're completely ineffective as far as I'm concerned... We used to have one personnel officer and a personnel clerk, who did a fairly adequate job for us. Now they have been taken away from us, given to the DOE centralized personnel services, and we have no service whatever. We might as well be a nondescript body floating in the atmosphere.'

A proud
accomplishment

Whatever his problems in 1974, Wally looked back on the Winnipeg experience warmly: 'The seven or eight years I spent in Winnipeg were perhaps the most productive of my life as an administrator, because I had this glorious opportunity of establishing what has become during this period perhaps the leading freshwater institute in the world. This was acknowledged as such by many at the Congress of Limnology held there this last summer.' Johnson had secured this congress for Canada: 'In 1968 I issued an invitation before the International Congress when it was meeting in Israel, without any blessing from anybody or not knowing where I could get one nickel to support it. I said that Canada would host the 1974 Congress in Canada. Most of the Canadians there said: "How can we commit ourselves to that?" I said: "We'll find a way."'

Wally concluded:

My years in Winnipeg were among the most satisfying. It was a golden opportunity for anyone to have a chance to start a new institution, to plan a new building, a magnificent new facility, to recruit people from 14 different countries: the best people we could get any place on earth to come there. Of course, one of the crowning accomplishments was recruiting Jack Vallentyne, then a full professor at Cornell University in the States, a native Canadian; to bring him back to lead this group on eutrophication and limnology, and later appoint him our senior scientist. You know, it was Jack Vallentyne who really carried the ball in organizing the Limnological Congress. He was elected president of the International Society for the next six years.

Another of Johnson's recruits was Richard A. Vollenweider, a well-known European limnologist who had had extensive field experience from Italy to Sweden. His specialty was the 'secondary' production in lakes – the transfer of energy from plant plankton to herbivorous water-fleas, rotifers, and copepods. He was placed in charge of the institute's Great Lakes unit near Burlington, Ontario. This was in the Canada Centre for Inland Waters, situated on the long bar that cuts Hamilton Bay off from the rest of Lake Ontario. The Board's program dealt with the biological aspects of the Great Lakes studies, particularly those related to eutrophication and pollution. In 1973 it became an independent operational unit called the Great Lakes Biolimnology Laboratory.

The Winnipeg and Burlington laboratories formed the spearhead of a determined fight to improve our freshwater environments. An early result was the demonstration of the predominant role of phosphate-based detergents in rapidly increasing the biological production in lakes to nuisance levels, as shown by surface water blooms and oxygen depletion at a depth – a syndrome that has been given the name eutrophication. After considerable opposition and obfuscation from the industry affected, restrictions were placed on the use of such detergents. The effectiveness of this and other measures is being measured by changes in the abundance of certain algae.

As a substitute for phosphates in detergents, extensive tests were conducted on nitrilotriacetates (NTA), which proved non-toxic and much less biogenic than phosphorus. They also presented promise as a sequestering agent to combat copper and zinc pollution from mining activity.

Studies of crustacean zooplankton in Lake Ontario demonstrated that thermal factors could be assessed independently of other pollution problems.

The persistence and effects of specific pollutants were tested in small experimental lakes that were low in total dissolved solids and could be readily manipulated. Studies were conducted with humic substances as part of a program on transfer of nutrients between sediments and water.

Research on microbial conversion of industrial wastes that pollute water was under way at the Vancouver laboratory, in cooperation with the University of British Columbia. The first completed phase concerned the synthesis of vitamins from waste-sulfate liquors.

Pesticides in water could be reliably detected from their effect on fish tissue cells cultured in suspension. This new method was found to be simpler than conventional ones for studying water pollution.

Engineering studies on clarifying waste-water from fish-processing plants showed that a combination of screening and air flotation was most effective. Costs could largely be offset by the products recovered.

Hayes must have known that time was running out for the Board. The Glassco Commission recommendations were suspended over its head like an axe and needed only the weight of a strong parliamentary majority to have them im-

269

The Great Lakes Biolimnology Laboratory is housed in the Canada Centre for Inland Waters, Burlington, beside the ship canal. Shown in the photograph is the Centre's seven-story Main Administration and Laboratory Building, behind the Research and Development Building, the Water Quality Pilot Plant, and the Hydraulics Laboratory. Behind the buildings are some of the trailers that originally housed the Centre's staff

Zooplankton sampling nets, Great Lakes

Lowering a Dietz-LaFond bottom grab in Howe Sound for a bottom sample to be analyzed for mercury. One of the pulpmills is shown in the background (Photo courtesy the Vancouver Sun)

plemented. This majority was to be provided in the elections of 1968 which saw 'Trudeaumania' sweep the country. Nevertheless Hayes remained unintimidated and in that same year a 178-page document, *The Fisheries Research Board of Canada and Its Place in Canada's Scientific Development, 1968–1978*, was prepared by the Board with the endorsation of its chairman. In it the Board laid bold claim to a sharply expanded role of research in the five areas of environment, resource, increasing the resource, harvesting and management, and commercial products. Carefully documented and well-reasoned, the report's summary concluded: 'To meet commitments in the proposed plan for development in the next decade, FRB annual operating expenditures will have to be expanded more than threefold, from 10.9 million dollars to about 40 million. The corresponding increase in research scientists would be from 261 to 470.'

Hayes may be charged with various errors of omission and commission in his five-year tenure as chairman of the Fisheries Research Board of Canada. One thing that he may not be charged with is lack of boldness and imagination. And perhaps some of the most enduring contributions that the Fisheries Research Board brought about, took part in, or inspired were made during that tenure.

19 International agreements and commissions

Starting in the last century, committees have been set up from time to time by Canada and the United States to consider fishery matters (including marine mammals). However, these have usually been temporary bodies that dealt with special problems, and if they initiated research programs, they were of short duration. The first continuing international agreement concerning a 'fishery' related to the fur seals of the North Pacific. In 1911 a convention was concluded between Japan, Great Britain, Russia, and the United States. Its object was to end pelagic sealing, which had taken about 1,000,000 skins up to 1909, mainly from female seals, and had killed perhaps as many more again that escaped wounded or sank before they could be recovered. As a result the seal herds had decreased disastrously. The convention provided that henceforth the only kill was to be on land, where surplus young males could be selected, and the skins were to be divided among the four countries according to an agreed formula. Canada had been represented at the negotiations by W.A. Found, who a little later became assistant deputy minister. In the latter capacity he had to deal with a multitude of claims for compensation from ex-sealers and their heirs, in spite of the fact that sealing had become unprofitable and would soon have disappeared entirely. The 1911 convention did not call for research work on the seal herds, even by the nations that owned the rookeries, and in fact very little was done.

During the Second World War this convention was abrogated by Japan. In 1957, following protracted negotiations, a new convention was established between Canada, Japan, the United States, and the USSR, which set up the North Pacific Fur Seal Commission. This called for research to be done by all four nations, including pelagic investigations to determine the effects of the seals upon various fisheries.

The Canadian contribution has been made by Board and university scientists, notably G.C. Pike, Michael Bigg, and H.D. Fisher. Whereas up to 1940 the Pribilof seal herds had been increasing, during the postwar period they became stabilized. Eventually they were purposely decreased somewhat because rookeries seemed overpopulated and breeding success had declined, perhaps partly because of intensive utilization of their food fishes by the new Bering Sea trawl fisheries. Tagging studies showed that the seals ranged widely: those born on the Pribilofs were found off the coast of Japan as well as down the North American coast to California, and a few tags from the much smaller Soviet herds were taken in our waters.

INTERNATIONAL PACIFIC HALIBUT COMMISSION (IPHC)

The Halibut Commission was set up in 1924, on the basis of a 1923 convention between Canada and the United States, which has been modified several times subsequently. The impetus for its formation came from a marked decline in success of fishing for halibut on the older banks, and consequent belief that conservation measures were necessary. In fact, a winter closed season and certain closed areas were established by the terms of the original convention. The commission's first name was simply the International Fisheries Commission, reflecting the expansive horizons of its director of investigations, Dr W.F. Thompson, a student of David Starr Jordan's at Stanford. Dr Thompson had already done several summers' work on halibut during the First World War, in the employ of the province of British Columbia. During this period the Nanaimo station, after the pioneering studies by J.P. McMurrich and Arthur Willey, discontinued halibut investigations so as to avoid duplication, and deployed its very limited resources into other areas.

The new Halibut Commission was empowered to hire its own research staff from among biologists of both nations, and several men who had worked at Nanaimo were so employed, notably two future directors, Harry A. Dunlop and F. Heward Bell. The commission's headquarters were in Seattle, but for field operations it had workers at several ports, particularly Prince Rupert, and contacts between Board and commission biologists were frequent.

The Halibut Commission's intensive study of a single species produced very happy results: its excellent statistical system and penetrating biological analyses soon provided models for work with other groundfishes on both the Pacific and the Atlantic coasts.

INTERNATIONAL PACIFIC SALMON FISHERIES COMMISSION (IPSFC)

Negotiations to establish international control over the Fraser River sockeye salmon fishery dragged along at a snail's pace throughout the 1920s, but in 1930 a draft convention between Canada and the United States was drawn up and signed. However, seven more years elapsed before it was ratified by the United States Senate. The delay both in the negotiations and in the ratification came about largely because the Puget Sound salmon industry was loath to surrender its geographical advantage: it had first crack at the fish coming in from the ocean, and it was taking more than two-thirds of the total catch, most of it in traps fixed at strategic points along the coast. In 1934, however, the Washington seiners and sport fishermen 'ganged up' on the trap owners and passed a public referendum that made the traps illegal. Immediately Canada started to take the larger share of the sockeye catch, so that a legally imposed 50:50 division suddenly became quite attractive south of the border.

In 1937 the convention was ratified. Like the halibut convention, it set up a commission with power to conduct its own research program. The first director of investigations of the new commission was again W.F. Thompson and, predictably, its name was also quite inclusive: the International Pacific Salmon Fisheries Commission. Three members of the commission's original staff, R.E. Foerster, J.L. Kask, and W.E. Ricker, were all to loom large in the later history of the Fisheries Research Board.

The Salmon Commission, as it is usually known, mounted a series of investigations that confirmed and extended the findings of the international tagging program of 1918 and Canada's later identification of difficult water stages at Hell's Gate. At the same time it corroborated the very high rate of exploitation that the early upriver sockeye stocks had been experiencing: more than 90 percent after the Washington traps were abolished and more than 95 percent before that event. The commission proceeded to design and install fishways of a revolutionary new type at Hell's Gate, and for a time it severely reduced the fishing toll on the early runs. As a result, many upriver stocks of sockeye and pink salmon were soon restored to a reasonable abundance, although some have not yet responded adequately.

Over the years the commission has undertaken a great variety of additional research programs. Of particular importance has been the identification, by F.J. Ward and P.A. Larkin, of a mechanism of interaction between sockeye and trout in Shuswap Lake that can explain the 'big-year' phenomenon, similar to but not identical with the competition postulated by Huntsman as the cause of Atlantic salmon cycles. W.E. Ricker had previously shown that big years must be self-generated within each lake, and the development of the big Adams River line during the 1920s was in fact accompanied by suppression of initially promising runs of the other three lines there. From 1950 onward the commission's management policy has been based on the premise that great inequalities in the four sockeye lines are inevitable in most lakes and that, contrary to the former facile assumption, it is futile to try to make every year a big year. A few of the commission's investigations have been done in cooperation with the Board, for example Idler and Clemens' analysis of energy expenditures of sockeye of different stocks during migration.

In 1957 the United States was persuaded to bring pink salmon also under the control of the Salmon Commission, after the Canadian industry had mounted a seine fishery in the Strait of Juan de Fuca that threatened to cut off most of the supply to Puget Sound waters. To establish the relationships between pinks inside and outside of the treaty area a major cooperative investigation was carried out in 1959, involving the commission, the Washington Department of Fisheries, the Canadian Department of Fisheries, and the Fisheries Research Board. Dr A.S. Hourston was the Board's key man in this operation. It determined the migration routes of the pink salmon, and their rates of travel and rates of exploitation, in the marine waters of southern British Columbia and northern Washington as well as throughout the Fraser River system and in numerous smaller streams.

INTERNATIONAL WHALING COMMISSION (IWC)

Canada signed the International Whaling Convention in 1946, and participated actively in the resulting commission from its first meeting in 1949. The commission's global approach to development of knowledge concerning wide-ranging large whales gradually resulted in international agreements to protect endangered species and to conserve those stocks capable of producing annual harvests. This international perspective has been helpful to Canada in understanding and manag-

K. Radway Allen, director of the Pacific
Biological Station, Nanaimo, 1967–72

ing her land-based whaling industry in British Columbia, Nova Scotia, and Newfoundland. Whaling in British Columbia ended for economic reasons in 1968, and
reduced quotas were making Atlantic whaling uneconomic when the government
terminated commercial and sport whaling in 1972, the year of the United Nations
Environmental Conference.

The Board's Atlantic whale research began with D.E. Sergeant's study of the
small pilot or pothead whales of Newfoundland. In the 1960s most marine mammal
research was consolidated at the Arctic Biological Station. Intensive work by
Sergeant and E.D. Mitchell provided a basis for quotas in a temporarily revived
Canadian Atlantic whaling industry, and assisted the commission's work of stock
evaluation. On the Pacific coast G.C. Pike worked with whales processed at the
Coal Harbour whaling station, which took mainly fin whales but a fair number of
other species. Whale abundance there decreased gradually, however, primarily
because of the much larger numbers being taken by catchers from pelagic mother-
ships of other nations.

K.R. Allen, who had been a member of the 'Committee of Three' that made a
convincing analysis of the decline of the great antarctic whale stocks, continued his
interest in this field after he joined the Board in 1964. Recently he has headed a
commission subcommittee that is evaluating the condition of sperm whale stocks.
A commission 'workshop' on small whales, convened by Mitchell, produced a
major compendium that summarizes international knowledge of these species.
Contributions by Board staff include papers on white whales and narwhal in the
Arctic, on killer whales in the Pacific, and on minke, pilot, and bottlenose whales in
the Atlantic.

INTERNATIONAL COMMISSION FOR THE NORTHWEST ATLANTIC FISHERIES
(ICNAF)

Although the supply picture for the western Atlantic pelagic and groundfish
fisheries was bright enough in the late 1940s, most of the stocks were wide open to

exploitation by any European nation that cared to cross the Atlantic, and several were already in the picture. Far-sighted men like Alfred Needler and Stewart Bates could see that this participation might increase very rapidly, so moved to assist in establishing an international organization that could get information about all these fisheries and eventually regulate them.

Needler has recalled the steps which led to the formation of ICNAF:

In 1943 the government of the United Kingdom, in spite of the war, called a meeting in London on the conservation of fisheries in the North Atlantic and I attended this. I was the lone scientific adviser, and there were two deputy ministers: Keenleyside, who was under-secretary of state for external affairs, and Finn, who was deputy minister of fisheries. The Canadian position at that time was that maybe we should join with a North Atlantic organization. This was opposed by the United States. There were meetings following that ..., in Boston and in Washington, in which it was put up to the United States that unless they could come up with something better than the proposals for a full overall North Atlantic body or commission, Canada would join such a body.

There was a lull, and then along about 1947 the United States had some more preliminary discussions and developed the proposal which was finally adopted in 1949 in the Northwest Atlantic Convention. This established a commission for the northwest Atlantic by itself, and divided it into five areas with the principal purpose of making it possible to carry out regulations in the areas that they – the United States and Canada – were most concerned with, that is, Georges Bank and so forth. This developed without waiting to get the support of any more than the two countries that were interested there, but nevertheless having it binding on all of the others.

Well, ICNAF's history is very interesting. It is a good example, I think maybe one of the best examples, of the value of international commissions and fishery regulations, because the northwest Atlantic problem was much more difficult and complicated than, say, [that which concerned] the two-nation Salmon Commission, the Great Lakes Commission, or the Halibut Commission.

A need for statistics The commission's first executive secretary was W.R. Martin of St Andrews. He has described the first year as follows:

The primary requirement, of course, in the commission was a good statistical and basic information system, and I would say that my major contribution was to move around among all of the countries that signed the convention in order to develop a system of fisheries statistics, which has continued and evolved to the present time. This has provided the basis for all the research, the assessment work of the status of stocks and the evaluation of just what is happening to the resource, to the fisheries, the basis for all the regulatory measures that have developed in the commission. Mechanisms were set up immediately whereby we could assess the status of some of the most important groundfish resources, such as the haddock on Georges Bank. This gradually led to the initial regulatory action which was a form of mesh size regulation, large enough to protect small fish while they were growing rapidly, with the hope that they'd be caught again at a larger size and thereby return larger catches.

Eventually a Danish scientist, Dr Erik M. Poulsen, was appointed as the continuing executive secretary of the commission, and the headquarters was moved to Halifax. In 1963 Dr Poulsen was succeeded by L.R. Day of the Board's St Andrews staff.

ICNAF has no research staff of its own, so obtains all its scientific advice from member nations. Being the nation most vitally affected, Canada has assumed a major responsibility for obtaining information about the state of the various stocks and related ocean conditions. All of the Board's eastern stations contribute to the work, but primarily St Andrews and St John's.

From mesh
regulation to quotas

The evolution of ICNAF's regulatory activities was described by Alfred Needler in 1972 as follows:

Nothing is as effective as it should be, but in the last year or so it [ICNAF] has become surprisingly effective. What happened was that everybody, the scientists themselves in the early days, believed that such measures as minimum mesh sizes to protect the small fish would be adequate. They didn't realize that this simply wasn't going to be enough for any number of years, and then about 1965 – between 1960 and 1965 – the realization became pretty general among the scientists that something more would be needed, some limitation of effort or some quotas. So the commission set up a standing committee on regulatory measures; it was a committee of scientists, but it was attended by all the heads of delegations. It met in 1968, 1969, and 1970, and it was agreed that it would be necessary to have quotas. But if they had quotas, it would be necessary to have national allocations: otherwise there would be a mad scramble and the quotas wouldn't be equally beneficial for all the countries.

So, step by step the convention was changed so as to make it possible to recommend national quotas and also to take economic considerations into account. This authority was received in the middle of December 1971, and within six weeks there were quotas and national allocations on the three major herring stocks, and within six months there were quotas established and national allocations on another dozen stocks. So the commission did work pretty quickly when it got the authority.

I always see some function for ICNAF. It depends, of course, on what the Law of the Sea Conference decides about the extent of national jurisdictions or – I'm not thinking just about territorial waters – about the authority of coastal states over fishery management. The more this extends, the more ICNAF's role will become advisory, and I think that one would always have to have some body where the information is digested and advice is given.

In 1973 Wilfred Templeman saw problems ahead for ICNAF:

Future problems

It is very difficult to say with ICNAF, when you have 14 or 15 nations trying to agree and to share catches among the various nations; this is not easy. Until recently, catches of cod have been rising, and while this is occurring, it is not easy to persuade nations that they have reached their peak. The fact is that biologists didn't and couldn't know it. You must have fishing in order to prove your population. But this year, at least, they have got down to quotas. A quota, of course, is not the end. All sorts of things may be done under quotas.

The trouble with quotas might be the policing of them. The catching of small fish and going into fish meal if this is not taken care of. The catching of small fish as our own trawlers did in the early haddock days, dumping 80 percent of the catch in order to get the 20 percent big ones and not counting the ones that went over. The establishment of a quota is just the beginning of trying to do what a quota might be able to do.

The present danger is that there are still fish without a quota, and once a country fills its quota on cod, it may turn to redfish or something else, and lower that species beyond the proper level.

Ideally, every international fishery should have quotas at the same time, so that when people catch their quotas, they just have to stop fishing.

There are so few [solutions]. One would like to cut fishing power. This would be the

Electroshocking lampreys in Ontario, 1958

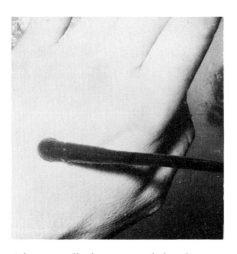

A lamprey clinging to a man's hand

Basket used to catch lampreys in streams

simplest thing. The nations are not very willing to cut fishing power, because the great nations with 3,000-ton factory ships want to be able to take them somewhere else after a quota in one place is used up: South Africa, South America, and so on.

GREAT LAKES FISHERY COMMISSION (GLFC)

The sea lamprey is not a fish; it is an eel-like creature belonging to an ancient vertebrate line that antedates the time when jaws were evolved. Instead of jaws it has a circular sucking mouth with horny teeth; these rasp their way through a fish's skin and drain its blood. Sea lampreys became acclimatized to fresh water in the basin of Lake Ontario probably during the postglacial stage when it was a brackish arm of the sea that contained both marine and freshwater creatures. As it grew fresher over the centuries, the lake's salmon, trout, and whitefish had time to develop some degree of physiological or behavioral defense against lampreys, and all three fish, as well as the lampreys, were abundant in the lake during the 19th century. Opening of the Welland Canal made it possible for lampreys to invade the upper lakes, but it was only during the 1940s that they established breeding populations there. When they did, they encountered lake fish that lacked any defense mechanisms against lamprey attacks, so that within 15 years the trout of lakes Huron and Michigan had almost completely disappeared and those of Lake Superior were about three-quarters gone.

The convention between the United States and Canada which established the Great Lakes Fishery Commission was signed in 1954, and the commission was organized in 1956. It had two purposes: to see if control of the lampreys would be possible, and to serve as a forum for improving the management of fish stocks of common concern, for example by recommending uniform fishing regulations where appropriate. Lampreys had the priority, and it was decided to work first on Lake Superior where considerable numbers of trout still survived. The initial approach was to establish electrical barriers that would prevent the adult lampreys from ascending rivers to spawn. The 1954 work on the Canadian side was under the auspices of a federal-provincial committee, and was directed by Drs F.E.J Fry of the University of Toronto and W.M. Sprules of the Department of Fisheries. In 1955 the Board's London station became involved. It established a branch at Sault Ste Marie under J.J. Tibbles, which later became a separate unit and eventually was taken over by Resource Development when most of the experimental phase was past. The electrified barriers proved less effective than had been hoped, and were abandoned in favor of a newly discovered poison that killed young lampreys in the rivers without harming fish. As of 1972, control had proved fairly successful, although expensive, and had been extended to the other lakes. The trout, however, were still far from completely recovered, even in Lake Superior, but coho and chinook salmon from the Pacific proved less susceptible to lamprey attacks, and good sport fisheries for these species were being maintained by annual introductions.

INTERNATIONAL NORTH PACIFIC FISHERIES COMMISSION (INPFC)

During the decade before the Second World War Japanese salmon fishermen began to make forays into the eastern part of the Pacific and the Bering Sea, and were able to make excellent catches. This move provoked great alarm in America, particularly in Alaska, where the Japanese drift nets captured concentrations of Bristol Bay sockeye in international waters just before they reached the traditional coastal fisheries. In an effort to eliminate this threat and other potential dangers, after the war the United States and Canada cooperated to get a reluctant Japan to agree in 1951 to a convention that included the 'abstention principle.' This stated that stocks of salmon, halibut, and herring that were currently fully exploited and under scientific study by any nation should not be subjected to additional fishing by another. Provisionally it limited Japanese salmon fishing to waters west of 175°w, while Canada was excluded from fishing salmon in the Bering Sea. A commission was set up to investigate the adequacy of this line of demarcation, to review annually the qualifications of stocks for 'abstention' under the criteria mentioned, and to make recommendations for management of certain stocks.

The need to learn the distribution and movements of salmon on the high seas sparked a major series of cooperative biological and oceanographic studies among the three nations. At the Nanaimo station those heavily involved included H.T. Bilton, A.J. Dodimead, Harold Godfrey, R.J. Lebrasseur, J.I. Manzer, Leo Margolis, Ferris Neave, and M.P. Shepard, and there were corresponding teams from Japan and the United States. At first salmon distribution, by area and depth, was mapped by systematic gill-netting. Later, in order to obtain fish lively enough to be tagged and released, Japanese-style longlines were used, and many tag recoveries were made inshore on both sides of the Pacific, sometimes far up spawning rivers. Scale patterns of different American and Asian stocks were compared and seen usually to exhibit average quantitative differences, but in the case of Rivers Inlet sockeye there was a very distinctive freshwater pattern. Most convincing of all, several parasites were identified that were characteristic of Asian and American salmon, respectively, and even one that was peculiar to a specific region – Bristol Bay in Alaska. Using these characters, it was found that the ocean range of North American salmon overlapped broadly with that of Asian salmon, both west and east of the 'abstention line.' Few fish from British Columbia's coastal rivers were found that far west, but salmon bred in rivers of northern British Columbia and the Yukon were of course involved. The seasonal distribution patterns found were compared with oceanographic conditions. Salmon were rarely found south of a temperature boundary at about 14°C, but efforts to relate their movements to prevailing ocean currents were unavailing: the fish could move either with, against, or across the current, as the requirements of their life-history demanded.

Of much greater importance in the long run was work begun on the initiative of the Nanaimo station, which showed that high-seas fishing is intrinsically a poor method of harvesting salmon. Drawing on the results of earlier marking experiments in British Columbia, Alaska, and Kamchatka, it was possible to demonstrate that in pelagic fishing salmon are caught while they are still making a large net gain in weight, that is, when their rate of growth greatly exceeds their natural mortality rate. In addition, studies by Japan and the United States demonstrated an important direct loss of salmon taken pelagically by nets and longlines, because many fall out or are removed by predators. Mainly for these reasons, the postwar development of an intensive high-seas fishery west of 175°w has been accompanied by a serious decline in the catches taken from Asian salmon stocks; these are now less than half of the prewar tonnage, and some once-numerous runs have been almost exterminated. Two other disadvantages of pelagic fishing are that it is more expensive than coastal fishing and that it does not permit the direct management of the size of spawning stocks required in individual rivers. With all this as background, Canada and the United States are making a determined effort to have high-seas salmon fishing abolished by international agreement.

The other pelagic fish under 'abstention' was the herring, although it is much less wide-ranging than is the salmon. Studies by F.H.C. Taylor and A.S. Hourston identified its ocean distribution and provided data indicating that the British Columbia stocks were being fully utilized. The same was not true of Alaskan stocks, which were eventually removed from the abstention list.

In recent years the spotlight at INPFC was shifted to halibut and other groundfish, particularly in the northern Gulf of Alaska and the Bering Sea. In that region very large catches of flounders and Alaska pollock are being made by Japan and the USSR, and the incidental catches of halibut have contributed to the recent steady decline in halibut abundance all along the coast. Studies by C.R. Forrester and K.S. Ketchen have provided important data bearing on this problem, and S.J. Westrheim has mapped the distribution and abundance of Pacific ocean perch from central Alaska to Washington.

Although effective within its terms of reference, INPFC could not become a general forum for research and conservation without the participation of other nations. The absence of the USSR was particularly critical because, quoting Needler:

A serious handicap

A far more important fishing nation in the Pacific, even in the northeast Pacific, than either the United States or Canada, is Russia ... So the commission was handicapped. It really couldn't, without the USSR, get into a serious and effective study of the resources that weren't covered by the convention, and the fact that the halibut in the northeast Pacific were already covered by the Halibut Commission didn't leave very much ...

The opposition towards Russia becoming a member ... came mainly from the United States, but partly from Japan, and not at all from Canada as far as the official position was concerned. I think the Japanese may have been influenced by the fact that they had a convention with Russia in the northwest Pacific, and they thought they might have been in a better position to work in both of these commissions than they would have been if they had developed into a single commission.

I was always in favor of ultimately a single commission for the North Pacific, and as a step towards that actually I did explore the possibility of Canada inviting the countries to establish a sort of ICES of the Pacific, a body which included everybody and all species of fish. I didn't hear any objections. I discussed this informally with people like Moiseev of the USSR and Walford of the United States. I think if I had stayed deputy minister much longer, I would have gotten around to pushing this harder. It still [1972] is a logical thing.

THE INTERNATIONAL JOINT COMMISSION (IJC)

The IJC is a permanent body that considers all kinds of problems that arise along the border between Canada and the United States. Two major problems have been addressed to it that have affected the Fisheries Research Board. The more important one is the increasing level of pollution and eutrophication in the Great Lakes, as described in the preceding chapter. The other is the question of Fundy tidal power.

Interest in the Passamaquoddy power project did not disappear with the debacle of the 1930s, and after the war it began to simmer once again. In 1956 it was referred to the IJC, which was asked, among other things, 'to study specifically the effects which ... tidal power structures might have upon the fisheries of the area.' The IJC set up an International Passamaquoddy Fisheries Board (IPFB) for this purpose; its Canadian members were J.L. Hart and A.L. Pritchard. Cooperative investigations were launched between the Fisheries Research Board and the United States Bureau of Commercial Fisheries, which drew upon a number of other government and university departments for special aspects of the problem. Oceanographers pre-

281

dicted the changed environment that would follow construction; biologists predicted the effect of the new environment upon the fisheries; and economists
translated the fishery changes into dollar values.

One breakthrough, by R.A. McKenzie of St Andrews, was the development of an
external tag for small herring, which for the first time gave direct information on
their movements in the region. The report on herring, by McKenzie and S.N.
Tibbo, predicted little effect upon the weir fisheries either inside or outside the
proposed dams, contrary to the prognosis of the 1930s. Changes there would
certainly be – for example, more ice in the bay in winter but warmer water in
summer, a bigger shipworm problem, and different relative abundances of several
groundfishes, as forecast by W.R. Martin and F.D. McCracken – but the IPFB
general report of 1959 indicated that no really serious effects upon the fishery
economy of the region were to be expected.

In the end, the power project has not gone ahead, at least not yet. Rumor is that an
improved cost-benefit analysis showed that in no way could it produce power as
cheaply as alternative hydroelectric or fossil fuel sources.

INTER-AMERICAN TROPICAL TUNA COMMISSION (IATTC)

In 1950 the first of the tuna commissions was launched by the United States and
Costa Rica, and an open invitation to join was extended to other countries fishing in
the eastern Pacific. It is concerned with a region from San Francisco to Valparaiso
and for up to 1,500 miles or so westward. Panama joined in 1953, and several other
countries subsequently; the present total is eight (in 1975). A strong element in the
motivation for establishing IATTC was a desire to ensure a supply of bait fish to the
California tuna fleet, which bait could be obtained most easily along the coasts of
the countries to the south. However, the commission's most important job was to
conduct research on and eventually to regulate tuna fishing in the eastern tropical
Pacific. Its first director, the perspicacious Dr M.B. Schaefer, soon developed a
model of the yellowfin tuna stock which indicated that sustainable yield from
traditionally fished waters was close at hand.

Canada's interest in Pacific tunas was for many years restricted to the albacore.
In summer these fish move north along the coast, reaching Alaska in some years,
and are taken by a Canadian troll fishery. By 1966, however, after the method of
tuna fishing in tropical waters had largely changed from pole-and-line to purse
seining, a significant number of Canadian seiners based in New Brunswick were
operating off Columbia and Ecuador, and Canada began sending observers to IATTC
meetings. In 1968 Canada joined the commission.

INTERNATIONAL COMMISSION FOR THE CONSERVATION OF ATLANTIC TUNAS
(ICCAT)

The convention that established this commission was concluded in 1966, and
Canada joined in 1968. In 1975 there were 13 signatory nations, mostly those having
an Atlantic coastline but including Japan and South Korea. Its purpose is to study
and to coordinate national studies of populations of tuna and allied fishes throughout the Atlantic Ocean, and also any other species that are not under investigation
by another international fishery organization.

Both the financial contributions and the research responsibilities of the various
members of the commission are keyed to the quantity of tunas and other oceanic
fishes that they capture, and to date Canada's involvement has not been very great.

282

Two oceanographic vessels, the *Hudson* (top) and *Quadra*

Measuring sharks on board the *Hudson* as part of a morphometric study of the Caribbean area, 1965

S.N. Tibbo (left) and W.B. Scott sorting plankton on board the *Hudson*, February 1965. Sargassum weed in foreground

However, Noel Tibbo of the Board's St Andrews station had already carried out important studies on bluefin tuna and swordfish, both of which are now being fished rather intensively internationally. The once-important Canadian swordfish fishery has been discontinued, however, because an unacceptably high natural level of mercury occurs in the flesh of this species.

ICCAT has not been long in operation, but has already improved the collection of statistics for the species concerned.

PERMANENT INTERNATIONAL COUNCIL FOR THE EXPLORATION OF THE SEA (ICES)

This senior international forum for fishery matters was established in 1902 with headquarters in Copenhagen. Although its area of concern includes 'the Atlantic Ocean and its adjacent seas,' it has in practice dealt mainly with the eastern North Atlantic. Research of the various nations in that region is reported annually to its committees, and some degree of coordination is achieved.

Although over the years Canada has not been directly concerned in most of the work reported to ICES, the discussions at its meetings are frequently of great general interest and applicability to other regions. Accordingly for many years one or more Canadian observers – usually Board employees – were sent to ICES meetings, and in 1966 Canada became a member. In addition to its annual meetings, ICES has sponsored and published a number of symposia, and its *Journal du Conseil* and *Rapport et Procès-Verbaux* are major periodicals in the fishery research field.

BILATERAL AGREEMENTS

The absence of an open commission to assess and make management recommendations for the North Pacific fisheries has made it necessary to have bilateral agreements. The most vexing problems, as far as Canada is concerned, are with the United States – particularly the interception of Canadian-bred salmon by United States fisheries in Alaska and Washington, and to a smaller degree the capture of American-bred salmon by Canadian fishermen. These and other problems are negotiated under the auspices of the Canada–United States Reciprocal Fishing Privileges Agreement, which also considers Atlantic problems.

With the appearance of the USSR as a major fishing power off our Pacific coast, negotiations to prevent interference between different types of gear were necessary, and also to protect stocks, of herring particularly, which Canada was already harvesting to capacity. Starting in 1971, representatives of the two countries have met annually to consider the state of the stocks in question. Poland has joined this trawl fishery, which today takes hake mainly, and a similar agreement with that country has just been concluded. All three agreements, but particularly that with the United States, require a major research contribution from the Board's Nanaimo station.

THE ROLE OF FAO IN FISHERIES

The Food and Agricultural Organization of the United Nations has a Division of Fisheries – added as an afterthought, it is said, by Raymond Gushue of Newfoundland during the organization meeting in Quebec. Canada is of course a member of FAO, and Canadians have been prominent on the staff of its Fisheries Division and as members of its committees. According to Needler:

FAO, and particularly its Committee on Fisheries, is the only world forum for discussing problems of fisheries development, fisheries management ... It is ... very active in gathering accurate statistics on the fisheries of the whole world. They get them through countries, of course, but they bring them together – they are critical about this – and they do it well. There is a considerable demand for FAO's activities, for help by FAO, in a lot of developing countries. I think that they are in a rather strong position.

I don't think that anybody associated with FAO really wants FAO to be involved in the actual carrying out of the regional management things. They would rather leave this to the bodies like ICNAF and so forth. They can provide quite a bit of technical assistance. In some regions they don't need technical assistance from FAO, but even in these regions, which are the northeast and northwest Atlantic, where they are pretty well advanced, they find the FAO's comments very useful.

So they have a role in helping technically, which can be pretty important in some of the less-developed regions, but I don't think they will ever get into the actual management, and I think that would be rather poor policy. Because, when you get into bargaining about quotas and the various things that affect people's livelihood like this, well, you tend to make enemies, and it is better for FAO to let the countries disagree with one another rather than have any group of countries say: 'Well, FAO is doing us dirt.'

Don Finn, retired from FAO, talked about the role of commissions in 1972:

The future of commissions

There are two kinds of commissions now. One is by treaty among nations with, of course, the cooperation of FAO. The other is an FAO initiative which creates a commission. Actually I think that perhaps in future there might be evolved a system which would require some enforcement on the question of, let's say, regulations on the high seas – perhaps an international approach to enforcement.

[This could come under the control of FAO] or of a subsidiary or an affiliate body of FAO. The trouble is, of course, that FAO encompasses the whole of food procurement with all the ramifications of agriculture, forestry, fisheries, economics, and so forth, and it is an unwieldy organization in that respect. Perhaps that is one of the things which makes FAO difficult to run, because the rules which govern forestry, agriculture, and other activities have to be so general – if we included all the possibilities – as to be somewhat difficult to apply.

Ideally, people should be much more selfish than they are, but it's got to be enlightened selfishness. We've got to know where our real interests lie, and these lie in conserving food materials, and encouraging growth, application of science, and so forth, to the question of getting enough food.

[Nations have] got to be more enlightened. We were concerned with the formation of the Whaling Commission in 1946. At one time, certain people in the United States thought that this sort of thing should be under FAO too. This was opposed by the whaling countries, Norway, Britain, and others. They agreed to the commission and its constitution, but they proceeded to have so many escape clauses that it just didn't work. This treaty just simply did not work, and it led almost to the extermination of the blue whale, and the fin whale following it, and the other whaling stocks are being exploited to a point where they are threatened.

Man the problem

We had a scientific examination of the situation from the point of view of population dynamics, and we formulated certain sets of regulations. FAO was asked to do this, but the commission itself has really never followed the recommendations of the objective scientists. We have come closer to it, but I've actually had government officials say to me – civil servants in Europe – that: 'Well, the industry knows very well that most of the whaling will lie between Japan and Russia, and you might as well let them make as much as they can out of it while the whales are there.' This was the attitude.

[FAO must work as a go-between among nations;] there's no question about it. There is this sovereignty, but the history of civilization shows that the giving up of sovereignty is proportional to civilization. You get the case of barbarism, where even two families won't live together because of sovereign rights, and you get communities having sovereign rights, and countries having sovereign rights, and this is the course of history. You've got to know where your real interests lie. You've got to be enlightenedly selfish. You can't do anything by altruism. You can't say that, well, you've got to be unselfish, and you have to do things for love. What we must really concentrate on is where our real interests lie; do we want to still keep on existing?

It's a funny thing, you know. Very often I've been asked in this FAO job: 'What would you say was perhaps the most urgent problem about fisheries? Where do the real difficulties lie?' And I would say: 'I can answer you very very quickly, in one word: Man.' You see, if Man would behave, the whole thing would be simple to handle.

20 The new university marine stations

Ecology comes
of age
The Board's discontinuance of support for volunteer investigators in 1935 created little immediate reaction in academic circles. There may have been talk of alternatives, but the depression was still in the land and soon the war absorbed everyone's energies. When the war was over, new apparatus like the electron microscope, and new techniques like electrophoresis and radioactive tracers, gave a strong impetus to studies of the structure, biochemistry, and physiology of the cell, and particularly the cell nucleus. The carriers of heredity were identified, and eventually even the detailed chemical and spatial structure of the chromosomes. For 20 years this is where the action was in biology, and for a while it seemed as though the entire science would disappear inside the cell wall.

At some universities there were delaying tactics – for example, the University of British Columbia founded its Institute of Fisheries in the face of the prevailing trend – but by and large ecology and field studies went into eclipse. During the middle 1960s, however, the tide abruptly changed. Almost overnight ecology became a household word, much to the astonishment of the surviving ecologists. In fact the term came to be applied so widely ('urban ecology,' even 'ghetto ecology') as to become almost meaningless. Be that as it may, the core of ecological knowledge and theory still pertained to communities of plants and animals in their natural environments, and to train a new generation of ecologists it was necessary that students should once again be put into close contact with nature. Such considerations were part of the incentive to establish in Canada three academically oriented and university controlled marine stations. For two of these the Fisheries Research Board did little more than offer moral encouragement; for the third it was the prime mover and also supplied substantial facilities.

MARINE SCIENCES RESEARCH LABORATORY OF THE MEMORIAL UNIVERSITY OF NEWFOUNDLAND (MSRL)

The first of the new university seaside stations was established in 1967 at Logy Bay – the 'place of habitation' of the 'lovely charmer fair' in the traditional Newfoundland ballad. The original building was a squat octagon perched on bare rock overlooking the pounding surf of the bay, an object incongruous enough to suggest a space ship recently arrived from Mars. In 1972 a more conventional building was added, rather spoiling the eerie effect but doubling the floor space available. The first director was Dr F.A. Aldrich, who divided his time with the biology department of the university, but in 1971 Dr D.R. Idler left the Board's Halifax station to become the full-time director.

The Logy Bay laboratory was built and is run with money from a great variety of construction grants, operation grants, individual research grants, and research contracts, from a number of Newfoundland and federal sources. It has a small research staff and several joint appointments with university departments; in addition, many non-staff researchers find it convenient to use its facilities. On any given day up to 30 scientists and graduate students may be at work in the laboratory, along with an equally numerous support staff. The Board's St John's station uses MSRL facilities for holding fish in uncontaminated sea water, and has contracted work to it.

Although it is an excellent research institution, the MSRL is not a biological field station in the usual sense. No courses are given, and there are no living quarters. Student training is only at the graduate and research level. Logy Bay itself is open to northeast gales, and the ever-present Atlantic surf makes work from small boats difficult. Thus for undergraduate training purposes Memorial University, after initial hesitation, was glad to become a member of the group that established the Huntsman Marine Laboratory a few years later.

THE BAMFIELD MARINE STATION (BMS)

The Bamfield station was founded by a consortium of western Canadian universities consisting of Alberta, British Columbia, Calgary, Simon Fraser, and Victoria. Their feasibility study considered needs, sources of financial support, and possible sites. Since the National Research Council had already contributed substantially to the Logy Bay laboratory it was felt that they could scarcely refuse to take similar action on the west coast, and so it proved. As usual in cooperative ventures, it was necessary to have one enthusiastic individual to push things along, and that person was Dr Norman J. Wilimovsky of the University of British Columbia.

The question of a site proved a thorny one. A number of university marine research and teaching units had been established since the war, including the institutes of oceanography at the University of British Columbia and at Dalhousie University, as well as the Marine Sciences Centre at McGill University, but these were all based on a university campus. Even the Logy Bay site is close enough to St John's that staff and visitors live in the city and commute out to it. The Pacific consortium did in fact consider following the same trend and locating the new facility at the University of Victoria. However, this was rejected, partly because the Victoria campus is not close to the sea, but also because the operation would tend to become lost in the stream of normal university activities. Departure Bay at

Nanaimo was considered, but it was 25 years too late to obtain land there at anything but wildly prohibitive prices. The Nanaimo station would have welcomed a new educational facility near by, but its own 'campus' was too small and too hilly to accommodate any major expansion; in fact it had been necessary to fill in the foreshore as foundation for a new wing during the 1960s. In any event the whole Departure Bay area was rapidly becoming urbanized, which made it unattractive for a field station.

In the end the consortium very wisely chose a rather isolated site situated on Bamfield Inlet, which opens into Barkley Sound near the exposed west coast of Vancouver Island and close to the new Pacific Rim National Park. Lots of land and sea frontage was available, as well as a brick building that had housed a former cable station. This was remodeled to provide excellent laboratory accommodations, and first-class saltwater and freshwater holding facilities were constructed. In the adjacent waters marine plants and animals are more abundant and varied than in the Strait of Georgia, and any threat of pollution is remote. The station began full-scale operations in 1972. A series of courses is given each summer for undergraduate and graduate students. Rustic accommodations for up to 80 staff and students are provided, and the whole atmosphere is that of a self-contained community, ideal for field biology and for undisturbed research. It is a return to the philosophy and the surroundings of the Minnesota Seaside Station of 1901, but with modern conveniences and equipment added.

There is no question that both Nanaimo scientists and Bamfield scientists lose something by not having their stations side by side. On the other hand, there is an overall advantage in having year-round research facilities adjacent to two different marine habitats. In any event there are considerable contacts between the two: for example, the present scientific director at Bamfield, Dr John E. McInerney, has spent a year at Nanaimo.

THE HUNTSMAN MARINE LABORATORY (HML)

In contrast to Logy Bay and Bamfield, the Board was intimately involved in the establishment of the Huntsman Marine Laboratory. While Board members approved in principle, the accomplishment required sustained effort by a deeply interested and persuasive individual, Dr John M. Anderson. Anderson joined the Fisheries Research Board in 1967, coming from Carleton University in Ottawa to become director of the St Andrews station. He had graduated from the University of Toronto in 1958 with a PH D in animal physiology, and developed his first interest in fish as an assistant professor in the biology department of the University of New Brunswick, where he supervised a research project for graduate students on Atlantic salmon. He approached John Hart, then director at St Andrews, for money to support the project. A research contract with the Board resulted, and was renewed on a yearly basis. When Anderson moved to Carleton in 1963 he continued his salmon research there on a yearly grant from the Board. Then he was recruited by Hayes to be director at St Andrews when Hart retired.

Anderson had accepted the appointment on two conditions: that he be permitted to introduce a program on fish behavior at St Andrews, and that he should endeavor to re-establish good relations between the station and the universities. Hayes agreed to both conditions enthusiastically: Anderson was talking his language.

A crucial first meeting   Anderson lost no time in making contact with the universities, with the idea of inviting them to use the facilities of St Andrews for their research projects. He has described what happened:

My very first St Andrews meeting was at the University of Guelph with the chairman of the biology department, Keith Ronald. Don Chant, then in his first year as head of Zoology in Toronto, was there too. I went there with the notion, well, I knew that my strength had to come from Ontario, because that's where most of the universities were. And I sensed ... difficulty with this new idea in the Maritimes. I know I was right. History has proved me right, too. You often have difficulty getting new ideas off the ground if you deal only with Atlantic Canada. So I decided to bring in Ontario first, then bring the Maritimes along afterwards. I knew they'd come in because they recognize a good thing when they see it.

So I attended this first meeting. It was in September of 1967. I'd only been in office a couple of weeks, and felt a bit guilty about going up to Ontario in my very first weeks ... [But] this was a golden opportunity. I invited myself to the meeting and gave them the pitch: We had this thing at St Andrews. I considered it to be an national asset. I happened to be director, and happened to belong to the Fisheries Research Board ... if it made any sense for university people to come down and work there, I'd be delighted to try to do everything I could to provide facilities and so forth ... When I made the point – which I felt was a devastating point, and was certain everyone would agree – that, of course, universities can't afford ships these days, they're too expensive, but I've got ships, I remember Don Chant ... pushed his chair back ... and said: 'How do they pay for them? By the Gods of War!' And he said that's the sort of talk that just drove him right up the wall; when the government scientists seemed to assume that they were the only agencies that could responsibly spend large sums of money: 'Of course the University of Toronto could run ships; we've had one. All you have to do is give us the money. We'll run one; we'll run a dozen of them; a whole bloody fleet. So don't tell me that we have to use your ships because we can't afford to run one.' Well, I thought: 'I'm dead. It was a good idea, but that's the end of that.' Then, after he had made his point, he calmed down, and said: 'Well, actually, that was a good idea,' and personally he would support it. But he just wanted to make the point. These government guys come around to the university assuming that the university is going to go hat in hand for everything.

In 1968 Anderson provided a large double trailer especially for university scientists, as well as the general facilities of the station. Ship time was made available, and new equipment had been ordered. The university men had their own programs, but as Anderson pointed out:

Interaction between Board and university scientists

Obviously it would be impossible for them to work on anything at St Andrews that wouldn't be of interest one way or the other to what we were doing. They wouldn't come unless the animal was there, or they could talk to some of our people; so you got built-in certainty that whatever was going on would be of some interest. Some cases were a bit remote, but nonetheless of interest to our people.

My concern was that ... it's all very well to have government labs invited to establish at universities, the assumption being that the government lab and its research would be better off for associating with university people. That's true; they will be. When you're doing practical work, as you must if you're spending taxpayers' dollars in the government lab, then your whole approach is different – different from that of a person who is working in the university environment, where the philosophy, the atmosphere, the background is basic research oriented ...

But there are different ways, different methods of approach, and they're both valid, and they both benefit enormously the individuals prosecuting these two kinds of research, by interacting one with the other. So it makes tremendous sense for the government person to be in a position of knowing where the recent action is, where the new ideas are, that may have application 10 years from now, or maybe next year; who knows?

... The same applies to universities. It's salutary for them to know what's happening in

The final draft of the Huntsman Marine Laboratory charter and by-laws is reviewed by M.J. Dunbar of McGill University (seated at left); Keith Ronald, University of Guelph (seated at right); D.C. Arnold, Mount Allison University (standing, at left); J.M. Anderson, director of the St Andrews Biological Station (centre); and C.C. Ferguson, University of Guelph

government labs, because that's where the real urgency in terms of national priority is. What better way, therefore, for the university researcher to find out what's happening in the applied field than ... going to a government campus, to do his own thing. By being there, of course, he's going to be associated with his colleagues working under the government umbrella. And, of course, now the emphasis is very much, as we all know, on the relevance of university research. Can we afford to have all university professors who are doing research doing whatever they want to do instead of what the nation wants them to do?

December meeting
in St Andrews

The first summer's response was sufficiently encouraging to Anderson to induce him to call a meeting of all the interested universities at St Andrews that December. The idea was to encourage them to formalize their interest in the project. He persuaded Hayes to endorse the project, though Hayes was initially skeptical. He doubted that anyone would wish to attend a meeting at St Andrews in December, and he felt that those who might come would spend their time arguing. Anderson went ahead with the plan anyway, though he knew that Hayes had good reason for his doubts:

Well, firstly, people came in great numbers, at which I was surprised. And, secondly, we had

292

Dr Huntsman cutting the ribbon at the opening of the Huntsman Marine Laboratory,
24 August 1970

The seal pool in the Museum-Aquarium jointly operated by the Fisheries Research Board of
Canada and the Huntsman Marine Laboratory, 1972

a two-day meeting. We spent the first hour and a half of the first day doing just what Ron Hayes said they would do: arguing. And then, I don't know what happened, all of a sudden; it was actually Memorial that was arguing the hardest, and so was Dalhousie, and then, somehow, it became apparent that the universities from Ontario weren't arguing so much. I think what concerned people, certainly Dalhousie, was that this was competition. And Memorial, because they just started Logy Bay, were frightened to death. No, I don't think they were; they were frightened of nothing at the moment because I don't think they believed anything would happen, that the whole thing would fall through. I guess Dalhousie and Memorial came because they couldn't afford not to come. The universities from Ontario came because they were really interested. So, after Dalhousie and Memorial tried ... pretty effectively, to scuttle the whole thing ... it became apparent they weren't getting much support anywhere else. The Ontario universities seemed to think it was a grand idea.

And, of course, one of the things which I predicted earlier on was that ... you can't get universities to support a thing like this if you put it on one university's campus ... I said: 'Look, there is absolutely no way you're going to get universities from Ontario, much less the Maritimes, to cooperate with a Maritime university. They won't do it if one of them has all the marbles!' However, in this case nobody suspected the government, because we had no ulterior motive ...; we weren't about to give degrees.

That's exactly what happened. The universities from Ontario realized that there was something here being offered that they couldn't get any other way. Our interest was genuine ... We would simply be the organizer; we'd have to keep everything going. Well, a steering committee was formed: myself, Keith Ronald, Max Dunbar of McGill, and Fred Aldrich from Memorial. Nobody from Toronto was included originally ... This was the first indication that it was going to work. Because Don Chant ... said: 'Well, look, in getting this thing organized, we'd better let Keith Ronald head it up, because if Toronto gets involved, people are going to suspect us, because we are so big. And they're going to say: Ah, well, there's Toronto again. We're not going to cooperate with them, or whatever.' So they kept a very low profile, and even the board of directors was formed ... we thought of Toronto but Don Chant said: 'No, Toronto shouldn't be on the board of directors. Again, you know, if you want to make this thing successful, you'd better make sure we don't seem to be in the driver's seat.' This was Toronto, really thinking big.

An unsung hero   Well, we got to work and the real driving force behind this is a very unsung hero. There are, I guess, three key people ... Keith Ronald was one, of course; he was chairman. Obviously, to bring the universities along it made good sense to put a university person in charge as chairman, because bringing along the government was the easiest thing to do.

Anderson himself was of course one of the three key people. The third was

a fellow called Charlie C. Ferguson, assistant to the vice-president of administration at Guelph. He was our honorary secretary for the first two years of operation, I guess, but he wasn't honorary at all; he was the secretary and treasurer. He did everything. He prepared the minutes. We got ourselves incorporated in Ottawa over a weekend. He was able to do that because of a lawyer friend in Guelph. He was able to push this through the Department of Consumer and Corporate Affairs. Just an absolutely marvelous fellow behind the scenes – working, driving, getting everything organized.

Then we went to the universities for membership. We deliberately did not go to the National Research Council at all. We decided to go to them after we had proved that there was a need and we could run our own show; now it was important to really make our organization truly effective. We wrote to all the presidents of universities from Manitoba eastwards, and ... we had two classes [of members]: full membership, and associate membership with a modest initiation fee initially. But the initiation fee, which went up to $1,500

294

after the first year, encouraged people to join as full members rather than associate members ... We now [1973] have 21 members, 11 of which are full members and the others are associate members, including the New Brunswick Department of Fisheries and Environment. The FRB is also a full member. The Woods Hole Oceanographic Institution joined last year ... The State of Maine is an associate member, but they're coming in as a full member. And the International Atlantic Salmon Foundation. We subsequently got grants from the NRC of $210,000 for two years.

Through some involved negotiations with the Board, the New Brunswick government, and the bank, Anderson was able to purchase property adjacent to the Atlantic Biological Station, including a magnificent house to serve as residence. A long-term lease was also obtained from the Board on other property. Thus the Huntsman Marine Laboratory took physical shape. Its name, of course, honors Dr A.G. Huntsman, concerning whom much has already been said in earlier chapters. Its first director was an equally well known Board figure, Dr A.W.H. Needler, who took the job when he retired from the position of deputy minister of fisheries. In 1976 he was succeeded by Dr W.B. Scott, formerly of the Royal Ontario Museum and co-author of two of the monographs of fishes published by the Board.

Although the main purpose of the Huntsman laboratory is to give university scientists access to a marine environment and to promote Board-university contacts, its founders envisaged an additional role for it. Quoting Anderson: 'We felt right from the start that there should be one or two things the HML does perhaps better than any other place – one or two areas of excellence in addition to the other things that go on. And so we looked around for something that the HML could take on as a major project, that could involve a lot of people who are members of HML, and give the HML some special place in the scientific galaxy. And we found it.'

HML's role in salmon research

What they found and promoted was a continuing study of the genetics and selective breeding of Atlantic salmon. The International Atlantic Salmon Foundation was interested and put up $2.4 million over a period of years for construction of facilities. The federal Department of the Environment promised to meet operating costs for six years, up to an annual maximum of $200,000. Anderson is extremely sanguine about the possibilities: 'There's no reason why we can't breed salmon that won't go to Greenland ... There's no reason why we can't get 13-pound grilse up the Saint John or the Miramichi, since they do it in a river in Ireland.' Field trials will be made in Maine as well as New Brunswick. Indeed it may well be that new strains of salmon will initially be more important in the New England states than in Canada, because there the native salmon are generally at a low ebb and ripe for major escalation. We can only hope that the results of the project will justify the tremendous enthusiasm of its chief promoter.

The Weir years, 1969–72

When Dr J.R. Weir's appointment as chairman of the Fisheries Research Board and also as adviser to the minister of fisheries and forestry on renewable resource development was announced by Prime Minister Trudeau in 1969, he pointed out that, with the Government Organization Act of 1969 now in force, the effective integration of the former departments of Fisheries and Foresty was a matter of major concern to the government. To attain this objective, it was essential to develop a strong organization for the government's scientific activities in the field of renewable resources.

He went on to note that the government had followed the numerous studies conducted in Canada by the Science Council and the Science Secretariat and these had given rise to general concepts of government science organizations in the field of renewable resources that were very promising. There remained the vital need for vigorous and imaginative application of these ideas.

In selecting a senior official and distinguished scientist for this task, the Prime Minister said he had been impressed by Dr Weir's extensive background in biological research as well as his experience as a consultant to the Glassco Commission and as deputy director, and more recently director, of the Science Secretariat. There he had had an opportunity to consider organizational patterns for science from a broad perspective, and it was anticipated that – with the department and the Privy Council Office – he would make recommendations for future organizational developments that could be speedily adopted.

The operation, he concluded, would be considerably facilitated by Dr Weir's appointment as chairman of the Fisheries Research Board, following the retirement of Dr F.R. Hayes, who had served with distinction as chairman since 1 August 1964 and reached his 65th birthday on 29 April 1969.

Weir was born on a farm near Wingham, Ontario, in 1912. He had a fondness for

The annual meeting of the Board, 1973. Seated (l. to r.): E.L. Harrison, H. Favre, J.R. Weir, J.B. Morrow, L.H. Omstead Jr. Standing: P. Smith, A.W.H. Needler, G. Filteau, D.R. Idler, K. Ronald, D.A. Chant, P.C. Trussell, P. Russell, S. Sinclair, R.R. Logie

mathematics but a wariness born of the depression years led him to the Ontario Agricultural College at Guelph with the idea of obtaining a teacher's diploma. Subsequently he took his M SC at the University of Alberta, where he came under the influence of the brilliant Ken Neatby, who persuaded him to take further graduate work in genetics and statistics at the University of Minnesota. There he obtained the PH D degree, majoring in genetics and minoring in physiology.

Weir returned to Guelph on the teaching staff and remained there for 12 years, finally leaving to accept the post of dean of the Faculty of Agriculture and Home Economics at the University of Manitoba. There he came in close contact with such farmer organizations as the Canadian Federation of Agriculture and the Farmers' Union, and also became involved in both provincial and federal farm activities. Then, he recounted in 1973,

along came a request to work on the Glassco Commission, so I took a year's leave from the university. This brought me in contact with the federal government, and I was interested in the operations and the research in it. And no sooner had I finished this when I became involved with the Ford Foundation in South America and the Rockefeller Foundation in Africa, doing reports on university organizations and renewable resources.

After 25 years, I felt I would like to have some experience in the federal field. Since I had been involved in the Glassco Commission work, it seemed interesting to me to go into the Science Secretariat, which was just forming and which, incidentally, was recommended in the Glassco Commission. To be in at the beginning of this and the development of the Science Council seemed important to me. Well, I was there for four years and I was interested in it and discussed all through this time the need to take a look at the organization of research in the renewable resources in Canada.

We talked in the Science Secretariat about the need to do a study on renewable resource organization. We thought about bringing people in to do it from the secretariat. But we felt that it probably wasn't the best place in the world to do studies of this kind, in an organization and a central agency that had some very serious constraints. So this seemed an opportunity to take a look at renewable resources within the operational environment of a department.

But the Board offered the opportunity of giving the freedom and the chance to look at this

297

J.R. Weir and W. Templeman on Templeman's retirement as director at St John's

issue, utilizing the Board's advantages to conduct a study on the resource itself ... This adds to my own thinking that this is an important part of the Board's function, probably its basic function.

Difficult days

The governmental atmosphere within which Weir was to lead the Board for the next four years would prove as difficult as any the Board had undergone since the days of Macallum. First of all, the new department went through further metamorphosis to become the Department of the Environment, with the inevitable organizational shake-up this involved, including shifts in emphasis on goals for all of its elements and, for Weir, the additional onerous responsibility of being assistant deputy minister of fisheries in the new catch-all department. Then the responsibilities of the Board expanded greatly as demands on its services created by the 'environmental thrust' increased. Similarly, expanding international responsibilities brought their increased demands. Yet for almost the entire period the strict limitations imposed by the Treasury Board on the Board's budget and the size of its staff were maintained. And throughout this time the ominous cloud of the Board's threatened demise as the research arm of the Fisheries Service hovered overhead, only to burst coincident with the Board's 75th anniversary in 1973.

Nevertheless, the Board continued to function effectively, though from time to time the pressures were reflected in the Chairman's annual reports.

The environmental thrust

Evidence of the new environmental thrust appeared in Dr Weir's first annual report when he outlined the three areas of activity involved in the Board's intramural program: commercial and recreational fisheries research, environmental research, and products and processing research. Support of university research, and administration and research services, completed the operations.

Weir expanded on these subjects:

298

Commercial and Recreational Fisheries Research includes investigations on harvesting – distribution, behavior, detection, and capture, and management – ecology, physiology, parasitology, abundance, population dynamics, and reaction to different levels of exploitation, and increasing the resource through aids to propagation, transplants, fish farming, and other methods. The objective of this research is to bring economic benefits or satisfaction to commercial and recreational fisheries and to Canadians affected by them ... Research to assist in the commercial exploitation of the fishery resource and related aquatic environment certainly must continue to be one of the major programs of the Fisheries Research Board of Canada. However, two additional aspects are now becoming of great importance: the rapidly increasing value to Canada of sport fishing and the need to consider the total environment ...

Environmental research includes studies on all the biological aspects of oceanography and limnology as they relate to productivity of marine and inland waters, and to fitness of the aquatic environment; detection, dispersion, and degradation of man-made pollutants in water; understanding and modification of natural ecological changes; and the resistance and tolerance of aquatic organisms to harmful substances ... The desirability of preserving a healthy and viable environment introduces the need for proper overall management of all renewable resources within the ecosystem so that its deterioration through such things as pollution can be avoided. This means long-range research and planning to ensure a more efficient, high quality, continuing resource for commercial and other uses that is compatible with demands for human health and enjoyment.

Weir emphasized the key role of the Board in this situation:

Because the Federal Government has the basic responsibility for living aquatic resources everywhere in Canada, the Minister of Fisheries and Forestry is the focus for management of Canada's fisheries and for protection of aquatic renewable resources from environmental changes. The Fisheries Research Board is the agency chiefly concerned with providing the research and technical background for management of aquatic renewable resources. The accelerating tempo of change has created an explosive demand for knowledge – knowledge needed as a base for technological advance in products and processing, knowledge required to assure high continuing yields from the resource, and to monitor the effects of management programs, and knowledge needed to detect deterioration of the environment and to develop remedial measures. The challenge of meeting this demand and of anticipating future demands forms the mission of the Fisheries Research Board of Canada. In the words of the Act of Parliament establishing the Board, FRB has 'charge and control of investigations of practical and economic problems connected with marine and freshwater fisheries, flora and fauna, and such other work as may be assigned to it by the Minister.'

Key role of the Board

With the organization of the new Department of the Environment in 1971, Weir noted that the challenges facing fisheries research and development had never been greater, and he expressed concern in his actual report:

A clear warning

There is an ever-present danger that public policies and goals may be guided in the future by the most pragmatic and expedient calculations. Scientists have been increasingly subjected to public criticism for being too abstract, and there is a technological backlash that cannot be ignored. We must respond to such views with stronger efforts to communicate an understanding of the value of our activities, and to nourish the public's interest. We must continue to plan and implement strong programs that are coordinated with others. We must also ensure a balance of projects that will utilize talents, manpower, facilities, and dollars more effectively.

299

The chances of success are greatest if goals and programs are set with some knowledge of science and its limitations. Scientists frequently complain that they are not consulted on contributions that can be made from the branch of science they know best. Avenues must be kept open for the collective views of the specialists. Assessment by peers has long been considered an integral part of the process in evaluating scientific results, but all too infrequently has it been used in assessing projects before they are launched. In the absence of any accepted mechanistic approach to the setting of goals and priorities, and to evaluate research, forums must be sought to use this expertise to capacity, and to bring them in concert with information users.

Freedom to plan and execute

Provided that objectives are in harmony with national goals, it is imperative that project managers and scientists on site have freedom to plan and execute their project operations so that their resourcefulness will not be constrained. Time must be allowed for new ideas to be developed sufficiently for their feasibility to be assessed. However, researchers cannot live in isolation, even though much of the work concerning living resources must be done near the problem site. Needed support from specialists in many disciplines can be provided only in large regional centers, which can also foster regionally coordinated programs, facilitate recruitment of competent and creative staff, and provide contact points for users in their areas. Such centers should also bring together the many functional approaches involved in fisheries science and its application, e.g. those depending on basic research, experimental development, applied research, innovation, technology transfers, surveys and data collection, scientific and technical consultations, and laboratory services; they should also help to avoid placing unnecessary emphasis on one approach. Administrative and support services, invariably an expensive component, should be responsive to the unique requirements of research and development, and must be supportive without being restrictive.

It was a strong plea for scientific integrity in the face of the approaching threat which was contained in a following paragraph: 'Question arises as to the future identity of the Fisheries Research Board of Canada, even though the Act still stands. Corporate structures change, but it would indeed be regrettable if the identity of the Fisheries Research Board of Canada, and the many things that it connotes, should pass into oblivion.'

With all flags flying

The final year (1972) of the Fisheries Research Board of Canada as the research arm for the Canadian Fisheries Service was a busy one, and the Chairman's report gave no hint of the changed role that was in store. The ship went down with all flags flying:

The past year was an eventful one for the Board. In both Renewable Resources and Environmental Quality areas, we have responded to challenging research needs. Our responsibilities for Renewable Resource research are increasing rapidly as a result of a series of actions which are increasing coastal state responsibilities and rights for the management of fisheries to the outer edge of the continental margin. These actions have already included national responsibilities for resources in exclusive fishing zones and territorial seas and in the establishment of quotas in the ICNAF convention area for several herring and groundfish stocks ... These responsibilities will be increased further as a result of recommendations from the 1973 Food and Agriculture Organization Conference on Fisheries Management and Development, and as a result of the actions taken in preparing for and originating at the forthcoming Law of the Sea Conference.

In the Environmental Quality field, we are playing a key role in responding to the need for a biological scientific input in the recently announced Canada–U.S. Great Lakes Water Quality Agreement. Partly in response to this pressure, we have changed the status of the FRB at the Canada Centre for Inland Waters at Burlington, Ontario, from that of a field detachment of

300

the Freshwater Institute in Winnipeg to a separate responsibility center. The new name of the FRB unit at Burlington is the Great Lakes Biolimnology Laboratory. Another urgent need has been providing bioassay procedures so that the Environmental Protection Service can set effluent regulations for various industries. An analysis of Department of the Environment activities has shown that the Board is doing approximately 75 per cent of the research in those areas which are of prime concern to the Environmental Advisory Council. There seems little doubt that our responsibilities in the Environmental Quality field will continue to grow.

A cynic might be tempted to suggest that the larger the Board's profile loomed in the environmental field, the more tempting a target if offered for assimilation. After all, in the early 1970s 'environment' was the name of the game in government circles, in Canada as elsewhere.

Weir announced, in his report, the inauguration of a new government policy which, to the layman, would appear to diminish the importance of such scientific research organizations as the Board. He wrote: 'A further challenge in 1972, of general importance to Board operations was the Government policy on Make or Buy; the intent of which is to encourage Government research units to contract out research and development problems rather than to solve them in house. The policy has much merit and is bound to affect the Board's operations. However, its full significance and implications for the Board are as yet unclear.' 'Make or buy'

It is difficult to avoid questioning the logic of this policy, at any rate as it applies to fisheries research. Over a period of 75 years the Board had developed as a superb scientific instrument, with a staff of more than 800, including scientists equipped with the most sophisticated technical equipment available, surely capable of handling any type of scientific problem that might develop in the fisheries. Millions of dollars had been spent and many careers had been dedicated over the years to this purpose. Now, apparently, it was the government intention to help develop parallel expertise in the private sector. Whatever the purpose of this maneuver, economy obviously did not play a role. Board stations had already encountered examples of private contractors or subcontractors who found themselves hopelessly at sea and came to them as the only source of information; this would then be regurgitated in a somewhat garbled form to their principals, who could easily have obtained an accurate story free by direct application. The point is that most ecological and environmental problems require something that no private contractor can obtain for himself. This is a backlog of observations over a period of time, which will reveal both year-to-year variability and the nature of any trends. The need for this information is well recognized in the study of the weather, tides, and river discharges, but it is equally important with living resources. Unfortunately far less provision is made for obtaining it, and thorough implementation of a Make or Buy policy would tend to dry up such sources as do exist.

STAFF CHANGES AND RETIREMENTS

Through the Weir years, a number of senior staff changes and retirements took place. In 1969 Idler was appointed Atlantic regional director (research), and J.P. Tully and M.W. Smith retired after fruitful careers. So, too, the following year, did the redoubtable Hugh Tarr, as well as A.W. Lantz at the Freshwater Institute and A.L. Wood at Halifax. Dr D.A. Munro came in from the Department of Indian Affairs and Northern Development to become deputy chairman, policy and planning, and W.E. Ricker was appointed research coordinator in the Pacific region. In

301

1971 Munro shifted to a similar post elsewhere in the department, W.R. Martin assumed responsibilities for strategic planning, Dr W.E. Razzell was appointed director of the Vancouver laboratory to replace Tarr, Dr E.G. Bligh became director at Halifax to replace Idler, and Dr A.W. Mansfield became director of the Arctic Biological Station, replacing C.J. Kerswill who was named acting director of the Program Planning and Coordination Branch. Later Idler moved to Memorial University and Dr T.R. Parsons moved to the University of British Columbia.

In the final year J.M. Anderson moved to Ottawa from St Andrews as director-general of research and development for the Fisheries Service, and he was succeeded as director at St Andrews by Dr R.O. Brinkhurst. However, Anderson soon left the federal service to become president of the University of New Brunswick. A Social Science Research Branch of the Research and Development Directorate was established, with T.C. Clarke as director. Templeman retired as director at St John's and was appointed first J.L. Paton Professor of Marine Biology and Fisheries at the Memorial University of Newfoundland. In September of that year he received a $2,500 Public Service merit award for an 'unusually high level of achievement over an extended period as a scientist, administrator, educator and adviser to government and industry.'

The colorful Wally Johnson moved from the directorship of the Freshwater Institute at Winnipeg to Nanaimo as director, succeeding K.R. Allen, who joined the Commonwealth Scientific and Industrial Research Organization in Australia, and the equally colorful Charlie Castell retired with honors at Halifax. The St John's Technological Unit, which had been under the direction of the Halifax laboratory, was joined with the St John's Biological Station.

PACIFIC ENVIRONMENT INSTITUTE

For some time the experimental station in Vancouver had been handicapped by the lack of a sea-water supply and by the high cost and the chlorine content of the fresh water that it used. Since early in the century a salmon cannery had existed on the seashore in West Vancouver, but the owners were preparing to close it and sell the site. Investigation showed that good salt water could be obtained at a depth just offshore, and also fresh water from a nearby creek via a gravel stratum and driven wells. However, to justify the purchase price some additional use for the property was desirable. Such a function surfaced when it was decided to establish a new center to handle most of the Board's environmental work in salt water and estuaries.

Accordingly in 1970 the Pacific Environment Institute was founded. To date it has operated in a series of trailers, plus two houses and a shed that had belonged to the cannery (the main cannery buildings were demolished). Dr Michael Waldichuk, formerly of the Pacific Oceanographic Group at Nanaimo, was named program head, and A.J. Dodimead accompanied him to the new institute. Other staff were recruited, and intensive work began on the estuary of the Squamish River – an environmentally sensitive area – and in the Fraser delta region. At the same time extensive saltwater and freshwater aquarium facilities were developed for the use of the staffs of the Vancouver station and local universities.

HARP SEALS

A natural resource that began to attract wide public interest during Weir's chairmanship was the stock of harp seals off our east coast. These seals spend the

The Pacific Environment Institute, West Vancouver, trailer laboratory complex and new dock, 1972

summer in Davis Strait, then come south in winter to bear their pups on the ice in the Gulf of St Lawrence and on the 'Front' off northern Newfoundland and southern Labrador. The annual seal hunt had been an institution in Newfoundland since early in the last century: the departure of the sealing vessels from St John's harbor in late winter was always a great public occasion. In addition, 'landsmen' from Newfoundland, the Magdelen Islands, and a few other points conducted local sealing operations from shore whenever possible, which provided an important supplement to their meager incomes.

Over the years the annual kill was apparently somewhat larger than replacement, but the respite provided by the Second World War had brought the herds up to a level of several million animals. Shortly after the war vessels from Norway joined the hunt, approximately doubling the former kill of pups and, what was worse, taking a rather large number of mature breeding seals. During the 1950s Dean Fisher, then at St Andrews, began reading the ages of the seals from layers in their tooth structure. In occasional years ice conditions made it impossible to take the usual number of pups, and Fisher observed that a brood having a low pup kill was represented in exceptionally large numbers among the juveniles and young adults of later years. He concluded that in most years the current level of hunting was killing a majority of the pups born.

Fisher's warning went unheeded for 10 years or so, during which time he made an attempt to get a complete census of the herds by aerial photography. Unfortunately this was not completely convincing: the white pups were difficult to distinguish against their snowy background, and it could always be argued that whelping patches somewhere or other had been missed by the aerial survey.

Seal studies were continued at the Arctic Biological Station, where David Sergeant developed Fisher's observation into a method of computing actual stock size from the pup kills each year. This, plus a gradual decrease in number of pups taken each year, finally made the decline of the herd blatantly obvious. It was then agreed that harp seals should come under review by ICNAF, and Canada also entered into bilateral negotiations with Norway by which kill quotas were agreed on for both nations. With prodding from Sergeant, from Fisher (who had moved to the University of British Columbia), and from Keith Ronald (an FRB Halifax scientist who had become dean of graduate studies at the University of Guelph), the quota has been reduced, and large sealing vessels have been excluded from the Gulf of St Lawrence. As of 1975 it was still a matter of contention whether the quota had been reduced sufficiently to reverse the downward trend of the adult stock. However, the herds are now under such close observation that if there is any further decrease it will bring a quick response.

HUNTSMAN'S LAST BOW

In the autumn of 1972 a symposium on Atlantic salmon was held in St Andrews, sponsored by the Huntsman Marine Laboratory. It brought together people from both sides of the Atlantic to review the current state of knowledge and to plan for the future. A few years previously Danish fishermen had begun to take large quantities of salmon off Greenland, although Greenland rivers have almost no salmon runs. It was soon established by extensive tagging and marking studies that these salmon came from the eastern Canadian provinces and Maine, and also from Iceland, the British Isles, and Norway. Salmon from all these areas converged on a rich feeding ground in Davis Strait, and later returned to their respective home waters to spawn. But Dr Huntsman had no use for the idea that a salmon could find its way home from such a distance. No known mechanism could explain *how* one could do it; therefore, he argued, it was impossible. He even wrote to a Danish colleague to congratulate Denmark for making good use of fish that had obviously got lost.

John Anderson, who had been a key figure in the development of the Huntsman Marine Laboratory, has told the story of the 1972 meeting:

In the symposium we had something like 400 people from 12 countries at St Andrews. It was an important symposium. Huntsman wanted to give a paper, [but], we deliberately didn't ask him because we knew perfectly well he was going to be controversial and cause all sorts of stir. So no one asked him. Well, this upset him. I might have known it would have. We hadn't really thought what we were going to do when he insisted on getting on the program.

So we had a hurried sort of emergency meeting and decided we'd do two things: we would insist that he write it out in advance, and when people read it, they could just read what he said and realize how wrong he was; and we'd let him have a little seminar at night, and we'd advertise it. If people wanted to go and listen to it, fine. We had, of course, the Press from all over the world there. I guess when you call a laboratory after a man, you must expect people to pay attention to him when he talks. You're telling the world that here is a famous man, and he is, of course, a living legend: A.G. Huntsman himself, 88 ...

Well, first we held a little trial thing in the FRB station's seminar room, got everyone down there. Of course, people didn't want to say 'no,' and they all came out. And then, of course, the Press, when they heard Huntsman was going to be there, were there in force, as were most of the Europeans, who'd never met Huntsman before. They assumed that they'd certainly have to hear what he had to say. I was chairman at one session when he got up, and tried to sidetrack a bit after he made his point about migration, and I was able, I think, to handle it reasonably well by pointing out that Dr Huntsman would be able to expound on his interesting theory more fully that evening in the FRB Biological Station lecture room. That was, of course, the wrong thing to say.

Well, it turned out, of course, he got all the headlines, the real headlines: HUNTSMAN DISCREDITS SCIENTISTS! It was no laughing matter because, as I'll mention in a minute, we're talking of a multimillion dollar project that darned near went down the drain as a consequence of A.G. Huntsman upstaging all of us. They went on to point out how Huntsman said that fish didn't migrate, they didn't go to Davis Strait and come back again, and therefore much of the theories of fisheries management were all so much hogwash ... It made a lot of us feel pretty foolish, because ... [we] looked to be devils, persecuting such a famous poor old scientist.

The reason that was rather serious was because we were then in the rather delicate ... negotiation stages for a very exciting selective breeding genetics program, for which the HML would be the administering body, involving 2.4 million dollars, to be supplied entirely by the International Atlantic Salmon Foundation. The fund-raising chairman for this was a man called Noel McLean, chairman of the Board of Trustees of the Woods Hole Oceanographic Institution, who was very impressed that we were naming this lab after Huntsman, because Huntsman had once been a trustee of the WHOL and had taken part in a famous oceanographic expedition with H.B. Bigelow. Not only that, but subsequent to the meeting, Huntsman began to write long letters to Noel McLean ... pointing out how stupid not only was the migration theory, but this whole business of genetic and selective breeding; he'd never seen such nonsense in his life before.

It took a little while, I assure you, to convince Noel McLean not to worry, that A.G. Huntsman wasn't the same now as in his heyday. Because genetics and selective breeding have come a long way, of course, since Huntsman was in it. For a while Noel McLean thought: 'My God. We're being taken to the cleaners by these Canadians.' Because here, after all, was Archibald Gowanlock Huntsman saying in effect that we were all crazy.

delicate situation (margin note)

Thus Dr Huntsman remained true to his principles, unconvinced by facts unless they would fit into an operational framework that he could accept. Being in a minority of one bothered him not a bit; quite the contrary, he was in his element as he maintained his role of gadfly at his last major public appearance. His career was remarkable.

A fourth-generation Canadian, Huntsman was born in 1883 on a farm near Tintern in the Niagara peninsula of Ontario. He entered the University of Toronto in 1901, receiving his BA in 1905; and in 1933 his BM degree, received in 1907, became an MD, although he never practiced medicine. Following his first season at Go Home Bay in 1904, Huntsman spent three additional summers there: 1905, 1907, and 1910. He was also active at the new stations at Nanaimo and St Andrews, which were opened in 1908. In 1911 he went to St Andrews as curator, and held this post between 1911 and 1913 and from 1916 to 1919, when he was appointed director. He remained director at St Andrews until 1934. He was also director of the newly established Fisheries Experimental Station at Halifax from 1924 to 1928. He served as editor of Fisheries Research Board publications from 1934 to 1949 and as consulting director from 1934 to 1953. Officially retired then, he remained an active and often controversial figure in fisheries research until his death in 1972.

Archibald Gowanlock Huntsman (margin note)

Huntsman also played an active role at the University of Toronto, serving as lecturer and later professor in the Department of Zoology from 1907 to 1954. He was a member of various special committees, notably the International Board of Enquiry for the Great Lakes Fisheries in 1940–2, of which he was secretary.

Huntsman's tremendous energy spilled over into learned societies and scientific organizations. He was president of the Royal Society of Canada in 1938 and Flavelle Medallist in 1952. He was first president and then Honorary President of the Ontario Society of Biologists, Honorary Member of the Canadian Society of Zoologists; Foreign Honorary Member of the American Academy of Arts and Sciences; Fellow of the American Association for the Advancement of Science; Honorary Member of the Marine Biological Association of the United Kingdom; Member of the British Association for the Advancement of Science; Member of the American Fisheries Society (president, 1936–7); Member of the Board of Trustees of the Bermuda Biological Station for Research, (vice-president, 1935–54, and Honorary Patron, 1963); and Honorary Trustee, Wood's Hole Oceanographic Institution.

But it was Huntsman's research work that represented his greatest achievement. He published more than 200 scientific papers and books, covering a wide range of zoological subjects. Although he was best known in biological circles throughout the world for his work on the Atlantic salmon, he made important contributions over a wide range of subjects, such as oceanography, marine invertebrates, marine ecology, growth and fatigue in fishes, migration, philosophy, the economics of fishing, and fish technology. He encouraged his students and colleagues to adopt the same critical attitude that he himself espoused, and did not take it amiss when they directed their criticism against his own work: rather, this initiated lively and sometimes protracted arguments. The breadth of his interests is perhaps best revealed in his late book *Man and the Universe*, which, interestingly enough, started out to be a history of the Fisheries Research Board!

W.B. Scott, appraising Huntsman in 1969, said:

In many of these areas of investigation, his works were truly pioneering and far in advance of contemporary thinking. His writings and discourses characteristically challenge accepted or established thought. Indeed this attitude of critical appraisal is not only most characteristic but also one of his most valuable contributions to science ... Officially retired since 1954, Dr Huntsman has remained active in scientific affairs, serving as Chairman for advisory committees on Great Lakes research and, by his writings, challenging the direction and accomplishments of current research in aquatic science. [FRB files, biographical material on Huntsman]

When the consortium of Canadian universities in 1969 undertook to establish a foundation to develop programs of teaching and research in marine science at St Andrews, it is not surprising that they named it the Huntsman Marine Laboratory Foundation. In many aspects, the story of the Fisheries Research Board is inextricably woven into that of Archibald Gowanlock Huntsman, as these pages have revealed.

A CHANGE LONG FORESHADOWED

The 75th anniversary of the Fisheries Research Board of Canada saw the conclusion of this unique organization's role as the active research arm of the Canadian Fisheries Service. In 1973 it was relieved of direct control over research programs and facilities, and became a wholly advisory body. The Board had built up a proud record, studded with numerous solid accomplishments. The milestones among these are given in 'Seventy-Five Years of Achievements,' an article by Dr W.E. Ricker published in the *Journal of the Fisheries Research Board* in 1975. It had

J.R. Weir presents a specially bound copy of his Russian-English dictionary to W.E. Ricker to mark Ricker's retirement during the celebrations of the 75th anniversary of the Board in Winnipeg, 1973

grown from a handful of dedicated volunteer scientists spending their summers at the movable laboratory on the Atlantic coast to a complex research organization with capital assets of some $90 million and an annual budget approaching $23 million. The Board's stations and laboratories, 10 in number, stretched from St John's, Newfoundland, to Vancouver Island: biological stations at St Andrews, Nanaimo, St John's, and Ste Anne de Bellevue; laboratories at Vancouver and Halifax; the Freshwater Institute at Winnipeg and the Great Lakes Biolimnology Laboratory at Burlington; the Pacific Environment Institute at West Vancouver and the Marine Ecology Laboratory at Dartmouth.

One explanation of the Board's changed status was offered by F.R. Hayes in his book *The Chaining of Prometheus* (p. 31): 'In a late 1972 restructuring of Environment, the FRB lost its independent status and was brought into line authority, reporting to the new assistant deputy minister for marine and fisheries. This will not immediately affect the work of the laboratories but the Honorary Board becomes a dead duck. While it is unlikely that a minority government [1973] would face the fuss of attempting to repeal the FRB Act, the transaction does offer a forecast about what is in store for other boards and councils. The government simply cannot contemplate the control of policy and funds by any but its own employees.'

Chairman Weir saw the development as a logical one and viewed it in terms of his own experience since coming to the Board in 1969. He had come to the Board in a twofold capacity: as chairman and also as adviser to the minister on renewable resources. The latter function required him to prepare a report for the government on how best to organize a research program on renewable resources.

When Weir looked at the research situation in fisheries, he found that parallel development had taken place since 1948. The Board's biological and technological stations engaged in their research programs, and 'quite sizeable research programs' were going on within the service itself, not only in the Resource Development Branch, but also in the Industrial Development Branch and in the Inspection Branch.

Marriage of research and development Like Kask before him, Weir felt strongly the need for linking research and development: 'I think that if you took all research out of operational activities and placed it in a separate branch you would ... sterilize and emasculate the operational activities, since the research input provides quite a needed stimulus to the people, and excitement, that is required for good work.' Conversely, Ricker has pointed out from his own experience that 'close contact with events at "production" facilities, and the routine monitoring that can easily be maintained there, can provide a valuable record of trends in the environment and in the fish stocks themselves, and can indicate new research needs. Also, any such facility may be in a particularly favorable situation for a particular experiment, and there should be the greatest flexibility in adapting to such opportunities.'

Thus integration of research and development offers real advantages. There are also disadvantages; in particular, research should never become restricted only to immediate problems of production and management. Kask's proposal was to bring resource development directly under the Board's control. What happened is that the Board stations became in fact, as well as in name, the research arm of the fisheries branch of the department, and the process of integration with parallel groups is proceeding as rapidly as personalities permit.

In the final analysis, after Board members had ceased to be active researchers at the stations it was inevitable that their role should become mainly advisory. They could suggest new programs or new techniques, phase out investigations that had revealed as much as they were likely to do, and if necessary they could veto a proposal – even from a high source – for work that promised little return in relation to the expense involved. Most important of all, they were usually able to protect and bring to completion research programs that require sustained effort, even after an initially enthusiastic minister or deputy had lost interest, or had been replaced by the vicissitudes of politics or the Public Service.

The loss of the Board's status as a separate employer has encumbered the stations with obstructive and expensive regulations and procedures, but to the Board members themselves it may have come as a relief. What remains to be seen is whether, with their new organization, they can as effectively initiate and encourage imaginative research, and also protect ongoing programs until they are brought to a satisfactory conclusion.

22   Publications and reports

Whatever political vicissitudes the Board may have suffered during its 75-year history as the scientific arm of Canadian federal fishery effort, its publication record has been a proud one, attesting visibly to the stature of the Board's scientific achievements. Today it is a simple fact that the *Journal of the Fisheries Research Board of Canada* is accepted as the largest and most widely respected publication on aquatic science in the world.

From its earliest days the Board had held to the view that scientific research is not complete until published, a practice underlined today by the great volume of research information gathered at considerable cost that languishes in filing cabinets throughout the world. Non-scientists often question the need for the publication of research results, and some even suggest that it is done only to inflate the ego of the writers. There are in fact two excellent reasons why publication is not only desirable but essential. One is to make information available to scientists throughout the world, particularly to those who are working in the same or related fields, and thereby to avoid costly duplication of research. Although we are justifiably proud of Canada's publication record in aquatic science, even the rosiest view cannot claim that it represents more than a fraction of the world output in that field. Any comprehensive work makes use of information obtained by researchers in several countries and published in several languages. Each paper contributes to a special aspect of a given problem and, especially in ecology, often it contains information that can be obtained only in one particular locality or from a particular organism living there. If a given research institution expects to arrive at far-reaching conclusions based to a considerable extent on work done elsewhere, it cannot escape the duty of publishing its own results in detail.

Vital role of publications in research

The other reason for publishing research work, and as promptly as possible, is to expose it to criticism from scientists everywhere who may be in a position to evaluate and correct it. Even the most knowledgeable scientist cannot possibly have a complete grasp of all the aspects of a research problem. Consequently every scientist sometimes makes mistakes and invalid interpretations, or fails to realize the full implications of a discovery. Through publication, mistakes can be corrected by others in the scientific community, and more penetrating analyses become possible.

The Board's first publication, *Contributions to Canadian Biology*, appeared as a supplement to the 32nd Annual Report of the Department of Marine and Fisheries, Fisheries Branch, in 1901. It was made up of seven separate contributions: a brief account by Professor Prince on the formation and the work of the Marine Biological Station (the movable station) during its first two years at St Andrews, and six research papers. The papers were on the effects of polluted waters on fish life, the clam fishery of Passamaquoddy Bay, the flora of St Andrews, the food of the sea urchin, the paired fins of the mackerel shark, and the sardine industry in relation to Canadian herring fisheries. So, from the first, the Board concerned itself with both applied and fundamental research.

It was not until 1907 that the second issue of *Contributions to Canadian Biology* appeared, as a supplement to the 39th Annual Report of the department. This covered research between 1902 and 1905, as the movable station moved along the Atlantic coast northward to the Gulf of St Lawrence, and 13 contributions made up the contents. With the appearance of the second issue of *Contributions* it seemed clear that publication was established as a major aspect of the Board's concept of managing research.

The Editors

Prince, the first editor
The importance that the Board attached to publication is shown by the fact that Prince, the first chairman, was also the first editor, receiving the reports of investigators, editing them tactfully, and following them through to press. This task was in addition to his other heavy duties, and he carried the load until 1918, when he requested that J.P. McMurrich be appointed to share the task. In effect McMurrich became editor, handling all papers intended for publication. By 1924, with the steadily increasing numbers of papers, McMurrich found the job too demanding, and Arthur Willey was appointed to succeed him, with A.G. Huntsman unofficially assisting. Willey filled the post for only two years, when he was succeeded by A.H. Leim in 1926. With Leim's appointment, an advisory committee was set up to assist the editor, but in 1929 Leim, too, resigned the post. Just a few months previously A.T. Cameron had been appointed associate editor for chemical papers, and McMurrich and Huntsman associate editors for biological papers. However, between 1929 and 1933 Huntsman was apparently the editor in effect, for in 1929 he reported to the Board on the state of its publications and in 1933 his position was regularized when the executive committee moved 'that Dr Huntsman be appointed Editor of the Board's publications with full power in cooperation with his Editorial Board to edit these publications, subject to such recommendations as may be made from time to time by the Executive of the Board.' Margaret Rigby served as managing editor throughout Dr Huntsman's editorship, which lasted until 1948.

Huntsman argues with authors
Huntsman would not have been Huntsman if during his long stint as editor he had not engendered strong feelings among the scientists who submitted manuscripts for publication. Many of his comments were pertinent and valuable but, especially in later years, some of them concerned quite trivial matters. The trouble was that he

310

tended to identify himself too closely with what was to be published, and his natural love of argument found a captive audience among the authors. The latter had no option but to respond to his repeated communications, each of which introduced a new set of objections. Most papers were finally approved, but meantime publication might have been delayed for a year or even two. For example, Brocklesby's important bulletin on fish oils was finally sent to press only after direct intervention by station director D.B. Finn. The result of this situation was that the flow of manuscripts to the Board's publications gradually slowed to a trickle; Board scientists increasingly looked to other publications, or became resigned to having their work appear only as Manuscript Reports.

In 1948 Huntsman answered his critics in a 'Final Report on Publications.' He stated bluntly that an editor's first duty is to the readers, not the authors. He also pointed out that he was in his final year of his responsibilities as editor, though it was not until 1950 that W.E. Ricker was named to replace him. The scientists were mollified and the storm died down. From 1948 to 1950 the chairman and vice-chairman, Reed and Dymond, carried out the editorial functions.

In 1950 Dr W.E. Ricker, a former Board scientist who had moved to Indiana University, became editor. Given a choice of universities and Board stations from which to operate, he felt he should be close to one of the major research units, and chose Nanaimo. Although Ricker regarded the editing, printing, and distribution of publications as his first responsibility, like all the editors before him he had time for scientific work as well. He was given assistance in 1955 when Dr N.M. Carter, director of the Vancouver laboratory, joined the Office of the Chairman in Ottawa as special assistant and associate editor. When Carter retired in 1962 the flow of manuscripts was becoming a flood, and a major change was urgently called for.

Ricker had a profound influence in shaping the future development of the *Journal of the Fisheries Research Board of Canada*, as the Board's primary publication was now called. He found that the Board, after a half-century of publishing, had done little to change the basic in-house nature of the *Journal*. Ricker made two important policy changes. The first was to actively encourage Board scientists to publish; also, because of his broad scientific connections, he received manuscripts from other Canadian scientists and from other countries. On occasion he spent a great deal of time with poorly prepared manuscripts in order to facilitate early publication, receiving the authors' gratitude for doing so. This was in sharp contrast to Huntsman's restrictive publishing policy.

Ricker's influence as editor

Ricker's second policy change was to extend greatly the distribution of the *Journal*. He set up a system of publication exchange agreements with research laboratories throughout the world. Also he offered complimentary subscriptions of the *Journal* to many scientists on whom he relied as referees.

These actions by Ricker, assisted by Carter, set the scene for the growth and international recognition that the *Journal* was to experience in the 1960s and early 1970s.

The 15 years previous to 1962 had witnessed a tremendous expansion of Board activities, and the amount and variety of scientific and popular reports had grown apace. It was no longer possible for a part-time editor, even with the assistance of an associate editor, to cope with the volume and determine effective editorial policies. The fact that the editorial function was physically separated in two locations made it increasingly difficult to handle the growing work load. Ricker handled biological papers at his office in Nanaimo, and Carter dealt with chemical and oceanographic contributions, looked after publication production, and supervised the Ottawa office.

The chairman of the Board, F.R. Hayes (right), presents a long-service award to J.C. Stevenson, 1967

When Carter retired the time was ripe to consolidate all editorial and production activities in Ottawa. At this time Ricker, though an excellent editor, was experiencing increasing demands for his other considerable talents, and he turned to full-time scientific work as biological consultant for the Board.

J.C. Stevenson, first full-time editor

It was J.L. Kask, the Board's first full-time chairman, who selected Dr J.C. Stevenson as the Board's first full-time editor. Stevenson, after 16 years as a scientist and assistant director at the Nanaimo station, had become assistant director of the Pacific Area headquarters of the Department of Fisheries in 1959. Kask supported the transfer, believing in the value of closer relations between the Board and the department. However, with the reorganization of the editorial functions, Kask 'thought it would be a good thing to have him back in a senior position with the Board. And as he, like Ricker, has patience with detail, patience with editing, and for reading long manuscripts critically, I thought he would make a good editor. The Board's publications are an important end product. Our publications tell the Canadian people and the world what we're doing, and how we're doing our job. So it is a very important post.'

Thus, in 1962, Stevenson came to Ottawa as editor and special assistant to the chairman (biology). Dr E.G. Bligh, a scientist at the Halifax laboratory, was named associate editor and special assistant to the chairman (technology). R.L. MacIntyre, publication production assistant, and Dorothy Gailus, secretary and editorial clerk, completed the staff of the newly consolidated Editorial Unit. Between 1962 and 1967 the publication output more than doubled and in 1965 R. Hazen Wigmore,

312

an editor with the Department of Agriculture, became the first appointment to the newly created position of assistant editor.

Wigmore, who retired in 1974, established himself as one of Canada's foremost scientific editors during his 25-year career in this field. His thorough approach to editing led to the coining of the word 'wigmorizing.' His dedication to editing had some similarity to that of Huntsman in the 1940s, but it was directed more toward clarity of presentation than to nature of content. In early 1974 Johanna M. Reinhart, the editor of the *Transactions of the American Fisheries Society*, was appointed to succeed Wigmore. 'Wigmorizing'

Until 1967 the Editorial Unit was an integral part of the Office of the Chairman, but in that year the growth of the latter required that the former be split off as a separate responsibility center. The Office of the Chairman was expanded to take over the special assistants' duties, and because of lack of contiguous accommodation the Editorial Unit was set up in another location in Ottawa and renamed the Office of the Editor.

Several changes have occurred in the associate editorship in recent years. Bligh became full-time special assistant in the Office of the Chairman in 1965 and Dr G.I. Pritchard, a nutritionist in the Department of Agriculture, became associate editor. In 1967 Pritchard accepted a senior position in the Office of the Chairman, and Dr L.W. Billingsley, a biochemist with broad editorial experience in the National Research Council and the Defence Research Board, became associate editor, serving until his retirement in 1974.

The increased demand for the services of the Office of the Editor reached a critical point in 1971, and a reorganization was made when Dr J. Watson, a scientist from the St. Andrews station, joined the editorial staff, becoming deputy editor in 1973. After Billingsley retired, the senior structure of the Office of the Editor comprised the director/editor, the deputy editor, two assistant editors – Reinhart (1974) and Dr D.G. Cook (1975) – and the chief of Publication Production and Documentation, J. Camp (1972). Growth and reorganization

Early evolution of the Journal

When Stevenson came to Ottawa he found a variety of publications bearing the Board's imprint. Undoubtedly the most important of these was the *Journal*, which had evolved over the years from the modest *Contributions to Canadian Biology* into a significant primary journal of aquatic science. The title *Contributions to Canadian Biology* had continued unchanged from 1901 until 1925, during which time 178 papers were published.

In 1922 a new series of *Contributions to Canadian Biology* was started, the volumes being numbered instead of designated by year. Only two volumes of the new series were published, with 25 papers appearing from 1922 to 1925. Contributions

In 1926 the title was expanded to *Contributions to Canadian Biology and Fisheries* (New Series), but the sequence of volumes was not altered. Beginning with volume 6 the articles were grouped into four series: A, General; B, Experimental; C, Industrial; and D, Hydrographic. However, this division tended to delay publication until enough papers for an issue accumulated. The various *Contributions* series existed from 1901 to 1934, during which period 355 papers were published.

The next change occurred in 1934 at an executive meeting of the Board, when 'the matter of the form of the series entitled *Contributions to Canadian Biology and Fisheries* was considered,' based upon a report submitted by the editorial commit-

313

tee, which stated in part: 'This series started as an occasional publication for results obtained in the Board's work, but has come to be a strictly scientific journal, still largely including the results of work under the Board's direction. In view of this it has seemed desirable to change the form to one better adapted to the new character.' On the basis of this consideration, it was decided that the title should be changed to *Journal of the Biological Board of Canada*, starting after the completion of volume 8. Three volumes were issued under this name in 1934–7. Numbering of individual papers was discontinued, as well as the grouping into the four series.

Finally, with the change in name from the Biological Board of Canada to the Fisheries Research Board of Canada, the name of the *Journal* was altered accordingly to *Journal of the Fisheries Research Board of Canada*. The first issue under the new name appeared in 1938.

Up to 1953 the issues of the *Journal* appeared irregularly; a volume was usually cut off when it reached about 500 pages. With the increase in number of manuscripts it became possible in 1954 to have one annual volume for each calendar year, each including six issues. By 1966 the amount of material was sufficient to put the publication on a monthly basis, an important event in the *Journal*'s attainment of international stature.

The printers
The cost of printing was borne originally by the department until after the passage of the Biological Board Act in 1912, when it was understood that the Board would be obliged to assume this responsibility.

For some years the printing continued to be done by the King's Printer, but in late 1917 the secretary of the publications committee of the government wrote to the Board requesting a substantial reduction in the publication of reports 'of a highly technical nature,' claiming that they were not widely read. At the executive committee meeting in early 1918, it was considered whether an alternative printer could be used. It was decided to try this out with a bulletin, and when there were no immediate repercussions, McMurrich was authorized to arrange the printing of *Contributions* by the University of Toronto Press. Thus the volume of *Contributions* for the period 1918–20 (issued in 1921) was the last to be printed by the King's Printer and issued by the Department of Naval Service, under which the Fisheries Branch had operated since the war years. The first two papers of the 1921 volume of *Contributions* were published in 1921, and the remaining 10 in 1922. Reporting as editor, McMurrich stated that the printing had been done economically and in a better style than formerly. It also made for efficiency to have the editor and the Press in the same city and only a five-minute walk apart.

In 1924 the assistant auditor general wrote the Chairman, objecting to having printing done outside the Government Printing Bureau. The Chairman replied by pointing out previous difficulties in getting work done by the King's Printer, and citing the numerous complaints from authors, but agreed to bring the matter to the attention of the executive committee. This was done and apparently a satisfactory agreement was concluded, for the arrangement with University of Toronto Press lasted into the 1950s. Possibly the King's Printer felt in 1925 that although the Board's work was inconsiderable in quantity it produced considerable headaches because of its highly technical content.

In 1955 the Board's publishing practices came under outside scrutiny once again, this time by a parliamentary committee empowered to review all government publications. H.A. Wilson, executive assistant to the Chairman, reported that members of the committee were astonished when they learned that the Board had for many years been contracting directly with its own printer. It was instructed to desist forthwith, and from January 1956 the Board's long, direct association with the University of Toronto Press came to an end. Printing henceforth was done or arranged by the Queen's Printer. With the best will in the world on everyone's part, this arrangement proved much more cumbersome, although it proved workable after the *Journal* contracts were let by the year rather than by the issue.

Since 1956 the *Journal* has been produced by commercial printers with varying success because of its highly scientific content. As a postscript, it should be noted that in 1975 the University of Toronto Press again became the printer because of its leading position in Canada as a printer of scientific journals.

The University of Toronto Press returns

An international journal

The success of a scientific journal is dependent on many factors. For the *Journal* a major factor was the decision in 1965 by the Board under F.R. Hayes to support a full-time editor. Previously editors traditionally held other responsibilities, and even today full-time editors are rare. This decision had several wide-reaching effects. It provided the editor with time to meet authors, to discuss research with them, and to encourage them to publish. Through these numerous contacts he recognized the difficulties inherent in the delicate author-referee-editor relationship. In his annual report for 1968 Stevenson wrote: 'A fundamental principle in primary scientific publication should be that the manuscript is the responsibility of the author and the journal that of the editor, which gives the former the right to withdraw and the latter the right to reject. The referee's responsibility is to his science. The author-referee-editor interrelations are sensitively balanced and leave no room for prima donnas in any of the three categories.' Development of this general principle proved useful in explaining to ruffled authors why their manuscripts had to be improved to meet the *Journal*'s standards, and this was a key factor that led to the *Journal*'s growth in subsequent years.

Author-referee-editor relationship

The phenomenal growth of the *Journal* and its predecessors can be shown by the numbers of articles published in three periods of editorship. Of a total of 4,007 articles published through 1974, 16 percent appeared between 1901 and 1949, 17 percent were published under Ricker's editorship (1950–62), and 67 per cent were published by Stevenson (1963–74). Put another way, an average of 13 articles per year appeared up to 1950, 51 articles per year between 1950 and 1962, and 224 articles per year since 1962.

But growth produces problems and bigness does not necessarily lead to excellence. From 1965 to 1969 the number of manuscripts published increased from 131 to 304, and it was clear that financial and manpower resources could not keep pace with continued growth at this rate. Hence, in 1969 standards of acceptance were raised (and progressively raised in subsequent years) and page charges, requiring authors to pay for publication, were instituted. Also, because of the large amount of editorial effort required to process poorly written papers, guides were prepared for authors, typists, and particularly for graduate students planning to submit theses for publication. The result has been that an average of 284 papers were published per year in the period 1970–74, and almost the same number of papers were published in 1970 as in 1974. This was in the face of an increasing number of manuscript submissions, which grew from 318 in 1970 to 421 in 1974 (460 in 1975). Also, from 1965 to 1974 the average length of articles decreased significantly.

Raised standards and new *Journal* sections

Another concern of the Office of the Editor in the late 1960s and early 1970s was whether the *Journal* should limit itself to publishing papers that were understood only by other scientists or whether the readership base should be broadened. Research findings concerning renewable resources such as fisheries or the aquatic environment have wide implications, and if interpreted or synthesized could be

Cover designs of the *Journal* and its predecessors (upper 6) with current design in centre, and cover designs of Board Reports, *Annual Report*, and *Review* (lower 3)

useful to many audiences. This thinking has led to a considerable change in the scope of the *Journal* in recent years. In 1974 two new sections of the *Journal* were initiated, Perspectives and Book Reviews, and in 1975 Letters to the Editor first made their appearance. These innovations have met with wide author and reader approval.

Perspectives, defined as 'essays of opinion or hypothesis that are of broad Canadian and international interest in fisheries management, ocean science, and the aquatic environment,' have significantly broadened the intellectual scope of the *Journal* by providing an influential forum for the expression of viewpoints on North American aquatic science. Twenty-four items in this section were published by the end of 1975, several of them being studies undertaken by the Board in its new role in advising on the future course to be charted for Canadian fisheries and marine science.

Special issues   From time to time the *Journal* has published special issues or collections of papers. During the 1950s it was usual for the *Journal* to publish the papers presented at the annual meeting of the Board's Committee on Biological Investiga-

316

tions, or the Canadian Committee for Freshwater Fisheries Research, or both. In 1958 two large special issues appeared, honoring the 50th anniversary of the St Andrews and Nanaimo stations, respectively. Both contained an illustrated résumé of the history and work of the station concerned. There have been five issues of the *Festschrift* type, honoring J.R. Dymond (1963), W.A. Clemens (1964), A.G. Huntsman (1965), board member T.W.M. Cameron (1969), and Cyril and Edith Berkeley (1971). Other large special issues have included the papers presented at a symposium on Salmonid Communities in Oligotrophic Lakes (1972), the FAO Technical Conference on Fisheries Management and Development in Vancouver (1973), a pre-conference series of Canadian limnological papers prepared for the 19th International Limnological Congress in Winnipeg (1974), and a series of papers on smaller cetaceans presented at a scientific subcommittee meeting of the International Whaling Commission (1975). Other series organized by federal government scientists included research on experimental lakes in northwestern Ontario, and biological and limnological studies on the Great Lakes.

The nature of the authorship of Board publications has changed markedly in the past decade, reflecting the growing international character and acceptance of the *Journal*. A comparison of the period 1962–4 with the period 1972–4 shows that the percentage of articles by federal government employees changed from 71 to 33, by other Canadian organizations from 17 to 28, by United States authors from 10 to 31, and by other foreign authors from 2 to 8. Another measure of the growth in the *Journal*'s international stature is the fact that the number of paid subscriptions increased from 319 in 1962 to 2,123 in 1974. Also, publication exchange agreements have ensured that the *Journal* is available in virtually every aquatic research laboratory in the world. By purchase or exchange it is received in over 100 countries, and the total monthly printing of over 4,000 copies is matched by few other primary scientific journals. *Authorship*

The *Journal*'s policy and practice has always been to accept papers in English or French, according to an author's choice. Since 1971, all papers have carried abstracts in both languages. This has been made possible by the bilingual expertise of Dr Yves Jean, who was formerly a scientist at the St Andrews station and the director general of fisheries for the province of Quebec.

Other indications of the pre-eminent stature of the *Journal* are reflected in a recent analysis, carried out by the Institute of Scientific Information in Philadelphia, on the number of times it has been cited in papers published in other scientific journals, and the fact that it has received six times the prestigious annual Fisheries Publication Award of the Wildlife Society for the best publication of original research for the year in which it was published. Awards

Professor Prince could never have anticipated the phenomenal evolution of his modest 'blue book' of 1901 over three-quarters of a century.

Bulletin series

While the *Journal* was growing in stature, another Board publication series was achieving a comparable success, and like the *Journal* its remarkable evolution could never have been anticipated when it was initiated in 1918.

The Bulletin series had its origin in the need that developed during the First World War to find new food sources. The Canada Food Board, appointed in 1917, cooperated with the Department of Fisheries in an effort to increase the utilization of fish. At its annual meeting in 1917, the Biological Board noted that any species of fish should be utilized if it was of comparable quality to the kinds generally marketed for food purposes. In early 1918 Huntsman proposed to the Board that History of new food fishes

317

Cover designs of recent Board Bulletins (upper 8) and of issues of the Miscellaneous Special Publications Series (lower 3)

popular accounts of the life histories of abundant but little-used food fishes be published and widely distributed to those directly interested in the fisheries. These would be written simply but accurately, omitting the technical details necessary for scientific papers. McMurrich and Huntsman, and later Macallum, made up a committee to arrange for the preparation and publication of these reports in a subseries to be called Histories of New Food Fishes.

Huntsman launched the series in 1918 with an issue on the Canadian plaice, and three more life histories of underutilized species, published in 1920 on the lumpfish, angler or monkfish, and muttonfish, ended this short subseries. In the 1920s emphasis in the Bulletins shifted to the processing of exploited species. The series became well established in the 1930s and its scope covered virtually every aspect of the Board's work. Many Bulletins have dealt with subjects of interest to the fishing industry, including reviews of problems and recommended improved methods of catching, handling, holding, processing, and distributing fish to markets. Methods of live storage of oysters and lobsters have been described, and detailed instructions for the culture of oysters on both coasts have been published. Also there have been numerous historical accounts of the development of important fisheries and the current state of their stocks, as related to life history, utilization, and economics. In addition many issues have been devoted to studies and reviews in oceanography, limnology, primary productivity, physiology, ecology, and pollution.

Up to the end of 1975, a total of 193 Bulletins were produced, and in recent years most have also been published in French editions. In general the original concept of writing Bulletins in non-technical language has been continued, but in recent years a number of scientific articles that were too long to be included in the *Journal* have been produced in the series. Standards of acceptance of contributions have increased over the years, and beginning in the 1930s many manuscripts of a type formerly included in the Bulletin series appeared in the newly instituted Circulars and Progress Report series. Starting in 1967 the Technical Report series served a similar purpose.

Over the years a major objective of the Bulletins has been to produce comprehensive reviews of the current state of knowledge on specific subjects. Usually these have included some research results not previously published. The monographic approach led to the production of many works that achieved national and international acclaim. The first major publication of this type in the series was the issue by Brocklesby and Denstedt on marine oils published in 1933, which was updated by Brocklesby in 1941 and by Bailey in 1952. Ricker's review of methodology in fish population dynamics originally published in 1958 was revised in 1975, and is a standard university textbook on this subject. A manual on sea-water analysis by Strickland and Parsons published in 1960, and revised in 1968, became one of the most frequently quoted works in aquatic science. The pioneering research of Roach, Harrison, Tarr, and Tomlinson on the use of refrigerated sea water for the storage and transport of fish appeared in the series in 1961 and was revised in 1968. Foerster's monographic work on the sockeye salmon was published in 1968.

The greatest impact of the Bulletin series has perhaps been through a group of books on Canadian fishes, beginning in 1946 with an issue on the Pacific fishes by Clemens and Wilby. Its success prompted Kask and Stevenson in 1963 to promote the publication of other Bulletins covering the rest of Canada's fish fauna. This resulted in the publication of Leim and Scott's *Fishes of the Atlantic Coast of Canada* in 1966, McPhail and Lindsey's *Freshwater Fishes of Northwestern Canada and Alaska* in 1970, Hart's *Pacific Fishes of Canada* in 1973, and Scott and Crossman's *Freshwater Fishes of Canada* in 1973. A Bulletin is being written on the arctic marine fishes to complete the national coverage, and a revision of the Atlantic fishes is planned.

The Bulletins have invariably had many favorable reviews in leading journals throughout the world, and four have received the Wildlife Society Award as the best book of the year: Ricker's *Handbook of Computations for Biological Statistics of Fish Populations* in 1960, Foerster's *The Sockeye Salmon* in 1970, and the previously mentioned regional fish books by McPhail and Lindsey and by Scott and Crossman in 1972 and 1975, respectively. The series of Bulletins on Canadian fishes was specially honored by numerous favorable reviews in leading scientific journals. For instance, Alwyne Wheeler wrote in *Nature*: 'The Fisheries Research Board of Canada has a most creditable publishing record on both fish and fisheries. By commissioning a series of excellent publications on the fishes of the major geographical regions of Canada, published in its Bulletin Series, it has made the Canadian fish fauna the best documented in the world ... [*Freshwater Fishes of Canada* is] one of the best books on a major fish fauna to be published this century.'

Other publication series

Over the years the Board has initiated several other types of publications, and although none have attained the lasting importance of the *Journal* and the Bulletin series, each has fulfilled an important function in disseminating scientific information. A major concern of the Board from the beginning was to develop and create publication media for the interpretation of its scientific findings, but there has been only limited success in this direction.

The following publication series are discussed in the chronological order of their initiation.

Fauna Series As early as 1912 Huntsman proposed to the Board that accounts be published of the local fauna and flora around St Andrews as a means of ready identification of species found in the Bay of Fundy, but it was not until 1921 that the Canadian Atlantic Fauna series was initiated by Huntsman in consultation with McMurrich. It was intended that the papers would give brief notes, illustrations, and keys to enable easy and accurate identification of species in each group of animals found on the Atlantic coast. A comprehensive plan covering all phyla was drawn up, and to ensure a more or less uniform style of presentation, a leaflet entitled 'Instructions to Contributors' was prepared in 1923. Copies were sent to various specialists, inviting them to contribute parts for their particular fields. Fraser's paper on the hydroids, which had been published in *Contributions*, was considered suitable, and was reprinted in 1921 as the first in the Canadian Atlantic Fauna series. Only seven parts in this series appeared at irregular intervals to 1948.

In spite of the poor response on the Atlantic, a Canadian Pacific Fauna series was started in 1937, but only 11 parts were published, the last in 1965. The failure of these series to become comprehensive had been a disappointment to all concerned. One reason was that qualified systematists were scarce. Another was that those qualified frequently felt that information was as yet too incomplete, and that their time would be better spent in improving knowledge of the species and its distribution before attempting a general review. More recently, comprehensive accounts of important groups of fishes, marine mammals, and aquatic invertebrates have appeared as Bulletins.

Progress Reports In 1929 the first issue of *Progress Reports of Pacific Coast Stations* was printed. Its purpose was explained as follows:

It has been felt for some time ... that the results of investigations being carried out on the Pacific coast should, in some way, be made available for those persons especially interested in and directly concerned with the various problems and their solutions. While conferences are excellent ... this method of reporting has its limitations ... a series of brief reports [is] to be distributed to those persons desirous of keeping in touch with the researches which are being carried out. In this way, it is hoped that a larger number of persons will become acquainted with the problems ... and the progress being made toward their solutions.

The *Pacific Progress Reports* aroused much interest and even enthusiasm, so that *Progress Reports of the Atlantic Coast Stations* was begun in 1931, which for a time included contributions from la Station biologique du St Laurent, Université Laval. In 1959 a short-lived series (two issues) was begun, called the *Progress Reports of the Biological Station and Technological Unit, London, Ontario*.

For 30 years (1929–59) the *Progress Reports* thrived; a total of 186 issues containing 1,111 articles was published. Virtually every Board scientist contributed

320

to the series, finding that suitable short articles could be quickly prepared and that the regional production of the issues led to speedy publication. Each issue typically contained 15 to 30 pages, an average of six articles, and a few brief news items. From two to four issues in each of the two major series (usually four from the Pacific series) appeared each year. Yet by 1959, after 72 issues had been published in the Atlantic series and 113 in the Pacific series, the *Progress Reports* were abandoned. The concluding issues in each of the three series were published in 1962 or 1963.

Two main reasons were advanced for this demise, one being that the original purpose was being served by other means. The argument was that scientists had available to them the Board's *Circulars*, departmental publications such as the *Canadian Fish Culturist* and *Trade News* (later *Fisheries of Canada*), and trade journals and newspapers. However, a critical factor in the success of the *Progress Reports* was the regional editors, who pressed scientists for contributions to fill the next issue; the pressure disappeared with the series. The scientist was now on his own to interpret his research for non-scientific audiences, and frequently he did not do so, lacking the encouragement provided by the *Progress Report* editors. Demise of progress reports

Another reason for discontinuing the series was an apparent tendency for some scientists to publish *scientific* articles in it, thereby absolving themselves of writing more definitive articles for the Board's *Journal* or other scientific publications. Although this view held some substance, most of the articles published in *Progress Reports* in later years adhered closely to the original concept of the series.

Circulars These were begun by three laboratories in 1935, and like the *Progress Reports* they were aimed at putting the Board's wealth of information in a form suitable for wide use. Unlike the *Progress Reports*, these series were under the control of individual laboratories without any surveillance at the regional (or headquarters) level. No central control was exercised on format, length of articles, printing process, or content. The result was that no less than 15 *Circular* series were initiated between 1935 and 1968, seven of which had been discontinued by 1972. One series consisted of a single issue, and the most prolific series reached 153. In all 561 *Circulars* were produced to 1974. Most have been addressed to government, to industry officials, or to fishermen (one series was directed to oyster farmers). Others have been used for data records or ship cruise reports from oceanographic operations, so that the various series have covered most areas of the Board's research activities.

In the late 1960s and early 1970s the number of *Circulars* produced dropped significantly. In spite of the large amount of information transmitted through the *Circulars* and *Progress Reports*, the Board recognized that it had not found an effective answer to the problem of transferring to users the knowledge it was generating.

Manuscript Report and Technical Report Series The Manuscript Report series were instituted for the purpose of recording field data, or preliminary notes and discussions on investigations in progress, or theses embodying Board work. With a few exceptions, they were intended principally for use within the Board. Three separately numbered series were initiated: (a) Biological Series, begun in about 1925, although reports written as early as 1910 were included in the series; (b) Experimental Series, begun in 1925 (active only until the early 1930s although the last report was dated 1963) and so called because the Board establishments contributing to it were originally known as experimental stations undertaking principally technological investigations on fishery products; and (c) Oceanographic and

Limnological Series, begun in 1957. These three series continued separately until 1966, when they comprised 899, 61, and 229 items respectively. By then it had become evident from the rapidly growing complexity of research disciplines that the original fairly distinct differentiation in the series no longer held and that the allocation of Reports to one or other of the series sometimes led to ambiguity. Also there was a growing tendency in the early 1960s to produce issues that were being more widely distributed than was the original intention for the series.

For these reasons the three series were concluded in 1966 and subsequent Manuscript Reports were issued as a single series (continuing the numbering of the Biological Series with no. 901; no. 900 was used for an index of the three series up to 1966) with distribution restricted, in keeping with the original concept of its in-house purpose. At that time (1967) a new series called Technical Reports was initiated for reports of unpublished stature which for many reasons should be distributed to research organizations outside the Board. One of many uses of this series has been as a repository for preliminary analyses and for data too extensive to be included in *Journal* papers.

By the end of 1974 the Technical Report series comprised 512 items and the reconstituted Manuscript Report series totalled 1,334 items (1,624 if all Manuscript Reports series are included).

Miscellaneous Special Publications Although the Board realized from the beginning the importance of recording all of its publications in numbered series for ready retrievability it was evident in 1970 that it did not always follow the practice. In the late 1960s several unnumbered publications had been produced for various purposes, which did not fit into any established series. This prompted the initiation of a new series in 1971 called Miscellaneous Special Publications.

By 1974 the series comprised 25 items dealing with a wide diversity of subjects, including a short history of the Board on its 75th anniversary, a brochure on the work of the Halifax Laboratory, a semipopular account of lake eutrophication (*The Algal Bowl*), a book on the fishes of the National Capital Region (a joint effort by four organizations), a manual for detection of fish diseases, and a handbook on the chemical analysis of fresh water.

STUDIES, INTERPRETIVE ARTICLES, AND TRANSLATIONS

Three other significant series reflect the Board's broad view of the publication function.

Studies Series
From the earliest years, Board scientists have published many papers in scientific journals other than those of the Board. To collect these for use in the station libraries and for distribution to a limited number of exchange organizations, reprints were obtained, numbered, given uniform covers, and distributed as work done under the Board's auspices. The series was formally started in 1919, and beginning in 1952 it was bound in annual volumes. By the mid 1960s the number of studies had grown such that the annual binding was produced in two parts. In 1970 the binding was discontinued because of cost, but the recording and numbering in the series has been continued. Additions to the series are listed in the *Journal*.

Many years after the series began it was found that a number of articles had been inadvertently omitted and a Studies Supplement Series was instituted to pick up these articles, the earliest of which was published in 1900. By the end of 1974 the

Studies Series consisted of 2,126 items and the Studies Supplement of 1,416 items. All have been included in Board publication indexes.

Interpretive Articles Series

In addition to scientific articles, there were many interpretive articles on Board work published in non-Board media. The number of items in this series reached 319 items by the end of 1974.

Translation Series

Translation of articles in other languages into English or French is required so that their information and ideas can be applied in Canadian research, particularly when the articles are written in Japanese or Russian, both of which are little known in Canada. From an early period translations had been made sporadically by Board employees and others, and in 1958 it was decided to keep a central record of these so that they could be made known and available to all Board establishments. The record also facilitated cooperation with similar activities in other countries, particularly the United States, so that costly duplication could be avoided. Many of the earlier translations were made by Board employees, particularly by Ricker during and after his period as editor, but as a result of the expansion of the government's Translation Bureau most of the work is now being done there.

An initial List of Translations was issued in 1958 and 14 supplements of titles have been provided to 1974 (on an annual basis since 1965). Translations of articles are made available at cost of reproduction to investigators on a world-wide basis. By the end of 1974 the number of items translated reached 3,325, and the series had become acknowledged as one of the major international translation series in aquatic science. Through a cooperative arrangement, the Marine Sciences Research Laboratory of Memorial University, St John's, has indexed most of the issues in this series; it plans to continue this project.

ANNUAL REPORTS AND BIENNIAL REVIEWS

A report on the investigations carried on under the Board was prepared by Prince and published yearly in the annual reports of the Department of Marine and Fisheries and the *Proceedings and Transactions of the Royal Society of Canada* more or less regularly from 1900 until 1913.

The first Annual Report to appear as such was issued as Appendix no. 19 to the *47th Annual Report of the Fisheries Branch*, and was entitled 'Annual Report of the Biological Stations of Canada for the year 1914,' by A.B. Macallum, then secretary-treasurer of the Board. From its contents and from the fact that it was published in Ottawa in 1914, this is considered to be a report of the work done in 1913. Three subsequent yearly reports appeared as appendixes to the report of the Fisheries Branch, but in 1917 the report of the Fisheries Branch was drastically condensed, apparently as a wartime economy, and the Board's report disappeared. It did not reappear until 1925, when it was signed by J.J. Cowie as secretary-treasurer. This practice continued until 1930, when an *Annual Report* on the work of the Biological Board of Canada for the year 1929–30 was published separately by the Board. Thenceforth the Board continued to publish its *Annual Report* separately in accordance with the statutory requirement in the Board's Act of Parliament, and it received wide distribution. A digest was published as well in the report of the Department of Fisheries for many years.

In the early 1960s the *Annual Report* had grown to between 150 and 200 pages because of its inclusion of extensive summaries of the research findings of all of the Board's establishments. It was felt that its length might deter non-scientists from reading it, and especially Members of Parliament to whom it was specifically directed. Therefore, in 1964 a shorter *Annual Report* was initiated, outlining the Chairman's views of the aspirations of the Board and the main research developments of the year. To supplement these, a new series entitled *Reviews* was instituted to provide a comprehensive view of the Board's research results for a scientifically oriented audience. The first *Review* was published for 1964, and four subsequent issues were produced on a biennial basis for 1965/66, 1967/68, 1969/70, and 1971/72. The *Review* was distributed internationally and was a significant medium for publicizing the content and scope of the Board's work.

DOCUMENTATION AND RETRIEVAL

Just as the Board believed that research was not complete until the results were published, it believed that publication of scientific information was not complete until it was indexed for ready retrievability. Several indexes were produced in early years, but the first comprehensive one appeared as a Bulletin in 1957, covering the period 1901–54. Beginning with volume 12 (1955) annual indexes of the *Journal*'s output were prepared by the editor and associate editor, and published in the concluding issue of each volume. By 1964 the amount of publication required outside indexing assistance, and former associate editor N.M. Carter undertook the task. Carter's 1901–64 Index, published in 1967, encompassed all Board output, except for the Manuscript Reports and the Translation series. This was followed by an amalgamation of the annual indexes of 1965 to 1972, prepared by Carter and published in 1973. These indexes proved to be valuable information tools to persons in all areas of the globe who wished to identify rapidly any area of work in which the Board had published.

In the late 1960s the burgeoning literature explosion resulted in attempts to develop mechanized rather than manual approaches to information handling. Not only was the production of material by Board scientists becoming difficult to handle manually, but the scientists themselves were experiencing difficulties in obtaining access to literature relevant to their work. In the United States computerized retrieval systems were developing rapidly because of the space program and in Canada the development of a national policy on scientific and technical information was being discussed at length. Economic restraints caused the Board's efforts in computerized information retrieval to be temporarily abandoned.

A major development in 1974 was the nomination of the Office of the Editor to formulate policy and coordinate Canadian input to the Aquatic Sciences and Fisheries Information System (ASFIS) of the Food and Agriculture Organization and the Intergovernmental Oceanographic Commission of the United Nations. The main product of this system was to be a computerized data base, Aquatic Sciences and Fisheries Abstracts, which would contain the relevant journal and report literature of the world. This stimulus created renewed interest in computerization as a more effective approach to information handling. In 1975 a consultant's study recommended that the Office of the Editor develop the capability for computerized retrieval of scientific and technical information, and become the aquatic sciences and fisheries center of excellence that would form one node of the national network of scientific and technical information services being developed by the National Research Council in response to a Cabinet directive.

The Board's system of managing research has two vital aspects. First, it can plan research for the long-term benefit of a nation without becoming too involved in the day-to-day distractions that beset operationally oriented organizations. Second, because of its long-range objectives, the Board system has a dedication to document and disseminate what it produces. This chapter has demonstrated the large measure of success that the Board achieved in publication over three-quarters of a century of high-quality research.

To the end of 1974 the Board had published 8,114 articles, books, and reports ranging in size from a 1,000-page monograph to a half-page *Journal* Note. It documented a further 7,186 articles, published as Studies and Interpretive articles by its staff in non-Board media or translated from foreign journals for the use of its staff. All 15,300 items have been documented by the Board and made widely available through its own indexes and through cooperation with international abstracting and indexing organizations.

The Board has developed many libraries that are acknowledged as among the world's finest in the field of marine and freshwater research, largely as a result of Ricker's farsighted decision in the early 1950s to enlarge greatly the exchange of publications on a worldwide basis.

Probably the greatest frustration of the Board in publication has been in seeking ways to make its research findings available to all who could use them. Certainly the Bulletin series has had a major impact in interpreting the Board's research for industry, management, and other non-scientific audiences. Also, as reported earlier, the *Progress Report* and *Circular* series had some success. But, as early as 1965 the editor commented in his annual report that 'the interpretation of research is not keeping pace with the amount of scientific information being generated,' and the situation has worsened since the demise of both *Progress Reports* and *Fisheries of Canada*. In its new advisory role the Board is now pressing the issue by advocating the establishment of an applied periodical to make research results broadly available.

During most of the Board's operational period it employed a large proportion of the federal government's aquatic scientists, and hence it was of little interest for other departmental groups to develop their own publishing media. In later years, however, as more and more scientists were employed outside the Board, various small non-Board publication units sprang up. This trend has now been reversed and the Office of the Editor has become the publication facility for virtually all federal groups in aquatic science, a tribute to the Board which established and nourished one of the most renowned scientific publication organizations in the world.

Appendixes

CHAIRMEN OF THE BOARD

E.E. Prince, BA, LLD, FLS, FRSC 1898–1921
A.P. Knight, MA, LLD, MD, FRSC 1921–5
J.P. McMurrich, MA, PHD, LLD, FRSC 1926–34
A.T. Cameron, MA, DSC, FIC, FCIC, FRSC 1934–47
G.B. Reed, OBE, MA, PHD, FCIC, FRSC 1947–63
J.L. Kask, BA, PHD 1953–63
W.E. Ricker, MA, PHD, FRSC 1963–4
F.R. Hayes, MSC, PHD, DSC, LLD, FRSC 1964–9
J.R. Weir, PHD, DSC, FAIC, FAAAS, FRSA 1969–

MEMBERS OF THE BOARD AND OF ITS PREDECESSORS

L.W. Bailey 1898–1923
A.B. Macallum 1898–1919
C.V.A. Huard 1898–1928
A.P. Knight 1898–1926
A.H. MacKay 1898–1926
E.W. MacBride 1898–1909
D.P. Penhallow 1898–1910
E.E. Prince 1898–1936
R.R. Wright 1901–12
A. Willey 1910–14
J.P. McMurrich 1912–38

A.H.R. Buller 1914–23
J.G. Adami 1915–19
R.F. Ruttan 1920–6
W.T. MacClement 1920–37
C.J. Connolly 1923–30
P. Cox 1923–35
A.H. Whitman 1923–41
J. Dybhavn 1923–43
C.H. O'Donoghue 1924–7
A.H. Hutchinson 1924–42
J.J. Cowie 1925–40

327

J.N. Gowanlock 1927–9
Marie-Victorin 1928–36
W.P. Thompson 1928–37
A.T. Cameron 1928–47
R.C. Wallace 1929–33
A. Vachon 1929–40
H.G. Perry 1929–42
J.A. Rodd 1929–47
A.F. Chaisson 1930–5
R.J. Bean 1930–7
D.L. Thomson 1933–53
R.R. Payne 1937–8
B.P. McInerney 1938–48
G. Préfontaine 1938–53
C.W. Argue 1938–65
G.B. Reed 1938–55
J.R. Dymond 1938–58
D.H. Sutherland 1940–51
R.E. Walker 1939–58
J-L. Tremblay 1940–9
O.F. MacKenzie 1941–55
S. Bates 1943–5
W.J.H. Deane 1943–7
W.A. Clemens 1943–56
J.H.L. Johnstone 1946–55
K.F. Harding 1947–56
I.M. Fraser 1948–56
J.H. MacKichan 1948–56
A.L. Pritchard 1948–63
P.-E. Gagnon 1951–60
J.L. Kask 1953–63
R. Gushue 1954–58
T.W.M. Cameron 1954–63
L. Piché 1954–61
D.B. DeLury 1956–65
C.J. Morrow 1956–60
F.R. Hayes 1956–69
W.L. Williamson 1956–68
R.B. Miller 1957–9
C.E. Desourdy 1957–62
L.R. Omstead 1957–62
E.S. Pretious 1957–62
I.McT. Cowan 1956–65

M.K. Eriksen 1958–67
D.S. Rawson 1959–61
D.F. Miller 1959–63
A.H. Monroe 1959–63
J.M.R. Beveridge 1959–68
Y. Desmarais 1962–3
S. Sinclair 1962–
R.G. Smith 1962–6
W.E. Ricker 1963–4
G. LeBlanc 1963–7
E. Pagé 1964–8
G.L. Pickard 1963–72
M. McLean 1964–8
R.L. Payne 1964–8
H.A. Russell 1964–7
G. Filteau 1964–73
O.F. Denstedt 1964–8
W.M. Sprules 1964–9
H. Favre 1964–
F.E.J. Fry 1965–9
L.E. Marion 1966–71
W.S. Hoar 1966–71
M.O. Morgan 1966–71
R.D. Connor 1967–
D.F. Corney 1967–71
B. Blais 1968–72
T.P. Pallant 1968–72
C.C. Pratt 1968–72
R.H. Common 1969–73
E.S. Deevey 1969–71
E.L. Harrison 1969–73
J.B. Morrow 1969–73
L.H. Omstead, Jr. 1969–73
J.R. Weir 1969–
D.A. Chant 1970–
R.R. Logie 1970–
A.W.H. Needler 1972–
P.A. Larkin 1972–
P.C. Trussell 1972–
D.R. Idler 1972–
K. Ronald 1973–
P. Russell 1973–
R. Pierce 1973–

PRINCIPAL ESTABLISHMENTS OF THE FISHERIES RESEARCH BOARD AND DIRECTORS
(including curators, program heads, and major 'acting' appointments)

Establishments are shown under 1972 titles, with original title beneath

MARINE BIOLOGICAL STATION (MOVABLE), 1899–1907

J. Stafford

GEORGIAN BAY BIOLOGICAL STATION, 1904–13

B.A. Bensley   J.W. Mavor
E.M. Walker

FRB BIOLOGICAL STATION, ST ANDREWS, NEW BRUNSWICK
Atlantic Biological Station (St Andrews), 1908–

J. Stafford   J.L. Hart
D.P. Penhallow   J.M. Anderson
A.H. Huntsman   F.D. McCracken
A.H. Leim   R.O. Brinkhurst
A.W.H. Needler

FRB BIOLOGICAL STATION, NANAIMO, BRITISH COLUMBIA
Pacific Biological Station (Nanaimo), 1908–

J. Stafford   A.W.H. Needler
G.W. Taylor   P.A. Larkin
C.McL. Fraser   W.E. Ricker
W.A. Clemens   K.R. Allen
R.E. Foerster   K.S. Ketchen
J.L. Hart   W.E. Johnson

FRB VANCOUVER LABORATORY, VANCOUVER, BRITISH COLUMBIA
Fisheries Experimental Station (Pacific), Prince Rupert, 1924–42
Pacific Fisheries Experimental Station, Vancouver, 1943–

W.A. Clemens   H.L.A. Tarr
D.B. Finn   N. Tomlinson
H.N. Brocklesby   W.E. Razzell
N.M. Carter

FRB HALIFAX LABORATORY, HALIFAX, NOVA SCOTIA
Fisheries Experimental Station (Atlantic), Halifax, 1925–

A.G. Huntsman   H. Fougère
A.H. Leim   D.R. Idler
D.B. Finn   J.E. Stewart
A. Labrie   E.G. Bligh
S.A. Beatty

Gaspé Fisheries Experimental Station, Grande Rivière, 1936–70

A. Labrie
A. Nadeau
H. Fougère

P. Dussault
F.W. VanKlaveren
R. Legendre

Technological Unit, St John's, Newfoundland, 1949–72

M.A. Foley
W.A. MacCallum

D.H. Shaw

FRB FRESHWATER INSTITUTE, WINNIPEG, MANITOBA, 1966–
Central Fisheries Research Station, Winnipeg, 1944–57
Biological Station, London, 1957–66

K.H. Doan
W.A. Kennedy

W.E. Johnson
G.H. Lawler

Technological Unit, London, Ontario, 1955–66

L.C. Dugal

FRB BIOLOGICAL STATION, ST JOHN'S, NEWFOUNDLAND
Newfoundland Biological Station, St John's, 1949–

W. Templeman
A.M. Fleming

A.W. May

FRB ARCTIC BIOLOGICAL STATION, STE ANNE DE BELLEVUE, QUEBEC,
1965–
Arctic Unit (Montreal), 1955–65

H.D. Fisher
J.G. Hunter

C.J. Kerswill
A.W. Mansfield

FRB MARINE ECOLOGY LABORATORY, DARTMOUTH, NOVA SCOTIA,
1965–
Atlantic Oceanographic Group, 1946–65

H.B. Hachey
N.J. Campbell

L.M. Dickie

PACIFIC ENVIRONMENT INSTITUTE, WEST VANCOUVER, BRITISH
COLUMBIA, 1971–
Pacific Oceanographic Group, Nanaimo, 1946–71

J.P. Tully
W.E. Johnson

M. Waldichuk

FRB GREAT LAKES BIOLIMNOLOGY LABORATORY, BURLINGTON,
ONTARIO, 1972–
Great Lakes Unit (Freshwater Institute), 1967–72

M.G. Johnson

EDITORIAL CONTROL OF BOARD PUBLICATIONS

EDITOR
E.E. Prince 1901–18
J.P. McMurrich 1918–24
A. Willey 1924–6
A.H. Leim 1926–9
   ASSOCIATE EDITORS
    A.T. Cameron (chemistry) ⎫
    J.P. McMurrich (biology) ⎬ 1929–33
    A.G. Huntsman (biology) ⎭
A.G. Huntsman 1933–48
   EDITORIAL BOARD 1933–49
   Chairman, B.P. Babkin
G.B. Reed   ⎫
J.R. Dymond ⎬ 1949–50
W.E. Ricker 1950–62
J.C. Stevenson 1962–
   ASSOCIATE EDITOR
   N.M. Carter 1955–62
   E.G. Bligh 1962–4
   G.I. Pritchard 1964–8
   L.W. Billingsley 1968–74
   DEPUTY EDITOR
   J. Watson 1973–

Sources

RECORDED INTERVIEWS

A basic source of information for the volume was a series of interviews conducted by J.C. Stevenson or the author with the following senior members and employees of the Fisheries Research Board: J.M. Anderson, 1973; S.A. Beatty, 1972; L.W. Billingsley, 1972; A.E. Calder, 1972; N.M. Carter, 1972; Orville Denstedt, 1972; Claire Duffy, 1973; D.B. Finn, 1972; R.E. Foerster, 1972; H.B. Hachey, 1972; J.L. Hart, 1972; F.R. Hayes, 1972; A.G. Huntsman, 1969 (interview by John Hart), 1972; W.E. Johnson, 1974; J.L. Kask, 1972; W.R. Martin, 1973; A.W.H. Needler, 1972; W.E. Ricker, 1973; Morden Smith, 1972; H.L.A. Tarr, 1973; W. Templeman, 1973; J.R. Weir, 1973; Harry Wilson, 1973; Otto Young, 1973. The tapes of these interviews are deposited in the Public Archives of Canada.

ANNUAL REPORTS

For the pre-Confederation period, two series of reports provided valuable material: those of M.H. Perley to the legislature of New Brunswick on the fisheries in 1849, 1850, 1851, later issued in one volume, *The Sea and River Fisheries of New Brunswick* (Queen's Printer, Fredericton, 1852), and the reports of Pierre Fortin on his operations in the Gulf of St Lawrence, which were issued as appendixes to the *Journal of the Legislative Assembly of Canada* for the years 1852 to 1857 (no reports were published for 1853 and 1854) and in the Report of the Commissioner for Crown Lands in *Sessional Papers* for 1858 to 1867.

For the post-Confederation period, pertinent material was gleaned from the annual reports of the following federal departments: Department of Marine and Fisheries, Fisheries Branch, 1868–1914 and 1921–30; Department of Naval Service,

Fisheries Branch, 1914–20; Department of Fisheries, 1930–68; Department of Fisheries and Forestry, 1969–70; Department of the Environment, Canadian Fisheries Service, 1971–4. From 1929 to 1936 the Biological Board of Canada issued its own annual reports, a practice which was followed by its successor, the Fisheries Research Board of Canada, from 1937 on. The reports of the latter were supplemented by the issue of a *Review* biennially beginning in 1964.

For chapter 4, the reports of the Ontario Department of Fisheries, Department of Fish and Game, for 1902–3 and 1905–10 were helpful.

SELECT TITLES OF THE FISHERIES RESEARCH BOARD

Carter, N.M. 1968. *Index and List of Titles, Fisheries Research Board of Canada and Associated Publications, 1900–1964*. Bull. Fish. Res. Board Can. 164
– 1973. *Index and List of Titles, Fisheries Research Board of Canada and Associated Publications, 1965–1972*. Fish. Res. Board Can. Misc. Spec. Publ. 18
Clemens, W.A. 1958. Reminiscences of a Director. *J. Fish. Res. Board Can.* 15 (5): 779–96
– 1968. Education and fish. An autobiography by Wilbert Amie Clemens. Fish. Res. Board Can., MS Rept. (Biol.) 974 (unpublished; in FRB Archives)
Contributions to Canadian Biology, 1901, 1902–5, 1906–10, 1911–14
Hachey, H.B. 1965. History of the Fisheries Research Board of Canada. Fish. Res. Board Can. MS Rept. (Biol.) 843 (unpublished; in FRB Archives)
Hart, J.L. 1958. Fisheries Research Board of Canada Biological Station, St. Andrews, N.B., 1908–1958. Fifty years of research in aquatic biology. *J. Fish. Res. Board Can.* 15 (6): 1127–61
Needler, A.W.H. 1958. Fisheries Research Board of Canada Biological Station, Nanaimo, B.C., 1908–1958. *J. Fish. Res. Board Can.* 15 (5): 759–77
Ricker, W.E. 1975. The Fisheries Research Board of Canada – Seventy-five years of achievements. *J. Fish. Res. Board Can.* 32 (8): 1465–90
Rigby, M.S., and A.G. Huntsman. 1958. Materials relating to the history of the Fisheries Research Board of Canada (formerly the Biological Board of Canada) for the period 1898–1924. Fish. Res. Board Can. MS Rept. (Biol.) 660 (unpublished; in FRB Archives)
Stevenson, J.C. 1973. *The Fisheries Research Board of Canada, 1898–1973: The First 75 Years*. Fish. Res. Board Can. Misc. Spec. Publ. 20

OTHER SOURCES

British Association for the Advancement of Science, London. Reports for 1897 (pp. xcvii, 683), 1898 (pp. xci, 582), 1899
Craigie, E. Horne. 1965. *A History of the Department of Zoology*. Toronto (Found, W.A.) Deputy Minister of Fisheries' Papers. PAC, RG 23, vols. 9–18
Hart, J.L. 1972. Correspondence with the author
Harvey, Rev. M. 1892. The artificial propagation of marine food fishes and edible crustaceans. *Trans. Roy. Soc. Can.* X (1893): 17–37
Hayes, F.R. 1973. *The Chaining of Prometheus: Evolution of a Power Structure for Canadian Science*. Toronto
House of Commons Debates 11 (1898): 7733
Huntsman, A.G. 1943. Fisheries research in Canada. *Science* 98 (2536): 117–22
– 1945. Edward Ernest Prince. *Canadian Field-Naturalist* 59 (1): 1–3

Innis, Harold A. 1940. *The Cod Fisheries – The History of an International Economy*. New Haven, Toronto

McAllister, D.E. 1965. *Bibliography of Dr. Wilbert Amie Clemens*. Vancouver: UBC Institute of Fisheries

McMurrich, J.P. 1884. Science in Canada. *The Week* I (49), Nov. 6

Prince, E.E. 1893. A marine scientific station for Canada. *26th Ann. Rept. Dept. Mar. Fish. Can.*, clxxxviii–cxcv (1894)

– 1897a. The fisheries of Canada. *Brit. Ass. Adv. Sci. Handbook of Canada* (Toronto), 264–74

– 1897b. The fisheries of Canada. *30th Ann. Rept. Dep. Mar. Fish. Can.*, xxxiii–xxxix (1898)

– 1901. Marine Biological Station of Canada. Introductory notes on its foundation, aims and work. Contributions to Canadian Biology, 1–2 Edward VII, *Sessional Paper* no. 22a (1902)

– 1907. The biological investigation of Canadian waters. *Trans. Roy. Soc. Can.*, 3rd ser. I (sect. IV): 71–92

– 1913. The Biological Board of Canada. *Fourth Annual Report of the Commission of Conservation*. Ottawa

– 1920. Fifty years of fishery administration in Canada. *Trans. Amer. Fish. Soc.* 50 (1921): 163–86

Reed, G.B. 1948. Obituary of A.T. Cameron. *J. Fish. Res. Board Can.* 7: 217

Royal Society, London. 1941. *Obituary Notices of Fellows* 3 (10): 747–59 (E.W. MacBride)

Royal Society of Canada. Biographical sketches of deceased Fellows published in the *Proceedings* (3rd ser.) for various years: V (1911): vii–x (D.P. Penhallow); VII (1913): xv–xix (G.W. Taylor); XIX (1925): xiv–xvii (L.W. Bailey); XXIV (1930): iii–vii (V.-A. Huard and A.H. MacKay); XXVIII (1934): iv–vi, xix–xxi (B.A. Bensley and A.B. Macallum); XXX (1936): iv–v (A.P. Knight); XXXI (1937): xx–xxiii (E.E. Prince); XXXIII (1939): 143–6 (J.P. McMurrich).

– Pertinent material leading up to the formation of the Board and relating to the early years of the biological stations published in the *Proceedings* for various years: X (1892): lviii–lx; 2nd ser. I (1895): xiii, cxi; II (1896): cv, cvii; IV (1898): iv, xiii; V (1899); VII (1901): xxv–xxviii; VIII (1902): xxiii–xxviii; IX (1903): lxi–lxvii; X (1904): xliii–xlviii; XI (1905): xci–xciii; XII (1906): lxxii–lxxiv; 3rd ser. II (1908): lxxi–lxxiv; III (1909): cxxxvii–cxxxviii; IV (1910): lxi–lxiv; VI (1912): lxxxiii–lxxxiv

Stevenson, J.C. 1971. Edith & Cyril Berkeley – An appreciation. *J. Fish. Res. Board Can.* 28 (10): 1359–64

Taylor, G.W. 1907. A plea for a biological station on the Pacific coast. *Trans. Roy. Soc. Can.*, 3rd ser. I (sect. IV): 203–7

Wilimovsky, N.J. 1974. Correspondence with the author.

Index

This selective index contains references to the people and places which have been prominent in the history of the Fisheries Research Board. Emphasis has been placed on chairmen, board members and managers, rather than on the scientists who were responsible for the achievements of the Board. The scientific accomplishments are well documented elsewhere, and are only superficially described in this history.

The index includes the principal research establishments, national and international agencies, and geographical locations associated with the Board's history. It does not profess to be complete.